THE EFFECTS OF
IRRADIATION ON THE SKELETON

The Effects of Irradiation on the Skeleton

JANET M. VAUGHAN, D.M., F.R.C.P.

Late Honorary Director,
Medical Research Council Unit for Research on
Bone Seeking Isotopes

CLARENDON PRESS · OXFORD
1973

Oxford University Press, Ely House, London W.1
GLASGOW NEW YORK TORONTO MELBOURNE WELLINGTON
CAPE TOWN IBADAN NAIROBI DAR ES SALAAM LUSAKA ADDIS ABABA
DELHI BOMBAY CALCUTTA MADRAS KARACHI LAHORE DACCA
KUALA LUMPUR SINGAPORE HONG KONG TOKYO

© OXFORD UNIVERSITY PRESS 1973

599.02
V367

FILMSET BY BAS PRINTERS LIMITED, WALLOP, HAMPSHIRE
AND PRINTED IN GREAT BRITAIN AT THE UNIVERSITY PRESS, OXFORD
BY VIVIAN RIDLER PRINTER TO THE UNIVERSITY

For D.G. and the Lands of Summer

Be well informed but always leave room for doubt

KO KU YAO LUN 1388

Preface

THE EFFECTS of skeletal irradiation are of interest to workers in many different fields of investigation: clinical medicine, radiotherapy, epidemiology, virology, cancer research, and environmental pollution, to mention a few. The following monograph attempts to collect some of the scattered literature concerned with the adverse biological effects of radiation on the skeleton. It is not concerned with the use of radiation in therapy or diagnosis. At the outset it must be emphasized that, though radiation may cause severe damage if used without due regard to the risks involved, it can be a safe and irreplaceable tool in clinical medicine, industry, and research.

It is difficult to write about the biological effects of radiation without becoming involved in problems of dosimetry and the physical differences between different types of radiation. Some of these are discussed very briefly and in extremely simple terms in the first chapter with a view to orientating the more biological reader. Reference however should be made by those who want further information to monographs concerned with the physical and radiobiological aspects of skeletal radiation such as *Radioisotopes in the human body* (Spiers 1968a), *Basic radiation biology* (Pizzarello and Witcofski 1967), *Manual on radiation dosimetry* (ed. N. W. Holm and R. J. Berry).

Acknowledgements

THE WORK involved in preparing this monograph was made possible by the generosity of the Leverhulme Trustees who awarded me an Emeritus Fellowship. I am deeply grateful to them for their generous help.

I am also grateful to many colleagues with whom I have worked over the last 45 years both in London and Oxford on problems of blood and bone diseases. I am particularly indebted for discussions on the problems of skeletal radiation to Professor F. W. Spiers, Professor Ray Oliver, and Betty Bleaney. Dr. John Loutit read an early draft and I owe much to his constructive criticism.

Thanks are due to the editors and publishers of the following journals, symposia, and books for permission to use figures and tables: Acta Radiologica; Academic Press; American Association for the Advancement of Science; Annals New York Academy of Sciences; The British Medical Bulletin; The British Medical Journal; The British Journal of Radiology; Blood; The British Journal of Experimental Pathology; The British Journal of Cancer; The British Journal of Haematology; Clarendon Press; Excerpta Medica Foundation; Health Physics; International Journal of Radiation Biology; International Atomic Energy Agency; International Review of Experimental Pathology; Laboratory Investigation; The Lancet; McGraw Hill Book Company; Medicine; Nature; Physics in Medicine and Biology; Journal of Bone and Joint Surgery; Journal of Pathology and Bacteriology; Pergamon Press; Radiology; Radiation Research; Proceedings Society of Experimental Biology and Medicine; University of Texas Press; University of Utah Press; Brookhaven National Laboratory; U.S. Atomic Energy Commission Division of Technical Information.

Special thanks must go to Janet Judge for her secretarial help and patience, to Miss Atkinson Smith for checking the bibliography, and to Margaret Williamson for general editorial care.

February 1972   J.M.V.

Contents

1. **Radiation and the skeleton**   1
   1. RADIOBIOLOGICAL EVENTS   1
      1.1. Character of the cell   1
      1.2. Character of the radiation   2
      1.3. Cofactors in initiation of malignant transformation   5
   2. COMPARISON OF DIFFERENT TYPES OF RADIATION   11
   3. DETERMINATION OF SAFE LEVELS OF RADIATION EXPOSURE   12
   4. GENERAL FACTORS INFLUENCING THE CHARACTER OF RESPONSE OF TISSUES TO RADIATION   14

2. **Structure and physiology of the skeleton relevant to irradiation**   16
   1. STRUCTURE   16
      1.1. Trabeculation   16
      1.2. Remodelling   20
   2. PHYSIOLOGY   23
      2.1. Osteogenic tissue   23
      2.2. Bone marrow   27
      2.3. Bone surfaces   29
         2.3.1. *Osteogenic tissue*   29
         2.3.2. *Bone matrix*   29
            2.3.2.1. *Collagen*   29
            2.3.2.2. *Carbohydrate protein complexes*   30
            2.3.2.3. *Lipids and peptides*   31
         2.3.3. *Calcium apatite*   31
         2.3.4. *Area of 'bone surfaces'*   32
         2.3.5. *Character of bone surfaces*   33
      2.4. Movement of ions into and out of the skeleton   34
         2.4.1. *Accretion*   34
         2.4.2. *Exchange*   34

3. **Skeletal tissues at risk from irradiation**   38
   1. OSTEOGENIC TISSUE   40
   2. MARROW TISSUE   41
   3. EPITHELIUM CLOSELY APPLIED TO BONE   41

4. **Histopathology of tumours induced by irradiation of the skeleton** ... 43
   1. TUMOURS OF OSTEOGENIC TISSUE ... 44
      1.1. Malignant tumours ... 44
         1.1.1. *Osteosarcoma* ... 44
         1.1.2. *Chondrosarcoma* ... 45
      1.2. Benign tumours ... 45
         1.2.1. *Cartilaginous exostosis* ... 45
   2. TUMOURS OF DOUBTFUL ORIGIN ... 45
      2.1. Fibrosarcoma ... 45
      2.2. Giant cell tumour or osteoclastoma ... 46
   3. TUMOURS OF MARROW ... 46
      3.1. Haemoproliferative lesions ... 47
      3.2. Other than haemoproliferative ... 49
   4. TUMOURS OF EPITHELIUM CLOSELY APPLIED TO BONE ... 50

5. **External irradiation** ... 51
   1. ENVIRONMENTAL EXPOSURE ... 51
      1.1. Natural ... 51
      1.2. Artificial ... 56
         1.2.1. *Fall-out* ... 56
         1.2.2. *Hiroshima and Nagasaki bomb survivors* ... 62
   2. OCCUPATIONAL ... 68
      2.1. Radiologists ... 68
   3. MEDICAL ... 69
      3.1. Diagnostic radiology ... 69
         3.1.1. *Foetal irradiation* ... 69
         3.1.2. *Adult* ... 73
      3.2. Therapeutic ... 74
         3.2.1. *Thymic enlargement in children* ... 74
         3.2.2. *Ankylosing spondylitis* ... 75
         3.2.3. *Pelvic irradiation* ... 77
            3.2.3.1. *Leukaemia* ... 77
            3.2.3.2. *Osteogenic sarcoma* ... 78
         3.2.4. *Other localized skeletal irradiation* ... 78
            3.2.4.1. *Osteogenic sarcoma* ... 78
               Normal bone ... 79
               Radiation dose ... 80
               Latent period ... 81
               Type of tumour ... 81
               Site of tumour ... 81
               Pathological bone ... 82
         3.2.5. *Dysplasia* ... 83
            3.2.5.1. *Retardation of growth* ... 83

CONTENTS

	Dose relationship	84
	Pathology	85
	3.2.5.2. *Fracture*	88
	Pathology	88
	3.2.5.3. *Osteomyelitis*	89
4. EXPERIMENTAL ANIMALS		89
4.1. Whole body radiation		90
	4.1.1. *Mice*	90
	4.1.2. *Rats*	90
	4.1.3. *Rabbits*	90
	4.1.4. *Dogs*	91
4.2. Partial body radiation		91
	4.2.1. *Mice*	91
	4.2.2. *Rats*	92
	4.2.3. *Rabbits*	92
	4.2.4. *Monkeys*	93

6. **Internal irradiation from the bone-seeking radionuclides** — 94
   1. PHYSICAL CHARACTERISTICS — 94
   2. CHEMICAL CHARACTERISTICS — 95
   3. BIOLOGICAL CHARACTERISTICS — 96
      3.1. Volume-seekers — 96
      3.2. Surface-seekers — 100
   4. RELATIVE TOXICITY OF BONE SEEKING RADIONUCLIDES — 101
   5. ^{32}P — 108

7. **Radium isotopes** — 109
   1. PHYSICAL CHARACTERISTICS — 109
      1.1. ^{226}Radium — 109
      1.2. ^{228}Radium — 110
      1.3. ^{224}Radium — 111
   2. BIOLOGICAL CHARACTERISTICS — 112
      2.1. ^{226}Radium — 113
      2.2. ^{228}Radium — 116
      2.3. ^{224}Radium — 116
   3. CLINICAL EFFECTS OF ^{226}RA AND ^{228}RA — 117
      3.1. The dial painters — 117
      3.2. Radium chemists — 117
      3.3. Therapeutic radium (iatrogenic cases) — 117
      3.4. Acute effects — 119
      3.5. Chronic effects — 120
         3.5.1. *Neoplasia* — 120

	3.5.1.1. *Osteogenic sarcoma*	121
	3.5.1.2. *Carcinomas of epithelium applied to bone*	122
	3.5.1.3. *Haemopoietic neoplasms*	123
3.5.2. *Dysplasia*		123
	3.5.2.1. *Bone*	123
	3.5.2.2. *Marrow*	129
3.6. Dosimetric considerations		129
3.6.1. *Macroscopic*		130
3.6.2. *Microscopic*		134
3.7. Effects on animals		138
3.7.1. *Neoplasia*		139
	3.7.1.1. *Mice*	139
	3.7.1.2. *Dogs*	144
3.7.2. *Dysplasia*		145
3.8. Histopathology of ^{226}Ra. ^{228}Ra induced tumours		148
3.9. Conclusion as to the hazard of ^{226}Ra		148
4. ^{224}RADIUM (THORIUM X)		149
4.1. Effects on man		149
4.1.1. *Neoplasia*		149
4.1.2. *Dysplasia*		150
4.1.3. *Dosimetric considerations*		151
4.2. Effects on animals		151

8. Strontium isotopes
1. STRONTIUM 90		154
1.1. Metabolism		154
1.2. Distribution		156
1.3. Effects of strontium 90 deposition		161
1.3.1. *Single administration*		162
	1.3.1.1. *Dogs*	162
	1.3.1.2. *Miniature swine*	165
	1.3.1.3. *Rabbits*	165
	Bone	165
	Marrow	169
	1.3.1.4. *Rats*	170
	1.3.1.5. *Mice*	172
1.3.2. *Continuous administration*		178
	1.3.2.1. *Man*	178
	1.3.2.2. *Monkey*	178
	1.3.2.3. *Dog*	178
	1.3.2.4. *Miniature swine*	180
	1.3.2.5. *Rabbits*	183
	1.3.2.6. *Rats*	183

CONTENTS

1.3.3. *Consideration of the effects of ^{90}Sr administration*	184
1.3.4. *Dosimetric measurements*	186
1.3.5. *Dose-response relationships following ^{90}Sr administration*	190
1.3.6. *Histopathology of skeletal tumours arising from ^{90}Sr irradiation*	194
1.3.6.1. *Osteogenic tissue*	194
1.3.6.2. *Marrow tissue other than haemopoietic tissue*	195
1.3.6.3. *Haemopoietic tissue*	196
1.3.7. *Dysplasia*	197

9. **Plutonium isotopes** . . . 201
   1. METABOLISM . . . 202
      1.1. Man . . . 203
      1.2. Animals . . . 205
   2. SKELETAL DISTRIBUTION . . . 207
      2.1. Bone surfaces . . . 211
      2.2. Marrow . . . 216
      2.3. Binding of plutonium . . . 220
         2.3.1. *Bone* . . . 220
         2.3.2. *Marrow* . . . 221
   3. EFFECTS ON MAN . . . 221
   4. EFFECTS ON ANIMALS . . . 222
      4.1. Neoplasia . . . 222
         4.1.1. *Osteogenic tissue* . . . 222
            4.1.1.1. *Site* . . . 222
            4.1.1.2. *Histopathology* . . . 222
         4.1.2. *Haemopoietic marrow* . . . 222
         4.1.3. *Epithelium closely applied to bone* . . . 224
      4.2. Dysplasia . . . 225
   5. DOSIMETRIC CONSIDERATIONS . . . 226

10. **Americium** . . . 228
    1. METABOLISM IN RELATION TO THE SKELETON . . . 228
    2. SKELETAL DISTRIBUTION . . . 231
    3. CLINICAL EFFECTS . . . 231

11. **Thorium isotopes** . . . 233
    1. ^{232}THORIUM . . . 233
       1.1. Distribution . . . 233
          1.1.1. *Bone* . . . 233
          1.1.2. *Marrow* . . . 235
       1.2. Clinical effects . . . 235

1.2.1. *Marrow*	235
1.2.2. *Bone*	236
1.3. Animals	236
1.4. Dosimetric considerations	236
2. ^{228}THORIUM	237
2.1. Animal experiments	238
2.1.1. *Metabolism*	238
2.1.2. *Distribution*	238
2.1.3. *Effects*	238
2.1.3.1. *Neoplasia*	238
2.1.3.2. *Dysplasia*	238
2.1.4. *Dosimetric considerations*	239

12. ^{32}P — 240

1. CLINICAL EXPERIENCE WITH ^{32}P — 240
2. ANIMAL EXPERIMENTS — 243
3. MARROW — 243
4. SKELETON — 244
5. DOSIMETRIC CONSIDERATIONS — 244

13. Other bone seeking radionuclides — 246

14. Conclusions — 248

Bibliography — 250

Subject index — 288

1. Radiation and the Skeleton

IT is useful, before discussing in detail the clinical effects of radiation on the skeleton to examine briefly certain more general questions. First, what is known of the underlying radiobiological events involved in skeletal injury? Secondly, how are the effects of different types of radiation compared? Thirdly, how are existing data used to determine safe levels of radiation exposure? Finally it must be recognized that there are a wide variety of factors independent of the radiation itself which will affect the response to radiation.

In broad terms radiation may cause degenerative changes, i.e. dysplasia, or malignancy. The available evidence suggests that, in the case of both experimental animals and man, induction of malignant change should be regarded as the limiting factor for radiation exposure of the skeleton since degenerative changes have not been seen when the dose is less than the carcinogenic dose (International Commission on Radiological Protection 1968).

It is not proposed to review the extensive literature dealing with the mechanism of carcinogenesis. As Huxley has said 'It is easy to waste a great deal of time in wholly unnecessary argument about the mechanism of cancer or leukaemia. We should recognize that multiple correlation is a characteristically biological phenomenon' (Huxley 1958).

The effect of an ionizing radiation passing through a cell will depend on both the character of the cell and the character of the radiation. A common term in discussing carcinogenesis is 'somatic mutation' suggesting that the carcinogen under discussion produces a change in a single somatic cell. As Curtis (1969) and many others have pointed out a simple mutation theory alone is not acceptable on the basis of experimental results. Several sequential steps are probably required to transform a cell, which may have mutated, to a malignant form. There is also increasing evidence that other determining factors lie in the interactions between a cell and its environment (Brues 1971). These are sometimes spoken of as cofactors or promoting factors.

1. Radiobiological events

1.1. *Character of the cell*

If an ionizing radiation passes through a fully differentiated cell, like an osteocyte, it may be killed or it may recover. If it passes through a cell capable of division, like a preosteoblast, it may be killed outright, it may recover sufficiently to divide for one or more generations and then die, or it may apparently recover and continue to divide or differentiate. In the latter case it will have suffered damage, which, on further division or following a further ionization event, may result in malignant transformation. Both clinical experience and experimental observations on animals suggest that the risk of carcinogenesis in cells is connected with their proliferative potential (International Commission on Radiological Protection 1968). It is not known which of the many organelles of a cell are affected by ionization or activation events leading to death, malignant transformation or injury capable of repair (Platzman 1967, Neary 1970). Chromosome changes may occur but their significance is still undetermined (Tough, Buckton, Baikie, and Court Brown 1960; Buckton, Jacobs, Court Brown, and Doll 1962; Bloom and Tjio 1964; Norman, Sasaki, Ottoman, and Veomett 1964; Miller 1964; Millard 1965; Nowell 1965; Warren and Meisner 1965; Bender and Gooch 1966; Awa, Honda, Sofuni, Neriishi, Yoshida, and Takashi 1971; Reiskin 1971).

1.2. *Character of the radiation*

The biological damage and chemical changes caused by radiation are initiated when the quanta or particles in the radiation beam collide with the molecules of the material and transfer to them large amounts of energy. These collisions lead to ionization (the ejection of one or more electrons from a molecule leaving behind a charged ion) or to excitation (the promotion of one or more electrons within a molecule to an abnormally energetic or excited state (Boag 1971)).

There are various types of ionizing radiation that may arise from sources outside the body or from radioactive elements deposited within the body. Charged particles such as alpha and beta particles, protons, mesons (present in cosmic rays), X-rays, and gamma rays all produce ionization in the medium in which they are released.

Neutrons are uncharged. They collide with nuclei and impart energy to them. The energy absorbed by the nuclei is highest for light nuclei such as hydrogen atoms. The recoiling nuclei produce ionization in the medium. Heavy charged particles of high energy such as alpha particles and protons produce dense ionization along the particle track which is almost a straight line. Secondary electrons produced by the interaction of the incident particles with the atoms of the absorbing medium produce further ionization but most of this ionization is confined to the region close to the track of the incident particle. These secondary electrons are called delta rays. (For 5 MeV alpha particles the range of 99 per cent of the secondary electrons is less than $0.2\ \mu m$ in tissue.)

A cell in the path of an incident particle undergoes intense ionization, whereas the adjacent cell may be missed by all the particles in the radiation and receive a negligible amount of ionization. Beta particles, which are electrons, because of their relatively small momentum produce much less ionization, and as they do not travel in straight lines the absorbed energy is scattered throughout the volume of the absorber. X-rays and gamma rays produce ionization which is more uniformly distributed over a volume the size of a cell. During the time intervals relevant to radiation damage, each cell will receive an amount of radiation similar to its neighbour.

The relationship between any ionizing or activation event and its radiobiological effect is extremely complicated and at present not completely understood. It involves a complex sequence of events initiated by the absorption of energy followed by biophysical, biochemical, and biological changes. It may well be that no ionization event, however small, is without some biological effect.

Since ionizing radiations produce their biological effects primarily through the agency of charged particles, the energy transfer to the biological system by the charged particle per unit path length is the property used to describe the quality of the radiation, i.e. the linear energy transfer commonly known as LET (International Commission on Radiation Units and Measurements, ICRU Report 16, 1970). Other properties such as the coefficient of absorption or the particle range determine the distribution of the absorbed dose on a macroscopic or even microscopic scale but the ultimate physical factor in the utilization of the absorbed dose in bringing about chemical and biological change is the submicroscopic distribution of the energy transfer along the particle track (Spiers 1968a). A list of types of radiation is given in Table 1.1 together with their energy, their range in

TABLE 1.1

Ranges and LET of ionizing particles
(Spiers 1968, by courtesy of author and publishers)

Particle	Energy (MeV)	Range in muscle	Mean LET (keV μm^{-1})*
Electron or β-particle	0.01	2.5 μm	4.0
	0.02	8.4 μm	2.4
	0.1	142.0 μm	0.70
	0.5	1.8 mm	0.29
	1.0	4.4 mm	0.23
	2.0	9.7 mm	0.21
Proton	1.0	22.0 μm	45.0
	2.0	72.0 μm	28.0
α-particle	5.0	35.0 μm	143.0

* Values calculated as mean energy per micron of track.

muscle, and their mean LET expressed as keV μm^{-2}. The same values for individual radionuclides are given later in the appropriate chapters. The present table shows at once how different both the linear energy transfer and the range in tissue may be for different types of ionizing particles. For instance the range of an alpha particle is of the order of $35\mu m$ and its mean LET is 143 keV while the range of a beta particle may be as high as 9·7 mm while its mean LET ranges only from about 2·4 to 0·23 keV.

Spiers (1971) has indicated that dose distribution and the average dose to tissues enclosed in bone cavities (i.e. marrow) are dependent, to a great degree, on the range of ionizing particles when, as is often the case, the particle range is less than, or of the same order, as the cavity dimensions. For example, secondary electrons released from a bone surface by 50 keV X-ray photons or an alpha particle from radium incorporated in the bone mineral both travel about 50 μm in soft tissue. These particles could therefore only reach a small fraction of bone marrow enclosed in a cavity

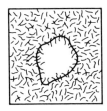

Low keV X-rays
Low energy β-rays

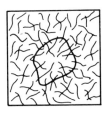

Medium keV X-rays
Medium energy β-rays

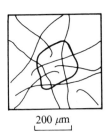

High keV X-rays
γ-rays
High energy β-rays

200 μm

FIG. 1.1. Qualitative picture of the irradiation of a marrow cavity by electrons released by X-rays or β-particles of various energies (Spiers 1971, by courtesy of author and publishers).

which might be typically 500 μm or more across. An energetic beta particle, on the other hand, could have a range 10–20 times greater than this cavity dimension. Shown diagrammatically in Fig. 1.1. is a qualitative picture of the irradiation of a marrow cavity by electrons released by X-rays or beta particles of various energies. This brief discussion may serve to indicate the sort of problems that arise in estimating radiation dose to skeletal tissues from different types of ionizing radiation.

1.3. *Cofactors in initiation of malignant transformation*

It is now recognized that probably no single molecular event necessarily determines the malignant transformation of a cell and that interactions between a cell and its environment either or both of which may be altered by radiation or chemical carcinogens may be significant. The cofactor or promoting factor which may effect the interaction between a cell and its environment and about which there is at present most discussion, are viruses. They are discussed below. There is also some evidence dependent on the work of Brues and his colleagues that when tissue cells are separated by chemically non-specific barriers so that they are no longer surrounded by other cells, the result is delayed tumour development (Brues, Auerbach, De Roche, and Grube 1969, Brues 1971).

Much of the evidence for the importance of viruses as cofactors has come from experimental studies of leukaemia and osteosarcoma apparently induced by different forms of radiation. Such studies must therefore be reviewed here. In the following discussion the term 'induction' is used to indicate any event or series of events which leads to recognizable cancer.

Leukaemia. The important part that radiation may play in experimental leukaemia (and other cancer) induction in mice was indicated by Furth and Furth in 1936. They suggested that X-rays so alter the constitution of certain cells that they, or some of their offspring, undergo malignant transformation several months after irradiation.

Gross in 1951 published evidence that cell-free filtrates and extracts prepared from spontaneous lymphomas of AKR mice could induce similar lymphatic leukaemias *de novo* after injection into newborn $C_3H/131$ mice. He postulated the presence of a *leukaemic agent* present in the filtrates and cell-free extracts. Subsequently it was shown that cells of a non-irradiated tissue could be induced to become neoplastic by virtue of their residence in an irradiated host environment (Kaplan and Brown 1954; Kaplan, Brown, Hirsch, and Carnes 1956). These observations stimulated both Kaplan and his colleagues and Gross to investigate further the possibility that an agent similar to the Gross leukaemia agent might play a role in the indirect development of thymic lymphomas in mice exposed to irradiation. Preliminary evidence that this was indeed the case was brought forward by both Gross in 1958 and by Lieberman and Kaplan in 1959. The whole complicated story was well reviewed by Kaplan in 1966 and 1967. Radiation could, they assumed, in some way activate a leukaemia agent in the host so that it would affect appropriate tissues, such as the thymic implant. They concluded that certain strains of mice with a low natural incidence of leukaemia may harbour a latent leukaemogenic virus, which is probably transmitted vertically prior to the time of intrauterine implantation. This virus can persist in the tissues of such naturally infected strains throughout life without producing any discernable ill effects, unless the animals are exposed to appropriate doses of irradiation or certain chemical agents which appear to trigger a change in the host-virus relationship leading to the development of leukaemia. In the strains with a high natural incidence of leukaemia, a similar situation exists with respect to the vertical transmission of virus but factors controlling the effect of the virus are different. This view of murine and also avian and feline leukaemia viruses has been further elaborated by Huebner and his colleagues (Huebner, Kelloff, Sarma, Lane, and Turner 1970) who characterize the virus type concerned as a C-type RNA virus. They suggest that, except in certain strains of mice which have high genetic susceptibilities for virion expression, and frequently also susceptibility to early tumour expression, the virus is generally transmitted in a repressed or 'switched off' state. Post-natal 'switch on' of virus can occur naturally, especially in old age, or it can be switched on by external agents such as radiation or specific chemicals such as methylcholanthrene. The cellular control mechanisms involved in the 'switch on' and 'switch off' described by Huebner and his colleagues (1970) are similar to the operons, regulators, and repressors postulated by Jacob and Monod (1961) involved in protein synthesis by bacterial cells.

The classic experiments on leukaemia induction and the part played by radiation and viruses have been carried out on mice. More recently a myeloproliferative dyscrasia has been described in miniature pigs fed ^{90}Sr continuously. In these animals also virus particles have been found in the plasma and in the affected tissues. These have a morphologic similarity to known leukaemic viruses (Howard, Frazier, and Jannke 1970; Howard, Ushijima, Hackett, and Frazier 1970). It is suggested by the workers concerned that this virus, in a latent state in the pig, is unmasked by the effect of ^{90}Sr administration on specific antibody synthesis since immunologic deficiency is observed in the swine. Such deficiency is known to result from radiation (Taliaferro, Taliaferro, and Jaroslow 1964). In a more recent report, Clarke and his colleagues (Clarke, Busch, Hackett, Howard, Frazier, McClanahan, Ragan, and Vogt 1972) describe the presence of two different viruses in the tissues obtained from the pigs treated with ^{90}Sr neither of which were present in controls. One is a C-type virus similar to that seen by Finkel and her colleagues in mice (Finkel, Biskis, and Jinkins 1966) and which they regard as the leukaemogenic agent, while the other is an adenovirus which they suggest plays the part of a helper virus.

Osteogenic sarcoma. It would appear that osteosarcoma induction in mice, like leukaemia induction, may also be facilitated by the interaction of viruses and radiation. The part played by viruses in various bone dyscrasias is now increasingly receiving attention. Evidence is accumulating to suggest that viruses are responsible for certain well recognized skeletal dyscrasias, such as avian petrosis, (Pugh 1927; Jungherr and Landauer 1938; Holmes 1961a, b, 1964; Simpson and Sanger 1968; Fritzsche and Bahnemann 1969). A wide variety of viruses, the Raucher, Molony, and Harvey sarcoma virus, and the parotid tumour agent and polyoma virus among others, are able to induce sarcoma in mesenchymal tissue adjacent to bone in mice, rats, and hamsters (Dawe, Law, and Dunn 1959; Kirsten, Anderson, Platz, and Crowell 1962; Sjögren and Ringertz 1962; Merwin and Redmon 1969; Soehner and Dmochowski 1969; Fujinaga, Poel, and Dmochowski 1970). These sarcomatous tumours are often described as osteosarcoma. It is still open to question, however, whether these lesions are true osteosarcomas. The evidence that bone is actually involved or that the tumours arise from osteogenic tissue, i.e. osteoblasts and their precursors, is still not entirely convincing. They are reactive lesions which may produce metaplastic cartilage and bone arising in mesenchymal tissue adjacent to the periosteum. Since in some cases these tumours can be transplanted into syngeneic mice their malignancy must be accepted but it is of low grade. In newborn rats the injection of polyoma virus has been reported to give rise to tumours associated with bone in 73 of 105 injected animals. The commonest histological pattern was said to be compatible with a diagnosis of osteochondrosarcoma (Kirsten, Anderson, Platz, and Crowell 1962). Sjögren and Ringertz (1962) have also reported bone tumours together with other tumours arising in mesenchymal tissues in mice following injection of polyoma virus. Recently it has become apparent that in at least one strain of mouse (CF 1/Anl) there is an endemic virus other than the recognized leukaemia viruses, which, when introduced into CF 1 mice, some other strains of mice, and hamsters, will induce connective tissue tumours of low grade malignancy in mesenchymal tissue adjacent to bone.* This virus was first isolated by Finkel and her colleagues from a spontaneous osteosarcoma in a CF 1/Anl male mouse and named the FBJ virus (Finkel, Biskis, and Jinkins 1966; Finkel, Jinkins, Tolle, and Biskis

---

* The CF 1 strain of mouse is supplied by Carworth Farms, New Jersey. The CF 1/Anl strain comes from a closed breeding colony of mice originally obtained from Carworth Farms maintained at the Argonne National Laboratory since 1960.

1966; Finkel and Biskis 1969). When injected into female mice of CF 1/Anl strain and into CF 1 strain the FBJ virus induces tumours which arise in soft tissue adjacent to the periosteal surface. These tumours may contain osteogenic, chondrogenic, and fibrosarcomatous tissue often in the same tumour. It is not yet altogether clear, however, that they are invasive nor that they can be classified as osteosarcomas though they are often described as such (Lisco, Rosenthal, and Vaughan 1971). It is of interest that two other strains of mice, the BALB/C and ICR strains, are unaffected by injections of FBJ virus (Yumoto, Poel, Kodama, and Dmochowski 1970). It is now generally agreed that the FBJ virus is serologically identical with a C-type RNA virus of the Gross serotype (Huebner, Hartley, Lane, Turner, and Kelloff 1969; Kelloff, Lane, Turner, and Huebner 1969; Rhim, Huebner, Lane, Turner, and Rabstein 1969) and that it is endemic in normal CF 1 mice. Finkel and her colleagues (Finkel, Biskis, and Farrell 1967, 1968, 1969; Finkel, Reilly, Biskis, and Camden 1972; Pritchard, Reilly, and Finkel 1971) also claim to have induced osteogenic sarcomas in hamsters with cell-free extracts of human osteosarcomas and to have demonstrated virus particles like those of the FBJ virus both in the original human tumour and in the subsequent hamster tumours. Viruses resembling murine leukaemia type C virus have also been observed in biopsy specimens and tissue culture from human osteosarcomas by Dmochowski (1969).

The evidence that radiation may potentiate viruses already existent in skeletal tissue, as it potentiates leukaemia viruses, comes from Finkel's studies with CF 1/Anl mice given both FBJ virus and ^{90}Sr. Virus-inoculated mice developed ^{90}Sr tumours sooner than uninoculated mice given ^{90}Sr only (Finkel and Biskis 1969). However, this picture in CF 1/Anl mice is complicated by the fact, already mentioned, that Huebner and his colleagues (Huebner, Hartley, Lane, Turner, and Kelloff 1969; Kelloff, Lane, Turner, and Huebner 1969) have shown that normal CF 1 mice harbour the FBJ virus. Indeed, it is possible that the shorter latent period for osteosarcoma production in this strain following ^{90}Sr injection compared with CBA mice (Finkel and Biskis 1968) is due to the natural presence of this virus. It has also been recognized for some years that the female CF 1 and the CF 1/Anl mouse have shown a 1–2 per cent incidence of spontaneous osteosarcoma (Finkel, Lisco, and Brues 1954; Brues 1949; Finkel and Biskis 1968).

A recent detailed study of 100, presumably normal, control CF 1/Anl female mice has enabled a more precise picture of the bone dyscrasia apparently common to this strain to be given (Lisco, Rosenthal, and Vaughan 1971). The degree of skeletal involvement was variable and tended to increase with age. There was a high incidence of osteomas (41 per cent), myelofibrosis (29 per cent), and malignant tumours arising from bone (4 per cent—3 osteosarcomas and 1 fibrosarcoma). In addition there was

79 per cent of lymphomas and 6 per cent of fibrosarcomas in subcutaneous tissue. On radiological examination, in addition to the osteomas, the majority of the mice showed areas of both increased and decreased density throughout the skeleton particularly in the long bones of the hind legs. This increased density was primarily due to endosteal thickening and was in some cases so severe as to suggest complete obliteration of the marrow cavity. There was no significant periosteal thickening except in the case of the localized typical osteoma which were often multiple. On histological examination the characteristic (non-osteoma) lesions were found to be endosteal in origin. The periosteal surface of the bones were rarely affected. The normal bone marrow and bone trabeculation was replaced to a variable extent by abnormal tissue showing actively proliferating connective tissue cells, osteoblasts, and osteoclasts associated with variable degrees of new bone formation in some areas and in others by focal and diffuse osteosclerotic changes made up of acellular compact and poorly organized bone with little or no evidence of osteoblastic or connective tissue cell proliferation. Both types of abnormal tissue sometimes appeared to invade the cartilage plate of the epiphysis, particularly in the upper end of the tibia and lower end of the femur, and the endosteal surface of cortical bone. The three bone sarcomas were large and characteristically osteoblastic, having grown to such extent when found at autopsy that it was impossible to determine their precise point of origin.

How far this dyscrasia in CF 1/Anl mice is dependent on the presence of FBJ or other viruses has not yet been determined. The finding of the dyscrasia and the virus in the same strain is at least suggestive, since a very similar dyscrasia has been induced in Balb/c mice by the injection of a filterable and serially transmissable agent isolated from sarcoma 37 (Merwin and Redmon 1969). It should be emphasized that the problem is further complicated in the CF 1/Anl mice by the fact that all experiments where the FBJ virus is injected into mice or hamsters result in tumours arising close to the periosteal surface but not necessarily being of osteogenic origin, while the undoubted osteogenic tumours potentiated by irradiation in CF 1/Anl mice are endosteal in origin (Finkel and Biskis 1969).

It is perhaps unfortunate that the only extensive comparative study of the effects of both external irradiation and of different radionuclides has been made by Finkel and her colleagues using the CF 1/Anl and the CF 1 female mouse, a mouse with a known bone dyscrasia and an endemic virus. These studies are discussed in more detail later. Inevitably the question arises as to how far the dose-response curves using the CF 1/Anl mouse are influenced by the presence of an endemic virus which is itself capable of sarcoma induction and therefore how far these curves can be used in determining radiation protection standards.

In CF 1/Anl mice osteosarcomas clearly occur under so-called normal conditions. Radiation increases the incidence. The same is true of leukaemia in certain strains of mice. Radiation alone is probably not responsible for either the osteosarcomas or the leukaemia. It acts as a co-factor. There is some evidence that radiation may act as a cofactor in the case of 'leukaemia' and osteosarcoma induction in the miniature pig (Chapter 8.1.3.2.4). It is known also that certain chemical agents like methylcholanthrene may act like radiation in potentiating the action of a virus (Huebner 1970a,b). Huebner would probably regard RNA tumour viruses as determinants of the generality of cancers (Huebner 1970a,b) since he describes them as being 'switched on' by a cofactor which may be radiation or a chemical. This view probably differs only semantically from that of Berenblum and Trainin (1963) who consider that radiation acts as the initiator by the formation of a precursor virus *de novo* and that promoting action, for instance by urethane in the case of chemical carcinogenesis, causes the precursor virus to be converted into an active virus. On the other hand very recent experimental work suggests that possibly, in the case of CF 1 ANL mice, the virus acts by inducing myelosclerosis so altering the pattern of bone trabeculation and therefore the pattern of radiation dose received by sensitive tissues from deposited radionuclides. Loutit and his colleagues (Bland, Loutit, Sansom, and Smith 1973) using CBA male mice, a strain which has no spontaneous incidence of osteosarcoma and no bone dyscrasia, have been unable to reproduce Finkel's results in female CF 1 ANL mice from deposited ^{226}Ra. ^{90}Sr had the same carcinogenic effects in both strains. The short range of the radium alpha particle would be affected by changes in trabeculation which would not necessarily affect the long range β particle of ^{90}Sr.

Further evidence of the possible part played by a virus in osteosarcoma incidence is derived from immunological studies (Morton, Eilbor, Mulmgren, and Cooke 1970). Immunofluorescent studies have yielded a high incidence of antibodies to osteosarcomas in the sera of patients with this disease. Radiation may also produce chromosome breaks. There is at present no evidence that such breaks alone may induce malignant transformation of the cell. On the other hand such breaks may theoretically facilitate the incorporation of tumour virus (Whitmore 1971, Reiskin 1971).

Emphasis in this discussion on the mechanism of radiation carcinogenesis has been laid on the possible part played by viruses, since experimental data involving the incidence of bone tumours and leukaemia have pointed to the importance of the interaction of the two agents. At a recent conference on the estimation of low-level radiation effects in human populations, it was made abundantly clear that radiation induced carcinogenesis 'may require or be greatly facilitated by cofactors with viruses being the most likely candidates' (Whitmore 1971).

For the sake of completeness mention must be made of a rather different view of the origin of malignant change dependent on radiation—namely that malignancy may arise in adjacent unirradiated repair tissue when radiation is localized. Glücksmann (1963), who holds this view, considers that induction of specific and progressive vascular changes results in unstable scar tissue, which, though unirradiated, becomes malignant. A symposium on the relationship of radiation damage to radiation dose in bone in 1960 (Vaughan, Lamerton, and Lisco 1960) concluded that gross skeletal damage was not a necessary condition for malignancy to arise. Experience with some of the radium-dial painters has shown the same. Late malignancy may develop in bone which is radiographically normal. On the other hand, Lisco (1956), in describing the histopathology of a case of radium poisoning that developed a bone tumour, suggests that in this case the fibrosarcoma arose from the preexisting extensive fibrosis of bone and bone marrow.

2. Comparison of different types of radiation

To enable a comparison to be made between the many different types of ionizing radiation use is made of the convention of *relative biological efficiency* (Report of RBE Committee 1963). The ability to make such comparisons is particularly important for protection purposes.

The term *relative biological efficiency* (RBE) used in radiobiology to express numerically the relative effect of two kinds of radiation, for instance A and B, is defined as the inverse ratio of the doses (D) that produce equal biological effects, i.e. it is the ratio D_B/D_A (for the same effect). Spiers (1968) emphasized that it is important to realize that the RBE is not the ratio of effects produced by a given dose. The basis of the comparison is the energy absorbed per unit mass of irradiated tissue; i.e. the dose (in rads) and factors, other than LET, which could influence the biological effect, are presumed to be the same. Thus the dose rate, fractionation of the dose, the dose distribution in the tissues, oxygenation in the tissues, and the nature of the biological effect should be the same (Report of the RBE Committee to the International Commissions on Radiobiological Protection and on Radiological Units and Measurements, 1963). This term RBE will be used frequently in the chapters concerned with internal radiation particularly, but it must be remembered that it is at best a useful convention, based on unproven radiobiological assumptions. It is further discussed in Chapter 6.4. in relation to internal emitters.

Even when conditions are not exactly comparable the RBE can still be usefully employed when the biological effects can be measured quantitatively. In the field of radiological protection, the biological effects are not always precisely defined, and the doses are far lower than those used in experiments to measure RBEs. The relative risk of different types of radiation is expressed by substituting a quality factor (QF) for RBE. The QF is an *estimate* of the effectiveness of a given radiation as evaluated from the LET. Values of QF recommended by the International Commission for Radiological Protection as a function of LET are shown in Table 1.2, taken

TABLE 1.2

Values of QF as a function of LET

LET in water keV μm^{-1}	QF
3·5 or less	1
7·0	2
23·0	5
53·0	10
175·0	20

from International Commission on Radiation Units and Measurements, ICRU Report 16, 1970

from the report of the International Commission on Radiation Units and Measurements (1970). In practice, QF for X-radiation and γ-radiation is close to unity, for beta particles it is greater than unity, at very low energies only, and it is usually 10 for fast neutrons and α-radiations.

For protection purposes the quantity obtained by multiplying the absorbed dose by the quality factor is called the dose equivalent (DE), and the unit in which the dose equivalent is then expressed is called the rem. The general definition of the dose equivalent is therefore the product of the absorbed dose D, the quality factor QF, and any necessary modifying factors, for example, one to allow for the geometry of the isotope distribution (DF) (Spiers 1968).

$$DE = D \times QF \times DF \times \cdots$$

3. Determination of safe levels of radiation exposure

It has already been said that it is probable that no ionization event occurring in a cell is without some biological effect even though these may be unrecognized. What is important from the point of view of human radiation protection is to determine what radiation dose will produce so little effect as to be acceptable on a cost/effect basis (Sinclair, Rowland, and Sacher 1971). Any analysis of the effects of low-level radiations involves the study of dose–response curves. Such dose–response curves may be plotted for a number of different endpoints, for instance life shortening, tumour incidence, microcateract, or chromosome aberrations, to mention a few. There is continuing discussion as to the significant form of a dose–response curve whatever the endpoint, i.e. whether it is linear or curvilinear, and whether there is a threshold below which no obvious injury occurs (Mole 1958). In general it has been said that the fact of a linear response is not in question—for all systems studied, thus far, yield a linear response in some part of the dose range (Sinclair, Rowland, and Sacher 1971). The essential question from the point of view of protection is whether all endpoints make a transition to a linear dose relationship at low doses. Further, does such a transition only occur at a point when it is unlikely that the recipient of the radiation will live long enough to show the effect of such a low-level irradiation? In other words, is there a threshold controlled by life span (Evans 1966)? For purposes of radiation protection the International Commission on Radiological Protection (ICRP Publication 8, 1966) have accepted the linear relationship at low levels. As Pochin (1969) says, in discussing the dose–response curve with cancer as the endpoint,

at present it is impossible to exclude a linear relationship, with the frequency of induced cancers directly proportional to dose between a few rads and a few hundred rads; it seems essential to base protection criteria upon this, the most pessimistic possibility—particularly as some tumours, such as the mammary tumours in rats, appear to follow such a relationship over a wide range (Bond, Shellabargar, Cronkite, and Fliedner 1960). If however the tumour frequency in any particular case was in fact proportional to the square of the dose rather than to the dose, then the estimated effect of a few rads, inferred on a linear basis from the observed effect of a few hundred rads would be too high by a factor of a hundred.

Experience of late effects in man and knowledge gained from animal experimentation are confined mainly to the higher dose ranges since it is impracticable or even impossible to carry out investigations on a scale sufficient to obtain results that are statistically significant at the incidence levels likely to occur at low dosage. The curves are only extrapolated to lower levels.

It will be noted in the following chapters, where the effects of internal radionuclides deposited in the skeleton are discussed, that dose–response curves with malignancy as their endpoint, obtained in animal experiments, are used by several workers to suggest that there is a threshold dose below which no malignancies are likely to occur and that this threshold is often imposed by life span. In looking at these results it must be noted that the number of animals used in the experiments on which such suggestions are based is often extremely small. It would appear unwise on statistical grounds to accept the conclusions. The most hopeful approach to an experimental determination of low-level risks must lie in attempting to understand the mechanisms by which radiation induces malignant transformation (Lamerton 1958, Mole 1958).

Another problem that must constantly be kept in mind is how far it is justifiable to transfer the results of dose–response curves obtained in mice, short-lived animals with tiny bones, to man, a long-lived animal with large bones in whom very different geometrical considerations must be involved in dosimetric calculations. Extrapolation from mice to man is far from satisfactory on theoretical grounds.

4. General factors influencing the character of response of tissues to radiation

It must always be remembered that the precise pattern of tissue injury that results from any radiation insult is dependent upon a large number of variables other than radiation dose, such as the time of exposure, whether exposure is acute, continuous, or fractionated, and the character of the tissue irradiated. In addition there are many other more general factors which may effect the response to radiation such as age, diet, general metabolic activity, species, and strain differences in sensitivity. There is also evidence, certainly in mice, that sex differences may be important (Hug, Gössner, Muller, Luz, and Hindringer 1969). Hug and his colleagues found that in mice treated with ^{224}Ra osteosarcoma were more frequent and occurred earlier in female than in male mice. The strain of mouse used was one raised in his laboratory. Further, the effect of a measured radiation dose to litter-mates of a known strain kept under identical conditions is not

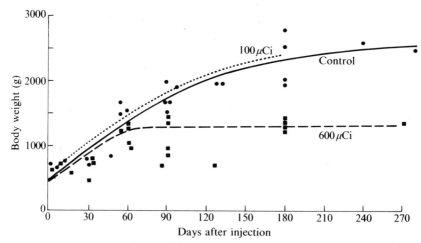

FIG. 1.2. Graph showing the increase in body weight plotted against time for control rabbits and for those which received 100 μCi kg^{-1} or 600 μCi kg^{-1} of ^{90}Sr. (Each point represents the mean of all animals at that time interval generally about 4.) (Macpherson 1961, by courtesy of the author and editors.)

necessarily identical (Carmon 1965). The same is true of human radiation exposure. Many individuals among those exposed to ^{226}Ra have as yet developed no skeletal lesion while others exposed for a shorter period and to lower radiation doses have already died, due to their ^{226}Ra body burden.

In discussing the pathological effects of radiation it must also be recognized that radiation may have a general effect on metabolism, possibly mediated through effects on the endocrine system. This general effect is well illustrated by the loss of weight seen for instance following the injection of a relatively high dose of ^{90}Sr, a bone-seeking radionuclide (Fig. 1.2). It is however possible to produce both neoplasia and dysplasia in bone by a point source of beta irradiation to the skeleton from ^{90}Sr which can produce few if any systematic effects (Jowsey and Rowland 1960). This suggests that a disturbance of metabolism is not necessarily fundamental to the development of either dysplasia or neoplasia.

2. The Structure and Physiology of Bone

THERE are certain points in the structure and physiology of bone which are of particular importance to the understanding of the effects of irradiation on the skeleton. It must be stressed that the following description of the structure of the skeleton is framed in general terms. The skeleton of the mouse is clearly different in many ways from that of man. These differences, for instance in the amount of trabecular bone and the fact that some growth may continue throughout life in small rodents, must affect the pattern of radiation dose that may be received by both osteogenic cells and marrow cells.

1. Structure

1.1. *Trabeculation*

The bones that make up the skeleton are extremely complex tissues. This complexity makes it more than usually difficult to interpret the effects of irradiation or to measure radiation dose. Great discontinuities of density and composition occur over spatial dimensions that may be of the same order of magnitude as the ranges of the ionizing particles from radioisotopes deposited in mineral bone. Further, the structure of this complex tissue changes with age. Two types of bone can be easily distinguished, particularly in man and the larger mammals—hard compact cortical bone which is found in the shaft of the long bones surrounding large marrow cavities and spongy or trabecular bone which is made up of a network of fine interlacing partitions (the trabeculae) enclosing much smaller marrow cavities. This latter bone is present in the vertebrae, the majority of the flat bones, and in the ends of the long bones. Figure 2.1 shows a longitudinal section of the head of an adult human femur from which the marrow has been removed. It shows the relation of both compact cortical bone and trabecular bone to the marrow cavities. Figure 2.2(a) shows a section of a human vertebra from a young adult illustrating again great variation in the size of the trabeculae and the marrow spaces within a single bone. The pattern of trabeculation changes appreciably with age as shown in Fig. 2.2(b) (Atkinson 1967). Such changes may be important in making estimates of

THE STRUCTURE AND PHYSIOLOGY OF BONE

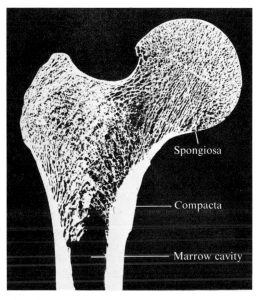

FIG. 2.1. Longitudinal section of head of human femur, from Moss 1966 by courtesy of author and publisher.

radiation dose. The vertebra in Fig. 2.2(b) is from an elderly but rather osteoporotic individual. It is even more important to appreciate that there are very great differences in trabeculation patterns in different species. The marrow spaces in pig vertebrae are so small and the trabeculae so fine that it is difficult to obtain a photograph that is meaningful. However, these differences are extremely important since they will inevitably lead to differences in radiation dose for the same body burden in different species and may therefore result in different pathology. Spiers and his colleagues (Spiers, Zanelli, Darley, Whitwell, and Goldman 1972) have recently made precise measurements of mean path-lengths in both cortical and trabecular bone of man, dog, and pig (see Table 2.1 and Fig. 2.3). The term *mean path-length* is based on their technique for describing the structure of trabecular bone quantitatively. Because the spaces between the trabeculae cannot be represented by any regular geometrical shapes a more general representation is required. It may reasonably be assumed that beta particles, for instance, will traverse the trabeculae and the small marrow spaces in approximately straight paths and that these paths will be randomly oriented in space. A special scanning device has been used to measure the possible path-lengths of a radiation from a deposited radionuclide in trabeculae and marrow cavities in a cross-section of bone (Darley 1968). By scanning in different directions across a bone section and by taking different sections through a bone, a combined 'omnidirectional' distribution can be

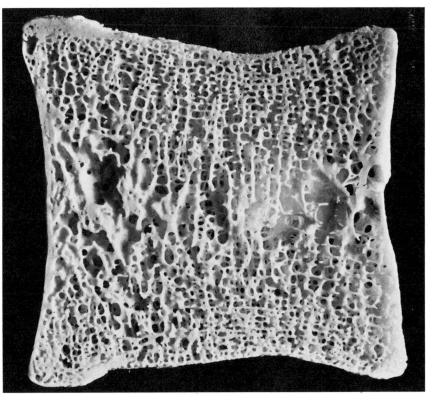

Fig. 2.2(a). Section of human vertebra of a young adult (courtesy of H. Sissons).

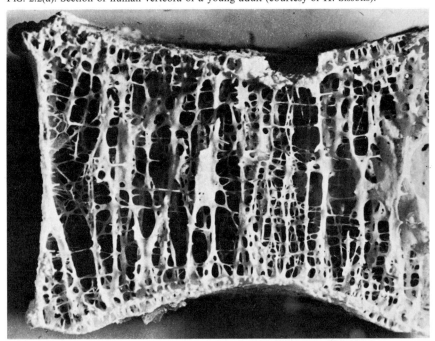

Fig. 2.2(b). Section of human vertebra of an elderly, rather osteoporotic individual.

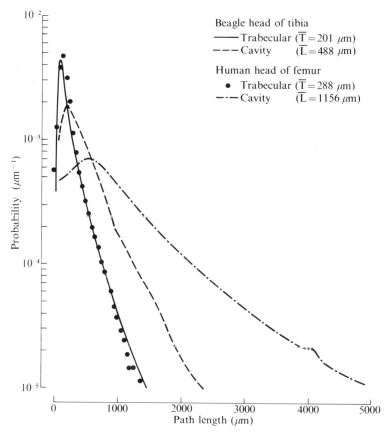

FIG. 2.3. Path-length measurements in the beagle head of tibia and human head of femur. (The figures in this diagram differ slightly from those in the Table but they are of the same order.) (Spiers et al. 1972, by courtesy of authors and publishers.)

obtained (Spiers 1968b; Spiers 1969; Spiers, Zanelli, Darley, Whitwell, and Goldman 1972). As shown in Table 2.1 it is found that the mean path-length across trabecular bone does not vary greatly between dog, pig, and human bone but that the pathway across the marrow cavities can vary by factors of three or four. For instance, the path length distribution across trabeculae in the head of the tibia of a dog bone was 201 μm and in the head of the femur of a human bone was 288 μm, but in the marrow for a dog it was 488 μm and for a human bone it was 1156 μm (see Fig. 2.3). These measurements must have serious implications in any attempt to apply data obtained on animals to man. Their significance will be discussed further when the identification and location of the cells at risk from radiation in the skeleton have been analysed, and also in relation to the different radionuclides that concentrate in bone. The only measurement available

TABLE 2.1

Mean path-lengths in human and animal trabecular bone
(*Spiers* et al. *1972, by courtesy of authors and publishers*)

Species	Bone	Mean path-length in μm (a) Marrow spaces	(b) Trabeculae	Path-length Ratio Trabeculae/ Marrow Spaces
Man[1]	Skull-parietal bone	390	510	1·31
	Rib	1705	265	0·16
	Iliac crest	905	240	0.27
	Cervical vertebra	910	275	0·30
	Lumbar vertebra	1230	245	0·20
	Femur-head	1155	230	0·20
	Femur-neck	1655	315	0·19
Beagle[2]	Tibia	490	200	0·41
	Radius	320	305	0·95
Dog[3]	Vertebra	425	180	0·42
Pig[3]	Vertebra	470	175	0·37
Cow[3]	Vertebra	470	375	0·80

[1] Adult male, aged 44 years.
[2] Beagle, UCRL, Davis.
[3] Origin unknown.

in mice for trabecular thickness is given by Lloyd and Hodges (1971) as 80 μm for trabeculae in the lumbar vertebra of an adult. These workers use a rather different measuring technique to that employed by Spiers and his colleagues but they state that the mouse trabecular thickness is much less than that in the dog.

1.2. *Remodelling*

The bones are constantly remodelled throughout life by processes of resorption and apposition. This remodelling is much more important during the period of skeletal growth and development, but internal remodelling (particularly within cortical bone) goes on continuously and must therefore affect the distribution of those radionuclides that as 'volume seekers' are distributed throughout bone (see Chapter 6.3.1). The fact of remodelling means that, in the young individual, removal of bone containing ingested radionuclides may be much more rapid and more extensive than in the old. Radionuclides may be removed by osteoclastic resorption or buried by the apposition of new bone. Figure 9.11 illustrates the removal of plutonium from the metaphysis as bone grows in length. In Fig. 9.4 plutonium originally taken up on the surface of a trabecula is buried by the apposition of new bone on the endosteal surface.

In older cortical bone remodelling is largely internal. Figure 2.4 shows microradiographs of cross-sections of human bone at different ages. There is great variation not only in the degree of calcification in different osteons but also in the whole structure of the cortex. If at any time the blood contains a radionuclide of the alkaline earth series, the alkaline earth will be taken up and retained in those osteons where active accretion of calcium is in progress but not in those which are fully calcified. This process accounts for the so called radium hot spots. Figure 6.4 shows a high power autoradiograph of a cross-section of cortical bone from a radium dial painter, who had ingested radium 36 years previously: one osteon contains numerous radium alpha-tracks, its neighbour, none. The latter was fully calcified at the time the radium level in the plasma was high while the osteon containing alpha-tracks was at that time in the process of laying down calcium. In trabecular bone, remodelling on the surfaces may remain more active than in cortical bone. It has been calculated that the mean surface/volume ratio of bone mineral in the third lumbar vertebrae of adult man is 110 cm^2/cm^3 with cortex and 120 cm^2/cm^3 without cortex. This is considerably less than is found in an adult dog, when the corresponding figures are 146 cm^2/cm^3 and 225 cm^2/cm^3. Measurements made by Sissons (Sissons, Holley, and Heighway 1967) and by Jowsey (Jowsey, Kelly, Riggs, Bianco, Scholz, and Gershon-Cohen 1965) as discussed in Section 2.3.4, Table 2.5, indicate that not more than 6–13 per cent of trabecular bone in the iliac crest is undergoing resorption at any one time from the age of 15. In 1-year old dogs, Lee, Marshall, and Sissons (1965) have reported high remodelling rates in lumbar vertebrae which they express as 230 per cent per year or 0·63 per cent per day. Amprino and Marotti (1964) estimate 0·5 per cent per day or 320 per cent per year in an 18-month old dog. In a 3-year old dog the rates have dropped abruptly to 0·11 per cent per day or 40 per cent per year for the femoral metaphysis and 0·26 per cent per day and 94 per cent per year for the vertebrae. It has been calculated that bone turnover in the vertebrae of a 1·5 year old dog is of an order of magnitude greater than that in the vertebrae of an adult man (Lloyd and Hodge 1971). This difference is extremely significant; it suggests that observations made on plutonium toxicity in dogs of this age group should not be transferred directly to man (Lloyd and Marshall 1972).

There are certain areas where remodelling is very low indeed, for instance in certain parts of the bones of the skull. This may result in long term retention of radionuclides deposited at the time of formation of such bone and in extremely high accumulated radiation dose even though the initial dose rate was very low (Vaughan 1970a).

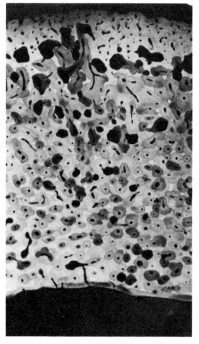

15 year old male

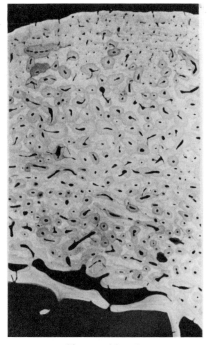

20 year old male

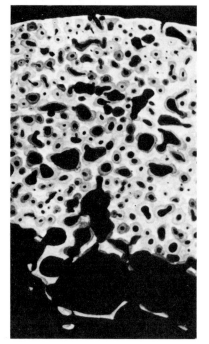

65 year old male

Fig. 2.4. Microradiographs of cross-sections through the mid-shaft of the femur of normal individuals of different ages to show large numbers of Haversian systems, variations in the size of the canals and in mineral density, and also the change with age (Jowsey 1963, by courtesy of author and publisher).

2. Physiology

It is common practice to divide the cellular elements of the skeleton into osteogenic tissue and marrow tissue. This is in many ways convenient but it must always be remembered that this is a geographical rather than a physiological differentiation. Osteogenic tissue lines bone surfaces and marrow tissue fills bone cavities, but both are of mesenchymal origin possessing protean possibilities of cell proliferation and differentiation and therefore of malignant change (Willis 1967).

2.1. *Osteogenic tissue*

All bone surfaces, endosteal, periosteal and Haversian, are covered with a layer of osteogenic tissue which varies in its thickness. It is concerned with the laying down of new bone and the remodelling of old bone. The characteristic cells are first the osteoblasts, which are fully differentiated cells that lay down the matrix; secondly, the osteoclasts which are also fully differentiated cells which resorb bone, and thirdly, the precursors of osteoclasts and osteoblasts which are proliferating cells capable of differentiation under appropriate stimulus. The osteocytes are osteoblasts surrounded by the matrix as it is laid down; each osteoblast as it were buries itself (Vaughan 1970a, Owen 1970). Whether fibroblasts are present in endosteal osteogenic tissue is at present uncertain. They are readily recognized in the outer layer of the periosteum. In 1968 Barnes and Khruschov, investigating sterile inflammatory tissue, concluded that the newly formed fibroblasts were derived from circulating precursor cells originating in bone marrow. More recently Barnes, Evans, and Loutit (1971) working with radiation chimaeras have produced evidence that the fibrosarcoma induced by implants of a pliable disc have arisen from the native host cells and not from the donated myeloid cells. It may well be that there are fibroblast precursors both in marrow and in connective tissue elsewhere in the body, including osteogenic tissue on bone surfaces.

It is extremely important to recognize that osteogenic tissue lining bone surfaces in young bone differs from that in adult bone. It is difficult to prepare good histological sections of adult bone for technical reasons; in thin (5 μm) paraffin sections of decalcified bone, separation between bone and adjacent structure, particularly bone marrow, is very apt to occur in the course of histological processing, as shown in Fig. 2.5. This makes it difficult to determine which tissue components are normally in contact with bone surfaces. Anatomical relationships are better preserved in celloidin sections of decalcified bone but these preparations are relatively thick (15 μm) with the result that it is difficult to see the details of thin layers of cells and to allow for the effects of obliquely sectioned surfaces. On the other hand it is very much easier to cut reasonably thin sections of young

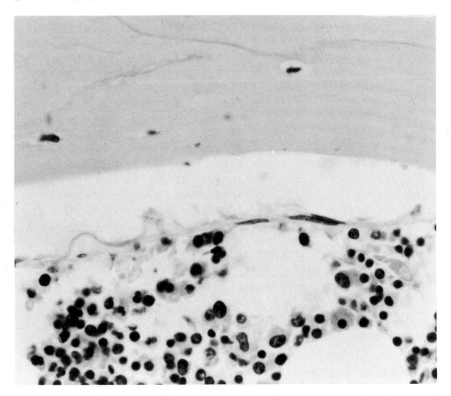

Fig. 2.5. 5μm paraffin section of iliac crest. Separation of bone marrow and investing layer of flattened cells from the bone surface. (Courtesy of H. Sissons 1970.)

bone, so that the character of osteogenic tissue in young bone has tended to be accepted as representative of all osteogenic tissue. This is not, however, true. In young bone processes of apposition and resorption are widespread owing to the activity of growth and remodelling; in older bone the greater part of the bone surface may be quiescent. Figure 2.6 shows a section of an active periosteal surface in young rabbit bone. This surface is covered with upstanding osteoblasts, external to which is a layer several cells thick of active preosteoblasts and finally a thin layer of fibroblasts. In Fig. 2.7 is an active resorbing surface again from young rabbit bone. There are many osteoclasts both on the bone surface and close to it and, as indicated, there are other cells taken to be preosteoclasts (Bingham, Brazell, and Owen 1969). On such surfaces the proliferating cell at carcinogenic risk from radiation will lie not on the mineral matrix but at least one cell diameter away.

In adult bone where about 90 per cent of the surface is engaged in neither apposition nor resorption the picture is different. Figure 2.8 shows a

THE STRUCTURE AND PHYSIOLOGY OF BONE

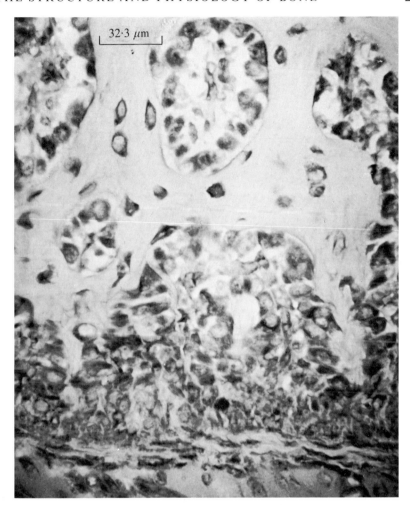

FIG. 2.6. Periosteal surface of the shaft of the femur of a young rabbit (10 days old) showing active osteoblasts lying adjacent to the mineral surface and within newly formed Haversian canals. Methyl green–pyronin. × 520. (Bingham 1968, by courtesy of author.)

section of the endosteal cortical surface of the human femur. Here there are thin flattened cells only on the surface. If the marrow is active it may be difficult to distinguish the osteogenic cells from the crowded marrow cells as shown in Fig. 2.9. The separation of this thin osteogenic tissue from the mineral matrix surface in many sections has already been illustrated in Fig. 2.5. These thin flattened cells have in the past been described as fibroblasts or resting osteoblasts (Pritchard 1956). It is now thought that they are mesenchyme cells, already differentiated to osteoprogenitor cells,

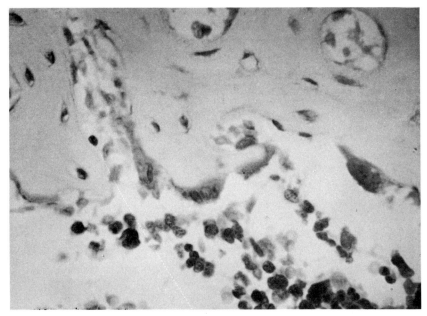

Fig. 2.7. Endosteal surface of the mid-shaft of the femur of young rabbit (10 days old) to show osteoclasts on the bone surface and scattered mononuclear cells (preosteoclasts) adjacent. Methyl green–pyronin. (Bingham et al. 1969, by courtesy of author and publishers.)

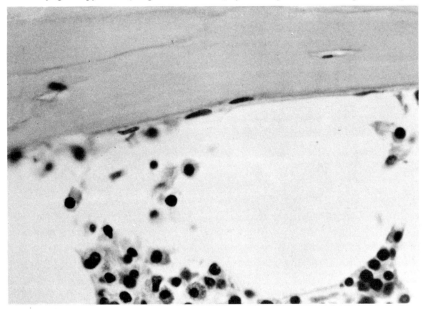

Fig. 2.8. 5 μm paraffin section of iliac crest. A thin layer of cells is present on the bone surface separating the bone from fatty and haemopoietic bone marrow. (Courtesy of H. Sissons 1970.)

THE STRUCTURE AND PHYSIOLOGY OF BONE

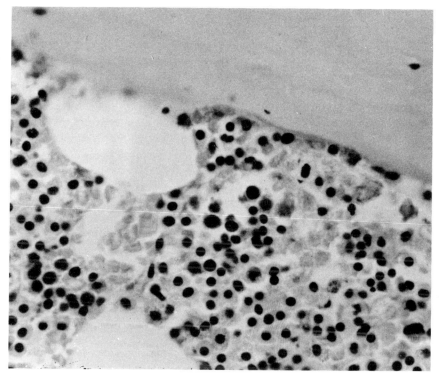

FIG. 2.9. 5 μm paraffin section of iliac crest. In this part of the section, marrow cells appear to be in contact with the bone surface. (Courtesy of H. Sissons 1970.)

capable of differentiating and proliferating as preosteoblasts or preosteoclasts with appropriate stimulation such as parathyroid hormone (Vaughan 1970a, Owen 1970) and are therefore probably at carcinogenic risk. Mathews (1971) emphasizes that though these cells are found on so called 'resting surfaces' they should not be regarded as resting cells. They may not be either laying down or resorbing bone but they are probably concerned with the exchange of both mineral and macromolecular elements of the bone matrix.

2.2. Bone marrow

Bone marrow is an extremely complex tissue containing in post-foetal life a wide variety of different cells. The majority of these cells are fully differentiated, but there are stem cells present and also cells which are in various stages of differentiation. The earliest stem cells cannot be recognized histologically but in some cases can be shown to be present by colony-forming experiments. In such experiments the injection of marrow into, for instance, the spleen of mice will result in the development of colonies of different haemopoietic cell types.

It is perhaps easiest to gain some understanding of the final complexity of the marrow by briefly reviewing its embryological development. Into the primitive osteogenic masses of cartilage and membrane bone, capillaries penetrate carrying with them primitive connective tissue precursors forming a loose network with specialized capillaries and sinusoidal vessels. These capillaries and sinusoids have extremely thin walls composed of endothelial cells and pericytes. The latter may or may not be related to smooth muscle cells and be contractile as first proposed by Rouget (1873) (Zimmerman 1923, Stout 1949, Marcial-Rojas 1960, Ham 1969, Backwinkel and Diddams 1970). Later these early capillaries become associated with venules and fully developed arterioles and arteries. Haemopoietic stem cells are carried into this network by the blood from the foetal liver (which has been similarly colonized earlier from blood islands in the yolk sac (Moore and Metcalf 1970). The stem cells leave the blood and colonize the interstices of the network formed by the invading capillaries and the associated loose connective tissue so forming the active haemopoietic marrow. As demand for circulating blood declines much of this active haemopoietic tissue is replaced by fat cells also of mesenchymal origin, probably differentiating from stem cells associated with the capillaries. Within the loose connective tissues also differentiate the large fixed tissue macrophages or histiocytes. Ham (1969) suggests that other phagocytic cells develop, namely the reticulo-endothelial cells which become arranged to form a somewhat imperfect lining of blood and lymph channels that are wider than capillaries: these cells phagocytise undesirable material that passes by. They are held in place by reticular fibres and serve as endothelial cells.

All these different cell types present in the marrow may be irradiated and those that have precursors capable of proliferation may undergo malignant transformation. Such are endothelial cells, pericytes, haemopoietic stem cells of the erythroid, granulocytic, and lymphoid series, reticulo-endothelial cells, and fat cells. There is also some evidence that marrow may contain two types of bone forming cells or osteoprogenitor cells (Friedenstein, Piatelzky-Shapiro, and Petrakova 1966; Friedenstein, Petrakova, Kuralesova, and Frolova 1968; Friedenstein 1968a,b; Owen 1970). One of these is called by Friedenstein a determined osteogenic precursor cell (DOPC) which will form bone without an inducing agent (transitional epithelium) being present; the other, which will only form and maintain bone when an inducing agent is present, he calls an inducible osteogenic precursor cell (IOPC). The question at once arises as to whether the determined osteogenic precursor cells in Friedenstein's preparation was not a piece of osteogenic tissue removed with the marrow rather than a true marrow cell. Such osteogenic cells remain attached to marrow rather than

THE STRUCTURE AND PHYSIOLOGY OF BONE 29

to the mineral matrix surface when marrow is removed (Bleaney and Vaughan 1971) and Friedenstein himself describes the DOPC cells as probably connected with endosteal bone (Friedenstein 1968a,b). He further states that 'the population of bone marrow cells is very heterogeneous including haematopoietic cells, reticular cells, and endosteum elements' (Friedenstein, Piatelzky-Shapiro, and Petrakova 1966). Whether this osteoprogenitor cell and the different haemopoietic progenitors are ultimately derived from a single progenitor cell present also in marrow has been disputed since the time of Maximov. Recent experimental results with radiation chimaeras obtained both by Friedenstein himself (Friedenstein and Kuralesova 1970) and by Barnes and his colleagues (Barnes, Carr, Evans, and Loutit 1970) suggest that in non-foetal marrow, at least in the mouse, there is only a defined osteogenic precursor cell or stem cell. There is no evidence for a precursor cell capable of giving rise both to osteogenic and haemopoietic cells (Vaughan 1970a). The nature of Friedenstein's inducible osteogenic precursor cell remains obscure.

2.3. Bone surfaces

The term 'bone surface' is in constant use particularly in relation to radiation injury to bone. It is therefore essential to define the constituents of a bone surface. It may be said to have at least three distinct elements, the osteogenic tissue, the matrix, and the mineral.

2.3.1. *Osteogenic tissue.* The character of the osteogenic tissue has already been discussed (Section 2.1). It may here be added that osteogenic tissue has recently been shown capable of metal binding (Bleaney and Vaughan 1971). Forty to fifty per cent of the plutonium on 'bone surfaces' may be bound by the osteogenic cells. This binding may be in the lysosomes, or mitochondria, or on the cell membrane which is rich in glycoprotein (Section 2.3.2.2). It remains to be determined whether other ions are also bound by osteogenic cells. Many are known to be concentrated on 'bone surfaces', as will be discussed.

2.3.2. *Bone matrix.* The organic matrix of bone is composed of collagen fibres associated with lipids (Wuthier 1968), carbohydrate protein complexes (Herring 1970, 1972), and peptides (Leaver and Shuttleworth 1968). In adult beef bone it is now recognized that only 88–9 per cent of the whole dry weight of the organic matter is collagen. A recent analysis of the composition of bovine cortical bone is shown in Table 2.2.

2.3.2.1. *Collagen.* The part played by collagen in nucleation of calcium apatite is still confused, though it is suggested that it may act as a nucleating agent binding calcium and phosphorus (Glimcher and Krane 1964, Bachra 1967). The mechanism of calcification is not for discussion here. Collagen may also bind plutonium though how it does this and to what extent is also

TABLE 2.2

The composition of bovine cortical bone
(from Herring 1972, by courtesy of author and publishers)

	Per cent by weight of whole dry bone	
	a	b
Inorganic matter	77·23	76·04
Organic matter	22·77	23·96

	Per cent by weight of organic matrix	
	a	b
Collagen	89·15	88·48
'Resistant protein'	4·87	0·98
'Osseomucoid'	1·14	
Chondroitin sulphate		0·81
Sialoprotein (BSP)		0·80–1·15
CP-S glycoprotein		0·31–0·44
Lipids (c)	0·42	
Peptides (d)	0·54	
Other glycoproteins, proteins, and matter not accounted for	3·88	7·19–7·67

(a) From the results of Eastoe and Eastoe (1954).
(b) From Oldroyd and Herring (1967) and unpublished results.
(c) Leach (1958).
(d) Leaver and Shuttleworth (1968).

uncertain (Borasky 1957, Taylor and Chipperfield 1970, Taylor 1970).

2.3.2.2. *Carbohydrate protein complexes.* The carbohydrate protein of the non-collagenous element in bone is extremely important from the point of view of metal binding and therefore probably affects the uptake and retention of many important radionuclides in the skeleton. It was noted in 1962 (Herring, Vaughan, and Williamson 1962) that sites in bone that concentrate yttrium, plutonium, and americium showed histological evidence of mucosubstances. Herring has since isolated from cortical bone at least five different glycoproteins which have been shown *in vitro* to have interesting and different capacities to bind a variety of metals (Peacocke and Williams 1966; Williams and Peacocke 1967; Chipperfield and Taylor 1968; Herring, Andrews, and Chipperfield 1971; Chipperfield and Taylor 1970). Table 2.3 shows the relevant binding capacity of these glycoproteins for plutonium, americium, and curium. The binding of plutonium to these glycoproteins in stronger than its binding to transferrin which carries it in the blood (Taylor and Chipperfield 1971). It has not yet been possible to demonstrate the presence of any of these glycoproteins on endosteal and

TABLE 2.3

Binding of plutonium, americium, and curium to proteins in vitro at pH 7·4
(Binding is expressed as the percentage of protein-bound metal recovered from Sephadex gel columns)
(Chipperfield and Taylor, 1971, by courtesy of authors)

Protein	Per cent of protein-bound metal		
	Pu(IV)	Am(III)	Cm(III)
Bone sialoprotein	54·7 ± 9·3	10·4 ± 2·9	12·2 ± 1·1
Chondroitin sulphate-protein	49·2 ± 3·0	14·9 ± 6·6	10·0 ± 1·6
CPC-soluble glycoprotein	36·6 ± 5·1	4·6 ± 2·9	8·7 ± 1·2
Glycoprotein I	30·0 ± 12·4	2·8 ± 0·9	37·5 ± 12·6
Glycoprotein II	50·3 ± 9·1	5·1 ± 2·0	8·2 ± 1·3
Soluble collagen	23·3 ± 4·3	0·6 ± 0·2	0·9 ± 0·3
Human transferrin	18·9 ± 6·9	0·2 ± 0·0	—
Bovine γ-globulins	13·8 ± 5·1	0·7	—
Polyglutamic acid	68·7 ± 0·5	8·0 ± 3·0	27·2 ± 5·5
Chondroitin sulphate	13·2 ± 5·7	0·3 ± 0·4	0·3 ± 0·4

periosteal bone surfaces. However, Williamson and Vaughan (1967) have obtained histochemical evidence of the presence of an uncharacterized sialomucin in the region of the hypertrophic cartilage cells where plutonium and americium are bound in high concentration.

2.3.2.3. *Lipids and peptides.* Little is known of the part played by lipids and peptides in the bone matrix though it has been suggested that the lipids may play some part in initiating calcification (Irving and Wuthier 1968) and may therefore be capable of metal binding. The strong anionic and cationic character of the bone peptides suggests that they may also be concerned in metal binding (Herring 1972).

2.3.3. *Calcium apatite.* The exact chemical character of bone mineral is still uncertain. The basic structure is in the form of hydroxy-apatite of which the prototype is $Ca_{10}(PO_4)_6(OH_2)$. Neuman and Neuman in 1958 published a table indicating that a wide group of ions might interact in one way or another with bone hydroxy-apatite. Radium is included in this table as being bound both on the crystal surface and within the crystal interior. Experiments with bone powder have led other workers to suggest that elements such as plutonium may also interact with crystals of calcium apatite. However such experiments should be treated with reserve. It must be remembered that Harrison and his colleagues (Harrison, Howells, and Pollard 1967) found that rat bone powder showed a higher retention of radium and barium than of strontium and calcium, the exact reverse of what happens *in vivo*. It may well be that bone powder contains material other than calcium apatite which serves to bind minerals. It is known, however, that the alkaline earths, other than calcium, if they gain access to the blood stream become distributed throughout bone probably associated with bone crystals and take part in the slow processes of long-term exchange (Marshall 1964, 1969b). These processes of long term exchange are therefore significant in the overall problem of radiation hazards from some internally deposited radionuclides. The importance of exchange is discussed further particularly in relation to radium and strontium toxicity.

2.3.4. *Area of 'bone surfaces'*. The area of endosteal surface is extensive compared with the area of periosteal surface. In man the trabecular surface area in a vertebral body of approximately 40 cm^3 is of the order of 1000 cm^2, in contrast with the periosteal area which is approximately 80 cm^2 (Dunnill, Anderson, and Whitehead 1967). The trabecular surface of approximately 40 cm^3 of iliac bone is about 1600 cm^2 (Sissons, Holley, and Heighway 1967). The periosteal area is not known precisely. The figure for cancellous bone in the head of the femur is of the same order (Dunnill 1970). These figures indicate that the area of endosteal surface is great compared with that of the periosteal surface of the skeleton. This becomes important in

TABLE 2.4

*Summary of published data on normal human lumbar vertebrae
(from Lloyd and Hodges 1971, by courtesy of authors and publishers)*

	% Bone volume*	Sample thickness	Perimeter area cm/cm^2	Surface area/ volume bone mineral cm^2/cm^3	Surface area/ volume bone tissue cm^2/cm^3
Dyson et al.	13·5 (1)	Surface only	165	207	28
Dunhill et al.	25 (30)	?	—	—	—
Amstutz and Sissons[10]	25 (average 4 areas in 1 vert.)	27 μm	—	147	33
Atkinson	69 (calc.) (28)	2·5 mm	—	—	—
Bromley et al.	15 (7)	8 μm	100	127	19
Lloyd (present study)	27 (1)	100 μm	94	120	32

* Number of cases in parentheses.

TABLE 2.5

Percentage of total surface-mean values for iliac crest (trabecular)
(taken from Sissons et al. 1967)

Age in years	20–9	30–9	40–9	50–9	60–9	70–9	80+
Formation surface	11	14	9	8	12	14	12
Resorption surface	8	12	9	12	13	10	13
Resting surface	81	74	82	80	75	70	75

(taken from Jowsey et al. 1965)

Age in years	15–20	20–9	30–9	40–9	50–9	60–9	70–9	80+
Formation surface	10	4	5	2	3·5	3·5	3·5	3·5
Resorption surface	9	4	6	6	6·0	6·0	9·0	9·0
Resting surface	81	92	89	92	90·5	90·5	87·5	87·5

radiation problems when it is apparent that some radionuclides are selectively concentrated on different surfaces. Extensive measurements of surface to volume ratios have recently been made by Lloyd and Hodges (1971). Their summary of published data on normal human lumbar vertebrae is shown in Table 2.4.

2.3.5. *Character of bone surfaces.* The figures obtained by a microradiographic technique (Jowsey, Kelly, Riggs, Bianco, Scholz, and Gershon-Cohen 1965) and by a histological technique (Sissons, Holley, and Heighway 1967) for the proportions of resting, resorbing, and growing bone surfaces in adult bone are shown in Table 2.5. They suggest that about 90 per cent of adult bone surface may be neither resorbing nor laying down new bone (see Table 2.5). They probably somewhat exaggerate the extent of growing surface since a layer of osteoid may have been laid down some time before the measurement was made and remained uncalcified; it will therefore still be estimated as a growing surface. The figures are important, again from the point of view of radiation hazards, since they indicate that in adult bone there is only a thin layer of osteogenic cells on the greater part of the surface. The reverse of course is true in children. However it must be pointed out that precise measurements of the amount of osteoid on bone surfaces may be greater than that indicated by the techniques used by Sissons, Holley, and Heighway (1967) and by Jowsey, Kelly, Riggs, Bianco, Scholz, and Gershon-Cohen (1965). This may be significant in the case of plutonium toxicity since this element is probably in part bound by the glycoproteins of the osteoid. Figure 2.10 shows the extent of the trabecular bone surface in the normal ilium that may be covered by osteoid rather than being fully

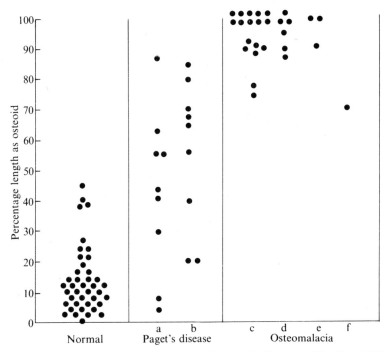

FIG. 2.10. Percentage of trabecular length covered by osteoid (Paterson, Woods, and Morgan 1968, by courtesy of author and publishers).

mineralized (Paterson, Woods, and Morgan 1968). In some healthy individuals over 40 per cent of the surface may have such an osteoid covering beneath the layer of thin osteoprogenitor cells. Chalmers and his colleagues (Chalmers, Barclay, Davison, Macleod, and Williams 1969) estimate, from biopsy studies, that 6 per cent of osteoid relative to total bone matrix is the upper limit of normal.

2.4. *Movement of ions into and out of the skeleton*

The complicated problems of calcium kinetics are not for discussion here (Rowland 1966; Marshall 1964, 1969b; Vaughan 1970a).

2.4.1. *Accretion.* When new bone is laid down there is an increase of calcium. Calcium ions form apatite crystals on the fresh osteoid tissue by processes which are not understood (Vaughan 1970a). If other alkaline earths are present in the blood stream they will be included in the new crystal formation though not to the same degree as calcium.

2.4.2. *Exchange.* Calcium ions are constantly moving into and out of the skeleton, with no loss or gain, by processes of exchange (Rowland 1966).

Exchange is a term much used in discussing the behaviour of mineral ions

THE STRUCTURE AND PHYSIOLOGY OF BONE

in relation to bone. Marshall and his colleagues (Marshall, Rowland, and Jowsey 1959; Rowland 1960a, 1966) have defined exchange 'as any process involving equal and opposite rates of transfer of calcium ions to and from a single microscopic volume of bone' (a microscopic volume of bone being a cube with sides of the order of 10 μm). 'Exchange' so defined may take two forms—*long-term exchange* with time constants longer than a week and *short-term exchange* with time constants less than a week.

The degree of movement of ions in exchange processes varies considerably from one part of the skeleton to another (Vaughan 1970a). This variation in 'turnover rate' has importance in the case of both strontium and radium toxicity, as will be discussed in Chapter 7.2.1 and 8.1.2 (Vaughan and Williamson 1967). It is essential however to remember that calcium plays an essential role in many physiological processes. The body does not handle the other alkaline earths, strontium, radium, and barium in exactly the same way since they have no known physiological role. However, they also move in and out of the mineral element of bone, exchanging, it is thought, with calcium atoms, but the rates of movement are different and their patterns of excretion are different both from calcium and from one another. Radium and barium are excreted much more rapidly than calcium

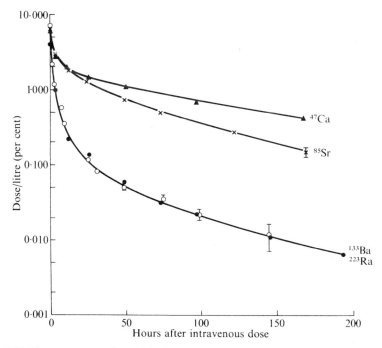

FIG. 2.11. Plasma concentrations of the four alkaline earths at various times after their respective intravenous dose (Harrison *et al.* 1966, by courtesy of authors and publishers).

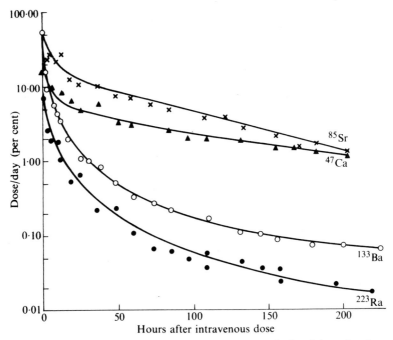

Fig. 2.12. Rate of urinary excretion of the alkaline earths over the first 8 days after the respective dose (Harrison *et al.* 1966, by courtesy of authors and publishers).

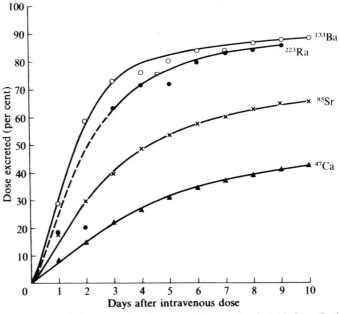

Fig. 2.13. Percentage of the respective dose excreted over the first 10 days. Ratio of total faecal to total urinary excretion after 8 days:

^{47}Ca	^{85}Sr	^{133}Ba	^{273}Ra
0·48	0·25	9·0	36

(from Harrison *et al.* 1966, by courtesy of authors and publishers.)

and strontium and their skeletal retention pattern is different (Harrison, Carr, Sutton, and Rundo 1966; Harrison, Carr, and Sutton 1967). Figure 2.11 shows the plasma concentration of the four alkaline earths after their respective intravenous injection into an adult man and Fig. 2.12 shows the rate of urinary excretion in the first 8 days. In Fig. 2.13 is the total faecal and urinary excretion (Harrison, Carr, Sutton, and Rundo 1966).

3. Skeletal Tissues at Risk from Irradiation

CRIPPLING degenerative changes in the skeleton have not been seen when the radiation dose is less than the carcinogenic dose, so it is now customary to regard malignant change as the most sensitive index of radiation damage to the skeleton (International Commission on Radiological Protection 1968). As already explained, both clinical experience and experimental observations suggest that the risk of malignant transformation is dependent on the proliferative potential of the cells irradiated. Therefore it is important to determine where in the skeleton cells with a proliferative capacity are present.

The account in Chapter 1 of the physiology of the skeleton will indicate where these proliferating cells are to be found, namely in the osteogenic tissue and in the marrow, but epidemiological studies and experiment would add epithelium closely applied to bone. In Tables 3.1 and 3.2 are shown the results of investigation of five groups of people exposed to both internal and external irradiation. In those exposed to external irradiation leukaemia is common while in those exposed to internal irradiation from radium osteogenic sarcomas and carcinomas of the sinus of the skull far exceed in number what might be expected to occur in a normal population. Carcinoma of the sinuses is a rare tumour. In the radium patients the ratio of observed to expected cases is 100:1. These epidemiological findings are confirmed (as will be discussed in the following chapters) by experimental observations on animals and in many detailed follow ups of individual human cases. It must be emphasized however that except in the case of the radium patients who have now in some cases been followed for 50 years the follow up period is relatively short. It is known that the latent period for osteosarcomas is very much longer than that for leukaemia (see Chapter 5.3.1.1) and there is still time for such osteosarcomas to occur both in the atom bomb casualties and the spondylitics. Such a table also only shows up the most common skeletal response to a carcinogen. Detailed analysis indicates that tumours other than those of haemopoietic origin may arise from marrow irradiation. These have in many cases in the past been classified as osteogenic, but probably arise from marrow elements (Chapter 4.3.2). It must also be made clear that total body irradiation will induce tumours in tissues other than the skeleton. These are not tabulated here

SKELETAL TISSUES AT RISK FROM IRRADIATION

TABLE 3.1

Five groups of people whose exposure to various types of radiation has been the subject of investigation

	Type radiation	Period exposure	Source	Approx. numbers	Follow up years
(1) Bomb survivors	Accidental	Minutes	External	100 000	15*
(2) American radiologists	Occupational	Years	External	3500	4–40
(3) British radiologists	Occupational	Years	External	1300	5–60
(4) Ankylosing spondylitics	Therapeutic	Months	External	14000	9–25
(5) Dial painters / Radium chemists / Therapeutic ^{226}Ra	Occupational therapeutic	Years	Internal	700	5–50

(1) I.C.R.P. Publication 8 (1966).
(2) Seltser and Sartwell (1965).
(3) Court Brown and Doll (1958).
(4) Court Brown and Doll (1957, 1965).
(5) Evans (1966); Hasterlik, Finkel, and Miller (1964).
* Table for years 1950–60—actual follow up 15 years.

TABLE 3.2

Ratio of observed to expected incidence of various diseases in the five groups of people described in Table 3.2

Disease	Bomb survivors	American radiologists	British radiologists	Ankylosing spondylitics	Dial painters, etc.
Leukaemia	10:1	2:1	2:1	9:1	1:1
Aplastic anaemia	2:1	n.r.	n.r.	30:1	1:1
Osteosarcoma	n.r.	n.r.	n.r.	2:1(b)	50:1
Carcinoma sinuses(c)	n.r.	n.r.	n.r.	n.r.	100:1

n.r. = nil recorded

(b) This figure possibly exaggerates the incidence of osteosarcoma. The reliability of the diagnosis of primary bone tumour on a death certificate is not high. Only three deaths have been confirmed as due to a bone sarcoma. The expected number of deaths estimated from data given by McKenzie *et al.* (1961) is 0·63. Difference between observed and expected is statistically significant ($P \sim 0.025$) on a one-tailed test.
(c) The Registrar-General's figures for the total numbers of deaths ascribed to carcinoma in the region of the skull for England and Wales in 1964 are shown below. The death rate for all these neoplasms together is 9 per million per annum. This expected rate has been used for comparison with the observed rates in American dial-painters, allowance being made, however, for a much longer period.

	Males	Females
Nose (internal) and nasal cavities	29	15
Eustachian tube and middle ear	10	18
Maxillary sinus	75	69
Other specified sinus (accessory)	13	6
Nasopharynx	94	49

(Seltser and Sartwell 1965; Court Brown and Doll 1965; Jablon, Tachikawa, Belsky, and Steer 1971; Stewart and Barber 1971).

1. Osteogenic tissue

The character of osteogenic tissue has already been discussed. Osteoblasts, osteoclasts, and osteocytes are fully differentiated cells which do not proliferate. There is no supporting experimental evidence for the view originally expressed by Young (1962) that these cells might revert back to the osteoprogenitor pool. The proliferating cells are the preosteoblast and the preosteoclast. In young bone these lie at least one cell away from the mineral/matrix surface but in adult bone they lie adjacent to the bone surface. Independent measurements made by Sissons (1970) and Vaughan (1970c) suggest that the proliferating cells lie almost entirely within 10 μm of the surface. Recently the International Commission on Radiological Protection has agreed that an appropriate convention for dose limits to endosteal cells would be that based on a dose rate of 15 rem per year as averaged over tissues with a range of 0 μm to 10 μm from bone surfaces. Clearly these cells are at risk from all forms of irradiation—fast neutrons, X-rays and gamma rays, short-range and long-range beta particles, and alpha particles emitted from radionuclides concentrated in the skeleton. Since it has been shown already (Chapter 2.2.3.4) that the area of endosteal surfaces is much greater (even excluding the surfaces of Haversian systems) than the area of periosteal surface it is not surprising that when careful

TABLE 3.3

Site of tumour origin in animals injected with ^{90}Sr or ^{45}Ca
(from Kshirsagar et al. 1965)

Author	Animal	Age at injection	Isotope	Endosteal*	Periosteal
Owen et al. (1957)	Rabbit	6–8 weeks	^{90}Sr	+++	0
Downie et al. (1959)	Rabbit	6–8 weeks	^{90}Sr	+++	0
Macpherson et al. (1962)	Rabbit	6–8 weeks	^{90}Sr	+++	0
Litvinov (1957)	Rat	3 months	^{90}Sr	+++	0
Kuzma and Zander (1957a,b)	Rat	?	^{89,90}Sr	+++	+
	Mouse	?	^{45}Ca	+++	+
Skoryna and Kahn (1959)	Rat	42 days	^{90}Sr	+++	+
Casarett et al. (1962)	Rat	40–117 days	^{90}Sr	+++	0
Nilsson (1962)	Mouse	75–85 days	^{90}Sr	+++	0
van Putten and de Vries (1962)	Mouse	70–105 days	^{90}Sr	+++	0

* Some few of these tumours probably arose from mesenchymal elements in the marrow rather than from endosteal osteogenic tissue.

histological studies using both internal irradiation and external irradiation have been made osteogenic sarcoma are found to arise from the endosteal rather than the periosteal surface. The results of such observations following internal irradiation from a single injection to small animals are shown in Table 3.3. It is probable, as discussed later (Chapter 4.3.2) that some of these tumours arise from mesenchymal marrow elements, though the majority are osteogenic in origin. The experiments using external irradiation, by Baserga and his colleagues, are discussed in detail elsewhere (Chapter 5.4.2.2) (Cater, Baserga, and Lisco 1959, 1960; Baserga, Lisco, and Cater 1961).

The one exception to the finding of an endosteal origin for the majority of osteogenic tumours arises in the case of the miniature swine fed ^{90}Sr continuously. It is stated clearly by Clarke and his colleagues (Howard, Clarke, Karagianes, and Palmer 1969) that these tumours arise from the periosteum. The authors suggest that the osteogenic tissue has received such a high radiation dose in these animals fed ^{90}Sr that the cells have been killed on the endosteal surface.

2. Marrow

It has already been said in discussing the physiology of the marrow that even in adult life it contains many precursor or haemopoietic stem cells together with a variety of other cell types of mesenchymal origin not involved in haemopoiesis. It has long been considered that the classical response to marrow irradiation is myeloid leukaemia. The high incidence of myeloid leukaemia may indeed be due, first, to the fact that cells of the granulocyte series predominate in the marrow, and therefore there are more cells at risk. Secondly, since the granulocytes are a population that is normally turning over rapidly, compared for instance to the lymphocytes, it is likely that there are more proliferating rather than fully differentiated cells in the population. However, there is increasing experimental evidence, which will be discussed, that varied forms of a myeloproliferative disorder may result from such irradiation (Chapter 4.3.1); further, that tumours may arise not only from haemopoietic marrow elements but from all the other mesenchyme elements present (Loutit and Vaughan 1971) (Chapter 4.3.2). These tumours are however uncommon because the number of involved stem cells at risk is relatively few.

3. Epithelium closely applied to bone

In the sinuses of the skull epithelium is closely applied to bone without, in the adult, intervening and obvious osteogenic tissue. At least in the rabbit the basement cells of the epithelium lining the external ear within the skull, where carcinoma following ^{90}Sr administration develops, (Chapter 8.1.3.1.3) has been shown by studies of tritiated thymidine uptake to contain actively proliferating cells and presumably this is the case in other epithelia (Kshirsagar, Vaughan, and Williamson 1965). It is not surprising therefore that malignant change may occur in those epithelial cells which are within range of both alpha and beta radiation from radionuclides in the underlying bone. Epithelial cells lining the sinuses are also exposed to radiation from decay products that may be retained in the air of the sinuses after escaping from the bone. Evans and his colleagues (1969*b*) have calculated in one human case of carcinoma of the frontal sinus that the radiation dose at a distance of 10 μm from the bone surface was 82 000 rad while the radon content of the air measured in the opposite sinus was at least 30 times that in the exhaled breath.

4. Histopathology of Tumours induced by Irradiation of the Skeleton

It has already been said that the tissues at risk from skeletal irradiation are the proliferating cells in
  (a) osteogenic connective tissue on bone surfaces,
  (b) marrow which contains all the haemopoietic cell types and their precursors, also blood vessels, connective tissue framework, and fat cells,
  (c) epithelium closely applied to bone.

Individual malignancies produced in these tissues by radiation do not differ from those which occur without radiation as a precipitating factor, but it may prove that certain malignancies which occur very rarely in a normal population are more common under the influence of radiation, and that their incidence may vary somewhat according to the character of the radiations involved. The nomenclature both of primary malignant tumours of bone and of haemoproliferative disorders has however always been confused and often controversial, and little attempt has been made, in the case particularly of the bone tumours, to pinpoint their site of origin.

Dosimetric considerations, particularly in the case of the radionuclides deposited in bone, should enable greater precision to be reached in determining the origin of radiation-induced malignancies and therefore in due course perhaps a more rational classification. A bone seeking radionuclide with an energy emission of short range like ^{45}Ca will mainly irradiate osteogenic cells on the bone surface while a radionuclide with an energy emission of long range like ^{90}Sr with its high energy daughter ^{90}Y will irradiate cells both on the bone surface and in the marrow. In the case of ^{45}Ca the radiation dose near the endosteal surface (D_s) of Spiers (1966) is important while in the case of ^{90}Sr both (D_s) and $\bar{D}_M$, the mean marrow dose, are likely to be significant.

Although the problem of nomenclature may be considered of academic importance only, the problem of the origin of different malignancies is of practical importance in relation to radiation protection. In the calculation of maximum permissible levels for any radionuclide it is important to know the tissue at risk with precision. An attempt is therefore made here to discuss tumours induced by radiation in relation to what is known of their cell of origin.

The most complete account of the pathology of tumours associated with bone is probably still that of Jaffe (Jaffe 1958). It is likely that the rarer tumours he describes, for instance the haemangioendothelioma, the angiosarcoma, and the primary reticulum cell sarcoma, do not necessarily arise from osteogenic tissue (i.e. tissue containing stem cells capable of differentiating into osteoblasts and osteoclasts) but rather from other mesenchymal cells in the marrow. As already discussed (Chapter 2.2.1), it was found that in radiation chimerae in mice there was no evidence that marrow contained a pluripotent stem cell capable of producing osteogenic sarcoma. All such tumours arise from the host tissue, presumed to be the osteogenic tissue on the bone surface (Barnes, Carr, Evans, and Loutit 1970).

The following account of the histopathology of tumours induced by radiation is not an account of bone tumours as such. There is for instance little to suggest that a malignancy with the characteristics of Ewing's tumour is induced by radiation. It is rather an attempt to pinpoint the origin of tumours known to result from skeletal radiation.

1. Tumours of osteogenic tissue

Two malignant tumours undoubtedly arise from the osteogenic tissue, as this tissue is defined in Chapter 2.2.3.1, namely the osteosarcoma and the chondrosarcoma. The origin of the fibrosarcoma and the giant cell tumour, as discussed below, is less certain. The benign cartilaginous exostosis also originates from osteogenic cells.

1.1. *Malignant tumours*

1.1.1. *Osteosarcoma.* The osteosarcoma is defined by Sissons (1966) as 'a malignant bone tumour having a predominantly osteoblastic pattern of differentiation as indicated by the formation of osteoid or bony intercellular material . . . the amount of ossified tumour tissue can vary greatly'. This variation applies not only to different tumours but to different areas of the same tumour. Tumour giant cells are often conspicuous. A tumour containing both cartilage and osteoid tissue is described as an osteosarcoma.

Jaffe warns that an osteosarcoma may be extremely pleomorphic in character and that it is essential to examine several parts of any tumour before reaching a diagnosis,

> . . . most osteogenic sarcomas are rather vascular. They are more so in some portions than in others, the still cellular stromal tissue being interspersed with smaller or larger sinuses and thin-walled blood channels bordered by tumour cells. In some parts in particular one may note engorged vascular channels lying close together. Viable tumour cells usually border them and some tumour cells may also be found admixed

with the blood in the spaces. Such telangiectatic areas may be near, or independent of, areas in which the tumour tissue is undergoing anaemic or even haemorrhagic necrosis.

1.1.2. *Chondrosarcoma.* Tumours containing abnormal cartilage elements only, chondrosarcomas, are much less common than osteosarcomas containing bone and osteoid tissue only or tumours containing both bone and cartilage. Tumours of all grades of malignancy occur and it is not always easy to distinguish between benign and malignant cartilaginous tumours. Large cells and large nuclei and the presence of a significant number of binucleate or multinucleate cells, particularly when these cells are large, are all features suggesting malignancy. It may be necessary to examine many sections before an unequivocal diagnosis can be made.

1.2. *Benign tumours*

1.2.1. *Cartilaginous exostosis.* The characteristic benign tumour induced in bone by radiation is the solitary cartilage-capped exostosis. This takes the form of a sessile and pedunculated bony mass arising usually from the metaphyseal end of a long bone but it may occur in the vertebrae. The exostosis appears to form as a result of cartilaginous metaplasia in the periosteum in the neighbourhood of the epiphyseal plate. As growth proceeds, the cartilage becomes replaced by cancellous bone as a result of endochondrial ossification. Such exostoses, whether occurring spontaneously or as a result of radiation treatment, become malignant only extremely rarely (Schajowicz and Gallano 1970). They are recorded in children who have received external irradiation for a wide variety of conditions (Hempelmann, Pifer, Burke, Terry, and Ames 1967) and in children treated with ^{224}Ra for tuberculosis of the knee joint (Spiess 1969).

2. Tumours of doubtful origin

2.1. *Fibrosarcoma*

It is not at present possible to determine the origin of the fibroblast as found in the skeleton. This is discussed in Chapter 2.2.1. It probably has a precursor both in endosteal osteogenic tissue and in marrow mesenchyme tissues.

A fibrosarcoma is a malignant tumour showing proliferation of fibroblasts and the intercellular formation of collagen. The presence of either bony or cartilaginous differentiation excludes a tumour from this group. The evidence as to whether fibrosarcomas arise from osteogenic tissue or from marrow reticulum is not at present clear. Nilsson (1970*a*), who has made an extensive study of the histopathology of tumours in mice induced by ^{90}Sr, regards fibroblastic sarcomas as arising from reticulum of the marrow on histological evidence. On the other hand the only fibrosarcoma reported by Barnes, Evans, and Loutit (1971) as occurring in mouse chimaerae was of host origin, i.e. like the osteoblastic sarcomas arising from osteogenic tissue. There is however evidence that a fibroblast precursor is present in the marrow (Barnes and Khruschov 1968) so it seems possible that a fibrosarcoma could arise from both marrow elements and osteogenic tissue.

2.2. *Giant cell tumour or osteoclastoma*

There is no agreement among pathologists or surgeons as to the origin and nature of the giant cell tumour of bone. It is listed by Sissons (1966) as a malignant tumour of bone. He defines it as a bone tumour showing conspicuous osteoclastic differentiation within a stroma-like tissue of ovoid or spindle shaped cells that do not show bony or cartilaginous differentiation. Histologically the tumour consists of two characteristic components, the giant cells and the intervening spindle-celled tissue. The giant cells are morphologically similar to osteoclasts. It is of course tempting to suggest that the tumour arises from osteoclastic precursor cells but there is no decisive evidence that this is the case. Some surgeons treat such tumours occurring in bone with radiation without consideration as to whether they are or are not malignant. Others consider that such treatment converts a benign into a malignant tumour. There are also surgeons who prefer not to irradiate a tumour which they regard as malignant at the time of first diagnosis (Chapter 5.3.2.4.1).

This type of tumour has been reported (Howard, Clarke, Karagianes, and Palmer 1969; Clarke, Busch, Hackett, Howard, Frazier, McClanahan, Ragan, and Vogt 1972) as occurring in miniature swine following continuous feeding of ^{90}Sr. It has not been reported following exposure to other radionuclides that do not, because of their short range in tissues, irradiate marrow.

3. Tumours of marrow

The marrow has already been described as an extremely complex mesenchyme tissue. It contains all the haemopoietic stem cells (Loutit 1967, Barnes and Loutit 1967) and possibly fibroblast precursors (Barnes and Khruschov 1967), also reticulum cells, fat cells and blood vessels lined with endothelial cells and associated externally with pericytes. Radiation of such a tissue might on *a priori* grounds alone be recognized as likely to give rise to a variety of malignancies.

3.1. Haemoproliferative lesions

Since bone marrow contains all the haemopoietic stem cells it is not surprising that its irradiation may result in malignant change in all the different cell lines, and that at times the resulting histological picture is extremely confusing. It is not possible at present to give a logical account of what is found, and why one type of haemopoietic neoplasm rather than another occurs in any particular individual.

The term leukaemia is often used to cover the many cytologically and clinically different conditions associated with proliferation of haemopoietic cells. There is considerable disagreement as to what constitutes a malignant and progressively fatal disease and what constitutes a non-malignant response to some agent such as severe infection.

Mole (1963) has given the following phenomenological definition of leukaemia:

Leukaemia, like many other indispensable terms in current use in medicine, and indeed in biology generally, is merely a descriptive word for a group of phenomena whose nature is not properly understood but which can readily be recognized by a trained observer. The gross symptoms of leukaemia may be highly variable and the underlying common feature is a progressive and ultimately lethal proliferation of cells which are morphologically not too unlike one or other (or more than one) of the different kinds of cells which make up the normal haematopoietic, lymphopoietic and reticular tissues of the body. Individual cases of leukaemia are usually named according to the predominant cell-type and it is a basic assumption that the particular cell (or cells) from which the leukaemia started is (or are) related to the predominant cell in the same way as stem-cells are related to their differentiated progeny, although in leukaemia, both generally and in specific cases, it is always uncertain just how far back in the cell lineage one has to go to find the cell (or cells) in which the leukaemia originated. The varieties of leukaemia are given specific names but, as in taxonomy in general, the boundaries between named species and even the grouping of particular sets of species into genera or families is often somewhat arbitrary. If this is forgotten it becomes all too easy to talk about leukaemia as if it were one 'thing' and thus to assume that there must necessarily be a common basic mechanism for every case of leukaemia. In this and many other ways the problems of leukaemia are just the same as the problems of cancer and it is generally, though not quite universally, accepted that leukaemia is in fact cancer of the haematopoietic tissues.

It is indeed possible to imagine that all leukaemias originate in totipotent cells which normally are capable of giving rise to fully differentiated daughter-cells as different as small lymphocytes, granulocytic leucocytes, erythrocytes, and macrophages. The predominant cell in different cases of leukaemia would then depend on the particular route of biological development of the daughter-cells of the 'leukaemic' stem-cell and how far along that route their mock-differentiation has proceeded. There might then be a single basis for all leukaemias, the particular morphological features of the predominant cells of a particular case being an 'accidental' consequence of some factor determining cell differentiation. On the other hand, most workers nowadays would believe that there are several really different kinds of leukaemia. The epidemiological evidence certainly suggests that the causes of acute leukaemia, of chronic myeloid leukaemia and of chronic lymphatic leukaemia in man are distinct (Court Brown and Doll 1959) and it is important to note that radiation has been shown to increase the first two kinds of leukaemia but not the third.

In the following pages the term 'leukaemia' is used when it is used by the workers concerned. In the present state of semantic confusion a wiser term is 'a myeloproliferative disorder' (Dameshek 1951, 1956, 1970). A myeloproliferative disorder has been defined (Vaughan 1970*b*) as

... abnormal hyperplasia of marrow cells with subsequent metastases to other soft tissues particularly the spleen, liver, and kidney. Clinically a myeloproliferative disorder so defined is associated with an anaemia of a leucoeythroblastic type but not necessarily a leucocytosis and almost invariably with enlargement of the spleen and liver. In adult man myeloid leukaemia, De Guglielmo's disease, myelosclerosis, megakarocytic leukaemia, and polycythaemia vera, all in both acute and chronic forms, are examples of a myeloproliferative disorder.

The diagnosis of aplastic anaemia may have been made erroneously in the past in patients who probably had an aleukaemic form of myeloid leukaemia. In infants the situation is complicated. Many of the conditions described as acute lymphatic or blast cell leukaemia probably fall into Bodian's type of non-teratomous tumour arising from embryonic cells (Bodian 1963).

In summary it may be said that external irradiation of the human foetus *in utero* gives rise to many different forms of leukaemia, while in adults external irradiation most commonly results in myeloid leukaemia. In young children all types of leukaemia are described many of which are difficult to classify, arising as they probably do, in many cases, from embryonic cells (Bodian 1963).

In the case of internal irradiation from radionuclides deposited in the skeleton myeloproliferative malignancy may develop both in man, from ^{32}P (Vaughan 1970*b*), and in certain experimental animals, particularly from continuous ingestion of ^{90}Sr. In pigs continuous ingestion of ^{90}Sr, a long-range beta-emitter, induces a spectrum of haemoproliferative disorders including myeloid metaplasia and myeloid leukaemia, lymphoma, and lymphocytic leukaemia and stem cell leukaemia (Dungworth, Goldman, Switzer, and McKelvie 1969; Howard and Clarke 1970). In dogs continuous feeding of ^{90}Sr has hitherto resulted in myeloid leukaemia and myeloid hyperplasia only. No lymphoproliferative response has been seen (Dungworth, Goldman, Switzer, and McKelvie 1969). Lymphosarcomas have been recorded however by Finkel and her colleagues following a single injection of ^{90}Sr (Finkel, Biskis, Greco, and Camden 1972).

3.2. Other than haemoproliferative lesions

It appears likely that the group of bone tumours variously described as angiosarcoma, haemangioendothelioma, reticulum cell sarcoma, and liposarcoma may arise from mesenchyme cells in the marrow rather than from osteogenic tissue. This group of tumours is usually osteolytic as seen in radiographs, and on section many of them are undifferentiated malignant tumours. Others, however, are remarkably vascular and show rosettes of ill-formed capillaries. They are rare in man but not uncommon in experimental animals given ^{90}Sr (Finkel, Biskis, and Scribner 1959; Finkel, Biskis, Greco, and Camden 1972; Dougherty and Mays 1969; Nilsson 1970a,b; Barnes, Evans, and Loutit 1971) (see Tables 8.2, 8.3).

It is of some interest that in the Utah dogs the predominating tumour is an osteosarcoma but in looking at the latest detailed table available, two dogs given ^{90}Sr died with haemangiosarcomas of the skeleton and one dog given ^{228}Th, a surface-seeker with a long-range alpha-radiation, is also recorded as having an haemangiosarcoma. Dogs receiving ^{226}Ra and ^{228}Ra and ^{239}Pu had only osteosarcoma (see Table 8.2). Finkel, Biskis, Greco, and Camden (1972) also report 3 angiosarcomas out of 7 bone tumours occurring in dogs given a single injection of ^{90}Sr, and McClellan (1972) 7 out of 17 bone tumours occurring in dogs inhaling ^{90}Sr chloride. In mice, Nilsson (1970a) also notes this type of tumour following ^{90}Sr but does not give precise figures. They are also well documented with good histological descriptions in rats (Kuzma and Zander 1957a,b; Skoryna, Kahn, and Webster 1958; Skoryna and Kahn 1959). Following administration of

TABLE 4.1

Some characteristics of certain bone-seeking radionuclides
(Bleaney 1971, personal communication by courtesy of author)

Radionuclide	Half-life	Radiation	Energy MeV	Range in soft tissue	Site
^{228}Th	1·9 yrs	α	5·4–8·8	39–88 μm	Surface
^{226}Ra	1620 yrs	α	4·6–7·7	31–70 μm	Volume
^{224}Ra	3·69 days	α	5·4–8·8	39–88 μm	Surface
^{239}Pu	$2·4 \times 10^4$ yrs	α	5·14	35 μm	Surface and marrow
^{90}Sr + ^{90}Y	28 yrs	β	2·18	10 mm	Volume
^{89}Sr	51 days	β	1·46	7 mm	Volume
^{45}Ca	1·65 days	β	0·254	900 μm	Volume

The energies and ranges of the first three nuclides include the alphas emitted by the daughters in the decay chains. The β energies are the maximum emitted energies.

other radionuclides to animals such tumours are not recorded in significant numbers.

Reticuloendotheliomas, reticulum cell sarcomas, and reticulosarcomas, which are reported rarely, are also probably of marrow origin. They occur most often in animals given ^{90}Sr. Ewing's tumour is rarely described as occurring as a result of any form of radiation. Table 4.1 briefly summarizes the character of the best known radionuclides and Table 4.2 the tumours described in experimental animals following their administration. The character of the radionuclides is discussed in greater detail in Chapter 6, Sections 1–4. It is at once apparent that the radiations with a short range in tissue (with the exception of plutonium, which deposits in marrow as well as on 'bone surfaces') characteristically induce tumours arising from osteogenic tissue, while those like ^{90}Sr + ^{90}Y with a long range in tissue may induce tumours arising from both osteogenic and marrow tissue.

TABLE 4.2

Radionuclides and skeletal tumour incidence animals

Radionuclide	Osteo	Chondro	Fibro	'Angio'[B]	'Leukaemia'[A]
^{228}Th	+++			+	
^{226}Ra	+++		+++		
^{224}Ra	+++				
^{239}Pu	++	+	+		+
^{90}Sr + ^{90}Y	++		+	++	++
^{89}Sr	++	+	+	+	+
^{45}Ca	++				

[A] The term leukaemia here is used to cover any myeloproliferative disorder.
[B] The term 'angio' is used to cover a miscellaneous group of tumours probably arising from marrow mesenchyme tissue: angiosarcoma, reticuloendothelioma, etc.

4. Tumours of epithelium closely applied to bone

Epithelium closely applied to bone occurs in the sinuses of the skull and in the mucous membranes of the mouth. A significant number of carcinomas are recorded particularly in the sinuses of the skull both in man and in experimental animals as the result of internal radiation from deposition of radionuclides. Unfortunately they are spoken of as carcinomas and no detailed description or classification is available at present. It is to be expected that they might vary in character as do the carcinomas of the same sites occurring in woodworkers. Squamous carcinomas, adenocarcinomas, and anaplastic transitional unclassified tumours are recorded in these workers (Hadfield 1970).

5. External Irradiation

EXTERNAL irradiation may induce either skeletal neoplasia or skeletal dysplasia in both man and experimental animals. In man such irradiation may result from sources of radiation in the environment, occupational exposure, or diagnostic and therapeutic medical exposure.

The evidence at present suggests that if a large part of the skeleton is exposed to an appropriate radiation dose the skeletal hazard is likely to be some form of myeloproliferative disorder. This conclusion may prove to be incorrect. The latent period for leukaemia depends on the cell type and the age at exposure. For myeloid leukaemia it varies from less than 1 year for cases of embryonic origin to over 8 years for later cases, for lymphatic leukaemia from less than 5 to over 20 years (Court Brown and Doll 1957, Stewart 1972ab). The latent period for osteosarcoma, even for children exposed *in utero*, is likely to exceed 10 years. The large scale surveys of populations whose skeletons have been exposed to irradiation which are discussed in the following pages have had a relatively short follow up period so it is possible that only the myeloid leukaemias have been fully recognized. In the case of thymic irradiation in children and in the patients with ankylosing spondylitis where the longest follow up data are available bone pathology was recorded. If exposure is strictly localized, as for instance following radical mastectomy, a bone malignancy is the more likely hazard. On the other hand leukaemia has not been looked for in such cases.

1. Environmental exposure

1.1. *Natural*

The characteristic of natural background radiation is that the *dose rate* is extremely low, as compared, for instance, to radiation given for therapeutic purposes. Even if a dose of 1 rad given in 1 second or 1 minute carries some leukaemogenic risk, there may not be any risks from exposure to the same dose spread over about 10 years, that is given at the average dose rate of natural background radiation. It has been argued that the present evidence of the relationship between dose and incidence suggests that background radiation may be responsible for 10–20 per cent of the cases of leukaemia that occur in America (Lewis 1957). This is not however generally accepted (Brues 1958, Lamerton 1958, Mole 1958).

Man has always been exposed to natural continuous external irradiation from his environment though this is variable depending upon differences in the character, particularly in the geology, of his surroundings. The major sources of such environmental radiation are cosmic rays and radioactive potassium, uranium, and thorium and their decay products. The actinium series ^{87}Rb, ^{147}Sm, and ^{14}C contribute little to the dose to man, either because of the character of their energy emission or because they are present in such small amounts (Spiers 1960). Uranium, thorium, and their decay products are present in the ground on which man walks and in the material of which he builds his home. Natural gamma radiation from the earth and from buildings makes up about one half of the total radiation dose out of doors. The dose rate ranges from about 50 millirad per year over sedimentary rocks to about 200 millirad per year in granite districts while the dose in houses has about the same range (Spiers 1960).

The possible effects of such external irradiation have been studied in Scotland by Court Brown and his colleagues (Court Brown, Spiers, Doll, Duffy, and McHugh 1960). They warn 'it is difficult to make direct observations of the effect of background radiation because any effect is likely to be small in comparison with that due to other causes. Populations of the order of hundreds of thousands would have to be studied and it is difficult

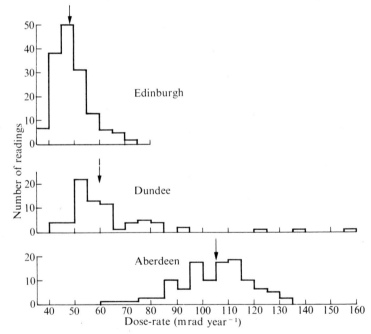

FIG. 5.1. Background radiation out of doors in Aberdeen, Dundee, and Edinburgh (Court Brown *et al.* 1960, by courtesy of authors and publishers).

EXTERNAL IRRADIATION

to maintain the same standards of diagnosis throughout such large populations'. Measurement of gamma ray dose rates (mrad year^{-1}) were made out of doors and in houses. Figure 5.1 shows the readings for roads in Edinburgh, Dundee, and Aberdeen, and Fig. 5.2 the readings within stone houses in the three cities. Aberdeen is a granite area where radiation is higher than in Edinburgh where the houses are of local stone from sedimentary rocks.

Estimates were made of the mortality from leukaemia for ten parts of Scotland over the years 1939 to 1956. The highest mortality (146 per cent of the expected) was recorded in Aberdeen, the second highest (124 per cent

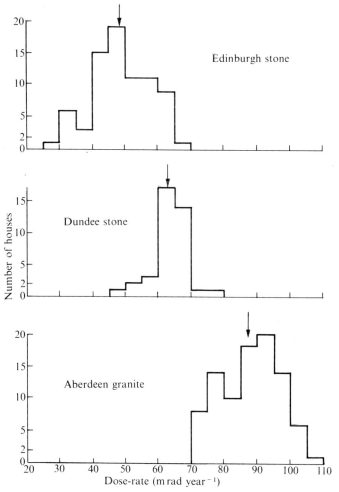

FIG. 5.2. Background radiation inside stone houses in Aberdeen, Dundee, and Edinburgh (Court Brown *et al.* 1960, by courtesy of authors and publishers).

of the expected) in Edinburgh, and the lowest (87 per cent of the expected) in Glasgow. The excess mortality in Aberdeen was attributed to acute leukaemia and myeloid leukaemia while in Edinburgh it was due to chronic lymphatic leukaemia. The excess noted in both towns could not be reasonably attributed to random fluctuation but possible explanations were thought to lie with social and economic factors such as better case finding and higher economic status. Estimates of the amount of radiation received from background sources out of doors indicated that the highest mean dose rate, 104 mrad year^{-1}, was received in Aberdeen and the lowest, 48 mrad year^{-1}, was received in Edinburgh (Court Brown, Spiers, Doll, Duffy, and McHugh 1960). The highest mean dose rate inside the house was 87 mrad year^{-1} in a granite house in Aberdeen and 48·5 mrad year^{-1} in a sandstone house in Edinburgh. Extrapolation from the effects of large doses given at very much greater intensities suggests that differences of this order were unlikely to account for more than 1 per cent of the observed differences in mortality—if they were capable of being leukaemogenic at all. The authors suggest that local differences in the average exposure to diagnostic radiography may be greater than the differences in background radiation but again are unlikely to be a major factor in its production. They conclude that it is insufficient to study leukaemia mortality in relation to background radiations alone. The proper interpretation of geographical variations can only be made when the effects of social and economic factors can also be assessed. There are parts of the world which have much more highly radioactive soils than are found in Scotland. The gamma dose rate measured 1 m above ground in the monosite area of Brazil has an average value of 500 mrad year^{-1} and a maximum of 1000. The weighted mean for the Kerala area of India is 1300 mrad year^{-1} (United Nations 1962, Seelentag 1968). Again there is no evidence at present that either man or animals in such areas are adversely affected (Penna Franca, Almeida, Becker, Johanne, Emmerich, Roser, Kegel, Hainsberger, Cullen, Petrow, Drew, and Eisenbud 1965; Gruneberg, Bains, Berry, Riles, Smith, and Weiss 1966). Neither dysplasia nor malignant lesions in the skeleton have been reported as the result of this high natural environmental radiation. The populations exposed are, however, for statistical purposes relatively small. The cosmic ray dose rates in high altitude areas may make an appreciable contribution to the tissue dose rate. At Cerro de Pasco, Peru, for instance, 330 mrem year^{-1} may be received from cosmic rays. The population exposed, however, is small, 19 187, and no skeletal abnormalities are recorded (United Nations 1962).

It has recently been calculated that the dose rate resulting from natural radiation to the osteocytes in man averages 137 mrem year^{-1} with a range of 80–300 mrem year^{-1}, and to the bone marrow 122 mrem year^{-1} with a range of 70–200 mrem year^{-1} (Schmier, Seelentag, and Waldeskog 1971).

This includes both external and internal radiation sources, and is rather higher than that given in the United Nations Report of 1962. The internal radiation sources arise from the ingestion of radioactive elements present in food and water. The question of ingestion from radioactive fall-out, particularly of ^{90}Sr (which contaminates the skeleton), is discussed in Section 1.2.1. It is recognized that the ingestion of naturally occurring radium isotopes (in particular) results in a constant, but negligible, level of alpha activity in all human skeletons (Turner, Radley, and Mayneord 1958).

There is no evidence that variations in the level of radium in water supplies have any effect on the incidence of leukaemia or osteosarcoma in the populations studied. The statistical problems and the necessity of studying very large populations however make it extremely difficult to determine very low incidences of any condition with certainty (Marinelli 1958, Lucas 1961). The observed level of alpha radiation in the skeleton is a factor of at least 1000 below the lowest body burden known to have resulted in a tumour (Turner, Radley, and Mayneord 1961). Dose commitments from nuclear explosions (1954–65) to cells lining bone surfaces and bone marrow are shown in Table 5.1 (Eisinbud 1968).

TABLE 5.1

Dose commitments from nuclear explosions
(*Eisinbud 1968, by courtesy of author and publishers*)

Tissue	Source of radiation	Dose commitments (mrad) for periods of testing 1954–65
Cells lining bone surfaces	External short-lived	23
	^{137}Co	25
	Internal ^{90}Sr	156
	^{137}Cs	15
	^{14}C	20
	^{89}Sr	0·3
	Total	240
Bone marrow	External short-lived	23
	^{137}Cs	25
	Internal ^{90}Sr	78
	^{137}Cs	15
	^{14}C	13
	^{89}Sr	0·15
	Total	150

1.2. *Artificial*

Artificial radioactivity may arise from the release of the products of nuclear fission or neutron activated material. This may occur by the explosion of thermonuclear devices in weapon testing or ploughshare projects (Gerber, Hamburger, and Hull 1966; Martell 1969).

The explosion of thermonuclear devices in the atmosphere gives immediate gamma and neutron irradiation and fall-out of radioactive products. Underground explosions which are used in weapon testing and may be used in ploughshare projects, if they vent, release radioactive products without the gamma and neutron primary radiation. Accidents at power reactors and ancillary plant (chemical) should give only special localized situations of contamination by radioactive material. In the case of conventional reactors, ^{131}I is likely to be the major hazard, but with fast breeder reactors ^{24}Na is a more likely risk. Other minor sources of contamination are the use of radionuclides like ^{90}Sr, ^{210}Po, and the transuranic elements in isotopic power supplies. The latter are used, for instance, to operate weather stations in remote areas, in coast-guard beacons, communication satellites in outer space, and in cardiac pace-makers.

The severity of the contamination in any instance will depend on the severity of the explosion. Industrial accidents are likely to contaminate only relatively small areas while the explosion of nuclear weapons has already led both to low-level contamination of the whole world and to severe local radioactive contamination and radiation injury.

1.2.1. *Fall out.* The most important of the fall-out radionuclides are ^{131}I, ^{137}Cs, and ^{89}Sr and ^{90}Sr, to which should be added ^{14}C and ^{3}H. Those that affect the skeleton particularly are the strontium isotopes. The amount of fall-out from nuclear explosions is highly dependent on the distance from the place of explosion, the height above ground, the type of explosion (whether fusion or fission), and the design and yield of the explosive.

It was soon recognized that contamination of food and water supplies by fall-out might involve particularly contamination by ^{90}Sr. It has recently been estimated that over 20 MCi ^{90}Sr have been produced by bomb explosions all over the world. Only 13 MCi have been accounted for in deposition (Bennett 1972). The difference is thought to be largely due to local deposition near the test sites and to decay of ^{90}Sr. The Northern Hemisphere has received about three times as much strontium as the Southern Hemisphere but current deposition rates are about equal. The pattern of fall-out in the two hemispheres and the cumulative deposition is shown in Fig. 5.3 (Bennett 1972). There is probably still about 0·05 MCi of ^{90}Sr in the stratosphere. This fall-out ^{90}Sr reaches the human skeleton first through direct contamination of food supplies and secondly through food chains leading to the consumption of contaminated milk by man.

EXTERNAL IRRADIATION

There has therefore been large-scale monitoring particularly of milk, vegetables, and human bone all over the world and much experimental work covering both the metabolism of ^{90}Sr and its toxicity. The daily intake of ^{90}Sr in the cities in the United States is shown in Fig. 5.4. and its distribution between different food stuffs in New York in Fig. 5.5. In Fig. 5.6 the ^{90}Sr content of human bone in the United Kingdom plotted in different years in the period 1959–68 is shown and in Fig. 5.7 the data are plotted according to age in any one year. Figure 5.8 shows the countrywide mean ratio of ^{90}Sr to calcium in milk in the United Kingdom. These figures all indicate that ^{90}Sr reached a peak in the environment in 1963–64, since

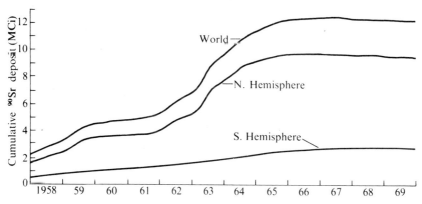

FIG. 5.3. Accumulated deposition of global ^{90}Sr fall-out (Bennett 1972, by courtesy of author and publishers).

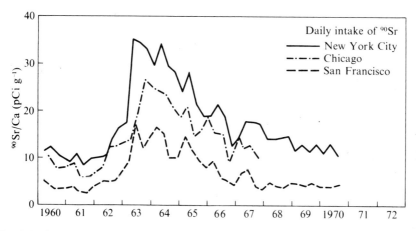

FIG. 5.4. Diet quarterly sampling results of ^{90}Sr intake in three cities (Bennett 1972, by courtesy of author and publishers).

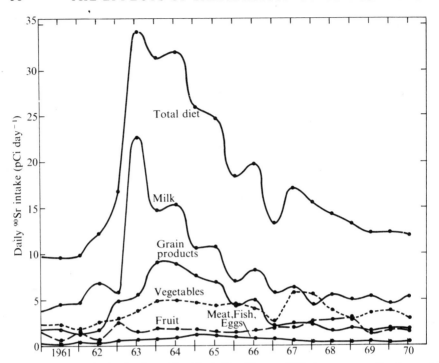

Fig. 5.5. ^{90}Sr intake in diet from New York City (Bennett 1972, by courtesy of author and publishers).

when it has been falling. This peak period in the environment was reflected by a peak in the ^{90}Sr content of bone in young children aged 1–6 months, as shown in Fig. 5.6. The fact that the strontium level is much higher in the young than the old is due to the fact that mineral accretion is much greater in young than in old bone. Figures obtained from similar monitoring studies in the United States are of the same order (Kulp, Schulert, and Hodges 1960; Kulp and Schulert 1962; Bryant and Loutit 1964; Bennett 1972). Since the present maximum permissible level of ^{90}Sr per kg of calcium for occupational workers is 2 μCi the figures for ^{90}Sr levels in the skeleton do not give cause for anxiety at present.

A medical survey of the peoples of Rongelap and Utirik Islands fifteen years after exposure to fall-out radiation following explosion of a nuclear device in 1954 showed no effects on the skeleton that could be clearly attributed to the strontium isotopes (Conard 1970). Defects of bone growth in the children proved to be dependent on thyroid deficiency due to ingestion or inhalation of radioactive iodine. A residual radiation effect on the bone marrow was, however, noted. There was a persistent slight depression of white cells and platelets in the peripheral blood and an increased number

EXTERNAL IRRADIATION

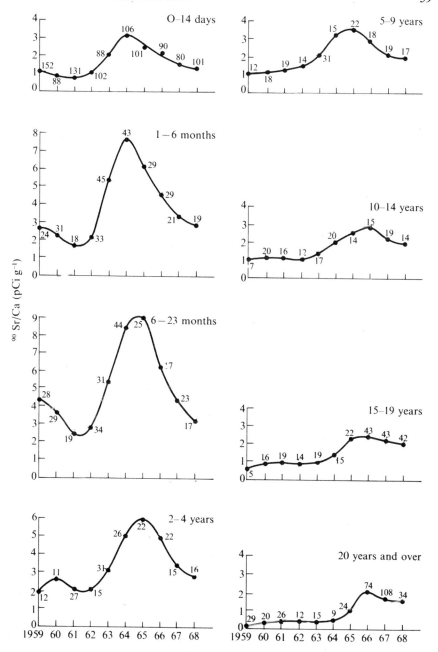

Fig. 5.6. ⁹⁰Sr in human bone plotted according to year (the figures adjacent to the dots indicate the numbers of samples on which the means are based) U.K.A.E.A. results 1959–68, Medical Research Council monitoring report no. 17 (1963).

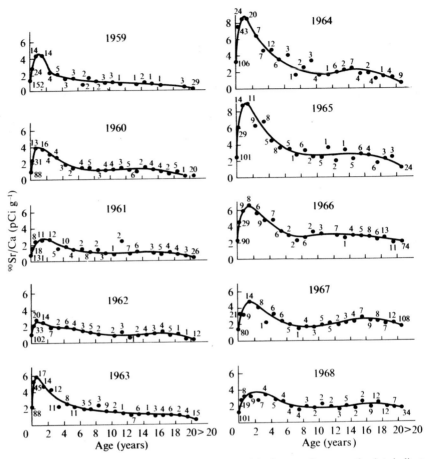

FIG. 5.7. ⁹⁰Sr in human bone plotted according to age (the figures adjacent to the dots indicate the numbers of samples on which the means are based) U.K.A.E.A. results 1959–68, Medical Research Council monitoring report no. 17 (1963).

of atypical lymphocytes in the exposed group. Examination of bone marrow in some exposed individuals revealed an alteration in the myeloid/erythroid ratio (increased red cell precursors), the presence of cells with abnormal chromatin material and double nuclei, and also an increased number of cells in mitosis (Conard 1970). Whether this effect on the marrow is due to ^{90}Sr in the skeleton or to the initial gamma irradiation is not at present clear. The estimated bone-marrow dose from natural and residual radiation sources both from external and internal sources was estimated in 1958, and is shown in Table 5.2. Cytogenetic studies of blood lymphocytes 10 years after exposure to radiation showed chromosome aberrations in 23 out of 43 exposed persons, half of which were of the

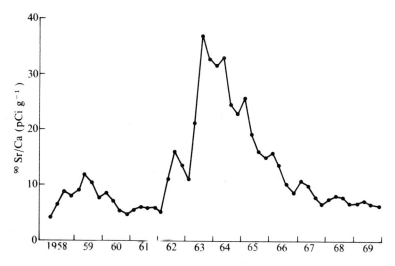

FIG. 5.8. Countrywide mean ratio of ^{90}Sr to calcium in milk. Medical Research Council monitoring report no. 17 (1963).

exchange type. An unexpectedly large number of acentric fragments, but no exchange type aberrations, appeared in a few unexposed people on the same island (Lisco and Conard 1967). Similar chromosome aberrations have been reported in the Japanese fisherman exposed to radiation from the same fall-out accident (Ishihara and Kumatori 1965).

Contamination with radionuclides in the islanders is thought to have been brought about largely through eating and drinking contaminated food and water and to a lesser extent through inhalation. The maximum

TABLE 5.2

Bone-marrow dose from natural and residual radiation sources in Rongelap islanders, estimated 1958

(Conard 1970, by courtesy of author and publishers)

	Dose to bone marrow mrad year^{-1}	
Internal		
^{90}Sr (11 nCi body burden)	11·3	
^{137}Cs + ^{65}Zn	120·0	
Natural (^{40}K etc.)	4·4	175
External		
Residual gamma (0·03 mR hour^{-1})	250	
Natural (cosmic etc.)	134	384
Total		559

permissible concentrations of radionuclides were approached or slightly exceeded during the first few days only in the case of ^{89}Sr and the isotopes of iodine. Body levels fell rapidly so that by 2–3 years post exposure they were far below the accepted maximum permissible level. Even 6 months after exposure activity in the urine was barely detectable. When the exposed population was returned to the island in 1957 body burdens of ^{90}Sr and ^{137}Cs rose slightly but by the early 1960s the body burden of ^{90}Sr in the Rongelap population had reached equilibrium with the environment. Little or none of the body burden in the exposed people at that time could be considered residual from the initial exposure. Since little difference was noted between the body burdens of the exposed and unexposed groups living on Rongelap Island, it can only have arisen from the slight continuing environmental contamination. It is of some interest that the local small Coconut Crab, a favourite food of the islanders, is still so rich in ^{90}Sr that it is a forbidden food. In 1961 it contained 1140 pCi per gram of calcium. In 1969 it contained 700 pCi/g^{-1} (Conard 1970). The detailed effects of ^{90}Sr on the skeleton are further discussed in Chapter 7.

1.2.2. *Hiroshima and Nagasaki bomb survivors.* An intensive follow up of the clinical effects of radiation caused by the bombs dropped on Hiroshima and Nagasaki is being carried out by the Atomic Bomb Casualty Commission (ABCC), together with attempts to measure the radiation received by survivors both within the cities and living outside (Folley, Borges, and Yamawaki 1952; Jablon, Ishida, and Yamasaki 1965; Bizzozero, Johnson, and Ciocco 1966; Miller 1969; Ishimaru, Hoshino, Ichimaru, Okada, Tomiyasu, Tsuchimoto, and Yamamoto 1971; Jablon, Tachikawa, Belsky, and Steer 1971).

Stewart (1972*a*) has argued that the fate of atomic bomb survivors has very little relevance to other groups exposed to external radiation and should as far as possible be avoided as a basis for estimating the population risks of small doses of radiation. She considers that the old and the very young may have succumbed to the acute effects of radiation leaving the healthy adolescents and young adults to survive long enough to develop malignancies later (Stewart and Barber 1971).

It is important not to forget that the explosions and the follow-up of survivors have allowed us to see, not the carcinogenic effect of varying doses of X-rays, but the effects of combining to a varying degree (i) the lethal effects of radiations which can be counted on to kill old people more easily than young people, (ii) the recuperative powers of the reticulo-endothelial system—which can be counted upon to save more young lives than old lives, and (iii) the carcinogenic effects of the radiations which can only be detected in long term survivors.

Certain facts of considerable importance do however arise from the surveys that have been made of the atomic bomb survivors though their precise interpretation may present difficulties.

It is not questioned that in the early years of the follow up the incidence of leukaemia, particularly of chronic granulocytic leukaemia, was greater in bomb survivors than in the normal Japanese population even though there has recently been a steep rise in the incidence of leukaemia throughout Japan (Segi, Kurihara, and Matsuyama 1965). Furthermore the longer the period of follow up the greater is the increase in other forms of malignancy (Wanebo, Johnson, Sato, and Thorslund 1968a,b; Wood, Tamagaki, Neriishi, Sato, Sheldon, Archer, Hamilton, and Johnson 1969). For instance, Jablon and his colleagues (Jablon, Tachikawa, Belsky, and Steer 1971) have followed up a group of children who were less than 10 years old in 1945 and who received a dose of 100 rad or more and who have in the ensuing 25 years developed an excessive number of malignancies which has not yet reached its peak (see Table 5.3). No osteosarcomas have yet been

TABLE 5.3
Number of deaths attributed to leukaemia and to other malignant neoplasms, l.s.s.[a] Cohort age<10 years a.t.b.[b] (all radiation categories)
(*Jablon et al. 1971, by courtesy of authors and publishers*)

Years	Leukaemia	Other malignant neoplasms
1950–54*	9	1
1955–59	5	2
1960–64	8	5
1965–69	4	15
Total	26	23

* From 1 Oct., 1950.
[a] l.s.s. = life span studies.
[b] a.t.b. = at time of bomb explosion.

reported but their latent period is known to be long. Factors that may contribute to the apparent high leukaemia incidence to date are first the relatively short latent period for myeloid leukaemia compared to that of other malignancies, and secondly the fact that marrow is, even in the adult, extremely rich in stem cells, particularly myeloid cells, so that probably more cells are at risk in this tissue than in any other. The myeloid cells will also be stimulated to divide to a greater extent than other cells if exposure involves tissue destruction elsewhere. The International Commission on Radiological Protection (1969) classifies the marrow as the most radiosensitive tissue in the body which is perhaps another way of saying that there are more cells at risk and more opportunities for division of mutant cells immediately after they are formed.

The most recent figures for the distribution of definite and probable leukaemia cases by type of leukaemia, exposure status, and city (as of 30th

June, 1967) are shown in Table 5.4 (Ishimaru, Hoshino, Ichimaru, Okada, Tomiyasu, Tsuchimoto, and Yamamoto 1971). Cases of definite leukaemia are defined as follows by the Atomic Bomb Casualty Commission: cases with good clinical information and a history of well studied and morphological documentation of the disease ...; cases with good clinical and

TABLE 5.4

Distribution of definite and probable leukemia cases by type of leukemia and exposure status and city (as of 30 June, 1967)

(Ishimaru et al. 1971, by courtesy of authors and publishers)

Type of leukemia	Total	Exposure status			
		A-bomb survivors		Other	
		2500 m	2500–9999 m	Not in city	Born after a.t.b.[b]
		distance from hypocentre			
Hiroshima					
Acute granulocytic	250 (36·8)[a]	45 (25·1)	25 (41·0)	114 (42·1)	66 (39·1)
Lymphocytic	123 (18·1)	27 (15·1)	9 (14·8)	39 (14·4)	48 (28·4)
Monocytic	35 (5·1)	8 (4·5)	3 (4·9)	18 (6·6)	6 (3·6)
Undifferentiated	36 (5·3)	10 (5·6)	4 (6·6)	8 (3·0)	14 (8·3)
Other type	25 (3·7)	6 (3·4)	4 (6·6)	12 (4·4)	3 (1·8)
Unknown	67 (9·9)	13 (7·3)	8 (13·1)	22 (8·1)	24 (14·2)
Chronic granulocytic	138 (20·3)	69 (38·5)	8 (13·1)	54 (19·9)	7 (4·1)
Lymphocytic	3 (0·4)	0 (0)	0 (0)	3 (1·1)	0 (0)
Other type	2 (0·3)	1 (0·6)	0 (0)	1 (0·4)	0 (0)
Chronicity unknown	1 (0·1)	0 (0)	0 (0)	0 (0)	1 (0·6)
Total	680 (100)	179 (100)	61 (100)	271 (100)	169 (100)
Nagasaki					
Acute granulocytic	235 (44·8)	19 (34·5)	34 (42·0)	117 (48·1)	65 (44·5)
Lymphocytic	108 (20·6)	16 (29·1)	9 (11·1)	28 (11·5)	55 (37·7)
Monocytic	33 (6·3)	1 (1·8)	10 (12·3)	13 (5·3)	9 (6·2)
Undifferentiated	14 (2·7)	2 (3·6)	6 (7·4)	4 (1·6)	2 (1·4)
Other type	32 (6·1)	2 (3·6)	4 (4·9)	25 (10·3)	1 (0·7)
Unknown	19 (3·6)	2 (3·6)	4 (4·9)	6 (2·5)	7 (4·8)
Chronic granulocytic	67 (12·8)	13 (23·6)	8 (9·9)	39 (16·0)	7 (4·8)
Lymphocytic	15 (2·9)	0 (0)	6 (7·4)	9 (3·7)	0 (0)
Other type	1 (0·2)	0 (0)	0 (0)	1 (0·4)	0 (0)
Chronicity unknown	1 (0·2)	0 (0)	0 (0)	1 (0·4)	0 (0)
Total	525 (100)	55 (100)	81 (100)	243 (100)	146 (100)

[a] Numbers in parentheses indicate percentage
[b] time of bomb explosion

haematological material that provide convincing evidence for the diagnosis of leukaemia and cases with morphological confirmation and a clinical history not inconsistent with leukaemia even though the clinical information is scanty. Cases of probable leukaemia are defined as follows: cases with convincing clinical information for the diagnosis of leukaemia but with little or no morphological material; cases with inadequate clinical information but with good morphologic material consistent with leukaemia ... cases with adequate clinical and morphologic material in which it is only remotely possible that some other clinical syndrome caused the haematologic abnormality (Bizzozero, Johnson, and Ciocco 1966).

The peak incidence for leukaemia was in 1951–52 and has since declined though the normal incidence level has not yet been reached. For both cities

TABLE 5.5

Period prevalence and incidence of definite and probable leukemia (all forms) among A-bomb survivors in the A.B.C.C. master sample by T65* total dose and city; Oct. 1950–Sep. 1966

(Ishimaru et al. *1971, by courtesy of author and publishers*)

T65* total dose range	T65* median dose (rad)			Period prevalence of leukemia			Person-years at risk (1000)	Incidence of leukemia	
	Gamma	Neu-tron	Total	Subjects	Leu-kemia	Pre-valence 1000		Cases	Annual incidence 100 000
Hiroshima									
300+	323	112	427	825	18	21·82	12·1	17	140·5
200–299	185	49	241	606	5	8·25	9·0	5	55·6
100–199	105	27	131	1652	11	6·66	24·1	10	41·5
50–99	56	13	68	2611	8	3·06	38·3	7	18·3
20–49	26	5	30	4555	15	3·29	67·0	14	20·9
5–19	8	2	10	10541	9	0·85	156·0	8	5·1
<5	0	0	0	62515	29	0·46	915·1	27	3·0
Total	—	—	—	83305	95	1·14	1222·7	88	7·2
Nagasaki									
300+	417	7	427	566	7	12·37	8·4	6	71·5
200–299	238	3	240	693	6	8·66	10·4	6	57·7
100–199	145	2	146	1174	3	2·56	17·7	3	16·9
50–99	69	0	69	1173	0	0	17·6	0	0
20–49	31	0	31	1354	0	0	20·0	0	0
5–19	5	0	10	4501	2	0·44	66·3	2	3·0
<5	0	0	0	20403	13	0·64	297·2	12	4·0
Total	—	—	—	29864	31	1·04	437·6	29	6·6

* T65 = latest calculation of dose

the granulocytic form is the most common in both acute and chronic leukaemia in the atomic bomb survivors and also in the 'not in city' group. In Table 5.5 is shown the period, prevalence, and incidence of definite and probable leukaemia (all forms) among atomic bomb survivors by T65 total dose* and city Oct. 1950–Sept. 1966 (Ishimaru, Hoshino, Ichimaru, Okada, Tomiyasu, Tsuchimoto, and Yamamoto 1971). The calculation of dose received (T65) differs from that used in some of the earlier surveys (T57) and is thought to be considerably more accurate (Brill, Tomonaga, and Heyssel 1962; Milton and Shokoji 1968). Inspection of these tables at once shows clearly that there are striking differences between the two cities both in the character of the leukaemias and of the radiation received. In Hiroshima the radiation was largely from fast neutrons while in Nagasaki it was largely from gamma irradiation.

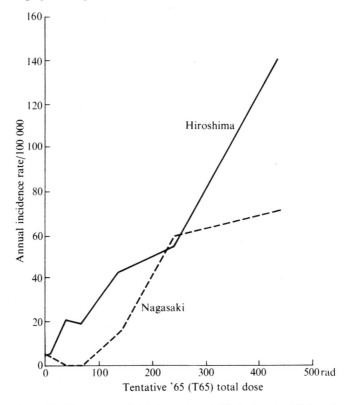

FIG. 5.9. Annual incidence rate of definite and probable leukaemia (all forms) per 100 000 population of A-bomb survivors in the A.B.C.C. master sample by T65 total dose and city; Oct. 1950–Sept. 1966 (Ishimaru et al. 1971, by courtesy of authors and publishers).

* T65 = latest calculation of dose
  T57 = early calculation of dose

Figure 5.9 shows the annual incidence rate of definite and probable leukaemia per 100 000 population of atomic bomb survivors in the Atomic Bomb Casualty Commission sample by T65 total dose and city from Oct. 1950–Sept. 1966. The incidence rate in Hiroshima is much greater than that in Nagasaki. From this evidence Ishimaru and his colleagues (1971) calculate that the RBE of fast neutrons compared to gamma rays is about 5, with no indication of a threshold. The risk of leukaemia in Hiroshima appeared to exist for those receiving less than 50 rad. The incidence of acute and chronic leukaemia in the two cities is shown in Fig. 5.10. Gamma

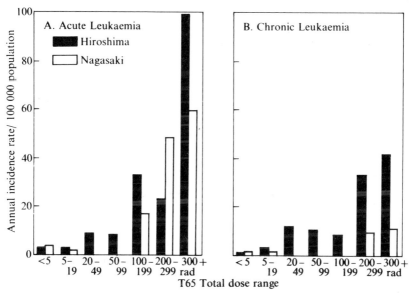

FIG. 5.10. Annual incidence rate of definite and probable leukaemia among A-bomb survivors in the A.B.C.C. master sample by chronicity of leukaemia, T65 total dose and city, Oct. 1950–Sept. 1966 (Ishimaru *et al.* 1971, by courtesy of authors and publishers).

irradiation in Nagasaki appears, from the data given, to cause more acute than chronic leukaemia while fast neutrons in Hiroshima induce both.

On the other hand six chronic lymphocytic leukaemia cases were seen in Nagasaki survivors who were beyond 3000 m from the hypocentre but no cases of this type were found in the Hiroshima survivors. Chronic lymphatic leukaemia incidence is however known to be low in Japan (Bizzozero, Johnson, and Ciocco 1966). The diagnostic criteria in the two cities are considered comparable, so these differences appear likely to be dependent upon differences in the character of the radiation received.

Ishimaru and his colleagues (1971) suggest that analysis of the atomic bomb data indicates that the leukaemogenic effect of the irradiation is stronger in males than in females and is greater in younger than in older survivors. These conclusions are open to question as Stewart and Barber (1971) have suggested. It may well be that more men and more young people survived the acute effects and therefore had time to develop leukaemia.

2. Occupational

Occupational exposure to ingested radionuclides which will result in internal radiation of the skeleton is discussed particularly in Chapters 7, 8, 9, 10, 11, 12, 13.

2.1. *Radiologists*

Information on the effects of occupational exposure to external radiation comes almost entirely from statistical studies on the health of radiologists. In discussing the effects on the skeleton it should be remembered that (1) external whole body irradiation will increase the risk of cancers in all tissues, (2) the modal induction period varies from one site to another and in adult life is longer for other cancers than for leukaemia, and (3) the total increase in cancer mortality within 20 years after whole body irradiation is approximately twice the increase due to leukaemia alone (International Commission of Radiological Protection 1966).

In the period when X-rays were first used for both diagnosis and therapy the possible risk of over-exposure of personnel who carried out the necessary procedures was not appreciated. In the first 20 years of this century exposure was largely due to X-rays of relatively low kilovoltage and therefore poor penetration. In the next 20 years the use of higher kV X-rays increased but at the same time there was increasing awareness of the hazards involved. However, following the report of the Adrian Committee (1959, 1960) it was recognized that as late as 1959 exposure to ionizing radiations, in many hospital departments, was still dangerously high. Regulations for protection have since been strengthened and it is unlikely that in the future radiologists and radiographers will suffer over-exposure except by accident.

In 1944, March first drew attention to the fact that radiologists dying in the 15-year period from 1929–43 had ten times as many deaths from leukaemia as did other physicians and again in the period 1944–48 he found somewhat comparable data (March 1950). Warren in 1956 pointed out that radiologists die on the average 5·2 years earlier than do other physicians and in 1957 Lewis pointed out the significant incidence of leukaemia among American radiologists. The most satisfactory statistical analysis came later, in 1965, from Seltser and Sartwell, who compared radiologists with other medical specialists over the period 1935–58. Their findings indicated that occupational exposure to ionizing radiations has in the past produced a non-specific life-shortening effect. Leukaemia showed the highest mortality rate expressed as a ratio of observed to expected deaths based on the age and time-specific mortality rates of the other specialities. The leukaemia excess and to a lesser extent the excess in other cancers, which was apparent but less significant, occurred in an earlier age group than the other diseases, thus supporting the existence of a relatively short latent period for leukaemia. Court Brown and Doll (1958) also assessed mortality rates in a group of British radiologists, as already shown in Tables 3.1 and 3.2, during a 60-year period 1897–1957. They noted a possible excess of leukaemia in those dying before 1921. There was no increase in leukaemia deaths in radiologists dying after 1921. Warren and Lombard (1966) have also noted a gratifying decrease in deaths from leukaemia among radiologists in recent years since better protection has been observed.

3. Medical

3.1. *Diagnostic radiology*

3.1.1. *Foetal irradiation.* In 1964 MacMahon and Hutchinson reviewed ten separate investigations of antenatal radiography and concluded that this form of foetal irradiation increased the overall risk of death from all forms of malignant disease between 0 and 10 years of age by about 40 per cent. Nine years earlier Hewitt had suggested that the post war increase in leukaemia mortality (observed particularly in the United States and Denmark) might be due in part to the increased use of drugs in medical practice (Hewitt 1955). This feature of modern medicine has since been shown to be playing a very minor role compared with the effect of antibiotics on the infection death-rate. This effect has played a key role by making it possible to restore a long standing comparison between acute leukaemia and acute infections as stated causes of death, and to show that as the death rate for pneumonia fell the rate for leukaemia increased (see Fig. 5.11). Meanwhile a retrospective and nationwide survey of childhood cancer (which allowed for the possible effects of radiation on the foetus) has succeeded in throwing new light on the oncogenic effects of radiation by identifying and pinpointing a 'cancer initiator' in the form of obstetric radiography. Unfortunately the continuing retrospective survey has not yet followed children from birth for more than 10 years (Stewart and Kneale 1968, Stewart and Kneale 1969, Stewart and Kneale 1970, Stewart 1972*b*, Stewart and Barber 1971), but it has covered the period during which most of the cancers which are initiated before birth were fatal (with the possible exception of bone sarcomas, cerebral tumours, and lymphomas).

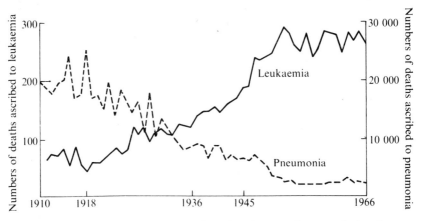

FIG. 5.11. Numbers of deaths of individuals under 10 years of age ascribed each year to pneumonia and leukaemia (official statistics of England and Wales) (Stewart and Neale 1969, by courtesy of authors and publishers).

Table 5.6 shows the proportion of juvenile cancers which can be attributed to obstetric radiography and to other causes. It is at once apparent that not more than 5 per cent of all childhood cancers have been caused by radiation shortly before birth and that although leukaemia is more common than other forms of juvenile cancer it is no more likely to be radiation induced than the other forms. It is also of interest to note that the latent period for embryonic osteosarcomas is much longer than for most forms of embryomas. This is well illustrated in Fig. 5.12. As the survey continues

TABLE 5.6

Proportions of juvenile cancers ascribed to obstetric radiography (radiogenic cases) and other causes (non-radiogenic cases)

(Stewart 1970, personal communication)

	Aetiological groups		No. of cases	Peak incidence (years)
	Radiogenic %	Non-radiogenic %		
Leukaemias	5·3	94·7	2974 ⎫	3–4
Lymphosarcomas	5·4	94·6	500 ⎭	
Cerebral tumours	4·8	95·2	1030	0; 4; 7.
Neuroblastomas	5·4	94·6	636 ⎫	1
Wilms' tumours	6·1	93·9	572 ⎭	
Osteosarcoma	5·4	94·6	131	Over 9
Total	5·3	94·7	5843	

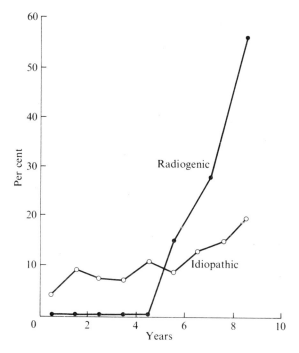

FIG. 5.12. Per cent distribution of radiogenic and idiopathic osteogenic sarcoma in the age range 0–9 years (Stewart 1970, by courtesy of author).

it is possible that the peak of radiation induced osteosarcomas will be found to occur during early adult life (i.e. after the adolescent peak which is probably caused by cases initiated during the first half of foetal life). The latent periods for other and lymphatic leukaemias induced by prenatal irradiation were also shown to be different as shown in Fig. 5.13. It is probable that myeloid leukaemia is a more acute disease than lymphatic leukaemia (Stewart and Kneale 1969). Only about $\frac{1}{6}$th of juvenile leukaemias are definitely myeloid, the others are described as acute lymphatic or blast cell leukaemias. Stewart considers that all juvenile cancers fall into the category of non-teratomatous tumours arising from embryonic cells (Bodian 1963) and that the only reason why leukaemia is so much more common than the other childhood cancers is because myeloid-lymphoid tissues form such a large component of the developing organism. A dose–response curve showing the relationship between exposure of the foetus to X-rays and the incidence of childhood cancer is shown in Fig. 5.14 (Stewart and Kneale 1968, Stewart and Kneale 1970, Stewart and Barber 1971). The lowest dose at which malignancies were noted was 200 mrad. The curve is linear and suggests that there is possibly no threshold. The extremely low

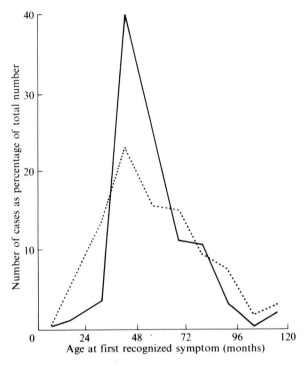

Fig. 5.13. Age distribution of 190 juvenile leukaemias initiated by obstetric X-rays shortly before birth (Kneale analysis of O.S.C.C. data). ——— lymphatic leukaemia (99 cases), ········ all varieties (190 cases), (Stewart and Kneale 1969, by courtesy of authors and publishers).

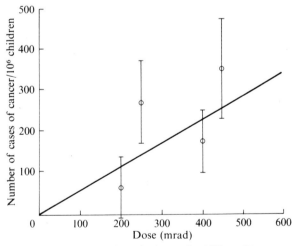

Fig. 5.14. Dose–response curve for radiogenic cancers in children. (Stewart and Barber 1971, by courtesy of authors and publishers).

TABLE 5.7

Leukaemias and lymphosarcomas
(Sample of adult deaths in England and Wales, 1954–58)
Estimated proportions of radiogenic and non-radiogenic cases
by cell types and X-ray history
(courtesy of Dr. A. M. Stewart)

Diagnostic groups	Aetiological groups[3]	Therapeutic X-rays %	Diagnostic X-rays %	Neither %	Total %	No. of Cases
M Series[1]	R	1·0	8·8	—	9·8	511
	Non-R	3·7	54·6	31·9	90·2	
L Series[2]	R	—	—	—	—	512
	Non-R	3·1	58·6	38·3	100	
Both	R	0·5	4·4	—	4·9	1023
	Non-R	3·4	56·6	35·1	95·1	
Total		3·9	61·0	35·1	100	

([1]) Comprising 88% myeloid, 8% monocytic, and 4% unspecified leukaemias.
([2]) Comprising 88% lymphatic leukaemias and 12% lymphosarcomas.
([3]) R = radiogenic or X-ray induced cases.

dose at which cancers have actually occurred should not be forgotten in considering all dose–response curves.

3.1.2. *Adult.* In 1962 Stewart and her colleagues published a study of the relation of diagnostic and therapeutic X-rays to the incidence of leukaemia and lymphosarcoma in adults (Stewart, Pennybacker, and Barber 1962). The population analysed consisted of 963 cases of leukaemia, 60 of lymphosarcoma, and 951 of other cancers, with 974 as controls. The results are shown in Table 5.7. The lymphatic leukaemias and the lymphosarcomas were combined to form a group of 512 cases (L series) for comparison with an M series consisting of 450 cases of myeloid leukaemia and 61 of other leukaemias. Following a detailed analysis of trunk and peripheral exposures to diagnostic and therapeutic X-rays at stated intervals, it was concluded that the radiogenic or X-ray-induced cases were confined to the M series and were caused only by X-rays of the chest or abdomen taken within 10 years of initial illness, usually within 5 years. The ratio of radiogenic to non-radiogenic cases was higher for therapeutic than for diagnostic exposures but the proportion of myeloid leukaemias ascribed to diagnostic X-rays (8·8 per cent) was higher than the proportion ascribed to radiotherapy (1.0 per cent).

3.2. *Therapeutic*

3.2.1. *Thymic enlargement in children.* A large number of children on the North American continent received irradiation for ablation of the thymus in the early part of this century. Treatment with X-rays (130–200 kV) with doses varying from 150–684 R was given. 6800 children were later traced (Latourette and Hodges 1959; Conti, Patton, Conti, and Hempelmann 1960; Saenger, Silverman, Sterling, and Turner 1960; Hempelmann, Pifer, Burke, Terry, and Ames 1967). In a group of 2878 with a mean follow-up period of 17·4 yrs Hempelmann and his colleagues found no osteosarcomas but ten benign osteochondromas in the irradiated area when 4·5 might have been expected and five in adjacent bone. There were six leukaemias when only two might have been expected. It should be remembered that in these children only a small part of the skeleton was irradiated which may account for the low incidence of leukaemia. It is particularly interesting to note the occurrence of benign rather than malignant bone tumours. This seems to be a feature of irradiation of the child's skeleton, for which at present there is no explanation, but Spiess has shown the same in children given ^{224}Ra (Chapter 7.4.1.2). The irradiated American children, it should be noted, also showed a high incidence of thyroid carcinomas which is not surprising since the neck area was involved in the radiation field.

3.2.2. *Ankylosing spondylitis.* In 1957 Court Brown and Doll published a follow up of 13 352 patients with ankylosing spondylitis who had received radiotherapy in the United Kingdom. The increase in mortality from leukaemia and aplastic anaemia in this group compared with the rest of the population was highly significant. They suggested that many of the patients diagnosed as aplastic anaemia were probably cases of leukaemia. In 1965 these authors reported a further follow up of 14 554 patients over a period which varied from 5 to 25 years (Court Brown and Doll 1965). The follow up, it must be noted, was in many cases considerably longer than in their earlier study. They estimated that, in an average follow-up period of 13 years after first treatment, the excess deaths from leukaemia and from other cancers arising in heavily irradiated tissues, which can be attributed to the effects of ionizing radiations, were 4 per 1000 patients and 6 per 1000 patients respectively. No excess was noted in lightly irradiated tissues. The other cancers arose in a wide variety of tissues.

Table 5.8 shows the observed deaths as a proportion of the number

TABLE 5.8

*number of observed deaths expressed as a proportion of the number expected,
by cause and period after first observation
(Court Brown and Doll 1965, by courtesy of authors and publishers)*

Cause of death	Years after First Observation						All periods
	0–2	3–5	6–8	9–11	12–14	15–24	
Leukaemia	6·4	12·8	10·6	7·0	11·1	3·7	9·5
Aplastic anaemia	27·3	50·0	33·3	14·3	0·0	0·0	29·4
Cancer of heavily irradiated sites	1·5	1·1	1·5	2·2	2·3	1·6	1·6
Cancer of colon	2·0	2·0	1·1	2·1	0·8	1·2	1·7
Cancer of lightly irradiated sites	1·3	1·1	1·0	1·5	0·5	1·7	1·1
All other causes	1·7	1·9	1·9	1·9	2·1	1·8	1·8
All causes	1·7	1·8	1·8	1·9	2·1	1·8	1·8

expected after first observation. Again it must be noted that patients diagnosed as aplastic anaemia on the death certificate on further enquiry often proved to have had leukaemia. Only 3 deaths were confirmed as due to bone sarcomas, while the expected number of deaths from this cause is estimated to be 0·63 (McKenzie, Court Brown, Doll, and Sissons 1961). However, as noted in the footnote to Tables 3.1 and 3.2, it must be reconized that this figure possibly exaggerates the incidence of osteosarcomas. The reliability of the diagnosis of primary bone tumour on a death certificate is not high, the difference between the observed and expected on a one-tailed test being $P \sim 0.025$.

The authors believe that it is reasonable from their results to adopt the working hypothesis that there is no threshold dose: 'This belief is based on the finding of a simple proportional relationship, over the range of dose observed, among those receiving only spinal irradiation and the fact that when the mean spinal dose to the whole population of patients is considered, in this case also, the incidence increases approximately in proportion to the dose at all but the highest levels' (Court Brown and Doll 1957). Figure 5.15 shows the incidence of leukaemia standardized for age in relation to the mean dose of radiation to the spinal marrow expressed in roentgens and in Fig. 5.16 the incidence of leukaemia, standardized for age in relation to the whole-body integral dose of radiation expressed in megagramme roentgens is plotted (Court Brown and Doll 1957). These results obtained in man, have been, to a large extent, the reason for the decision of the International Commission for Radiological Protection to accept a linear

76 THE EFFECTS OF IRRADIATION ON THE SKELETON

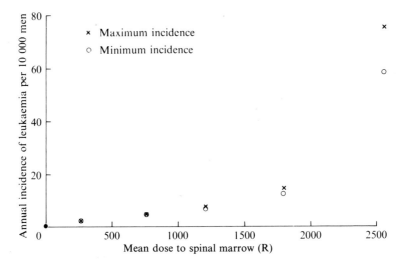

FIG. 5.15. The incidence of leukaemia, standardized for age, in relation to the mean dose of radiation to the spinal marrow. The two sets of points indicate the upper and lower limits of incidence calculated on the basis of the extreme assumptions made about the extent of the systematic errors in the investigation (Court Brown and Doll 1957, by courtesy of authors and publishers).

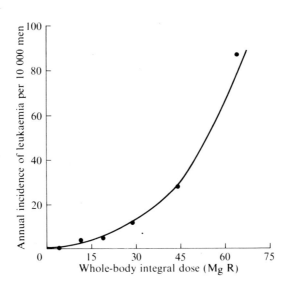

FIG. 5.16. The incidence of leukaemia, standardized for age, in relation to the whole-body integral dose of radiation (Court Brown and Doll 1957, by courtesy of authors and publishers).

EXTERNAL IRRADIATION

relationship between radiation dose and radiation malignancy as a working hypothesis, an hypothesis which is, however, questioned by some observers.

3.2.3. *Pelvic irradiation*

3.2.3.1. *Leukaemia.* Irradiation of the pelvic organs either for malignant or benign conditions inevitably involves irradiation of appreciable parts of the skeleton containing active haemopoietic marrow. Several surveys are reported of the effects of such irradiation. The results are far from clear cut, though on the basis of single case reports or analysis of small groups of patients, to be discussed later, such pelvic irradiation will result, particularly, in severe bone dysplasia. It has proved difficult in practice to estimate with any precision the bone or marrow dose received for many reasons. In some large series of patients followed up patients had received intracavity radium for carcinoma of the cervix together with external X-rays. In others, X-rays only had been given for benign gynaecological conditions such as metropathia haemorrhagica.

The results of the large scale follow-up of patients receiving pelvic irradiation will first be discussed. No clear answer as to the leukaemogenic effect of such irradiation has emerged. The reports are contradictory. In 1960 Simon, Brucer, and Hayes published the result of a follow-up by questionnaire of 71 000 patients treated for carcinoma of the cervix by radium, with or without external X-ray therapy in addition. They estimated a radiation dose of 2000–3000 rad was received throughout the pelvis. The marrow of the pelvis, they concluded, received a dose measured in thousands of rad, that of the head and necks of the femurs a dose measured in hundreds of rad and the thorax and legs measured in tens of rad. The follow-up covered five to fifteen years. The authors concluded that the incidence of leukaemia was no greater than that seen in the general female population in the same age group. This conclusion has been criticized however on the grounds that errors both of over and under estimation were possible in the way the survey was carried out and further that follow-up may not have been complete (Hutchison 1968, Doll and Smith 1968). Hutchison, in 1968, published an international prospective study of 29 493 patients with cervical cancer who were followed up for at least 5 years. Of these, 14 per cent had been treated with intracavity radium alone, 8 per cent with external therapy alone, and 69 per cent with a combination of the two, leaving 9 per cent who had had no radiotherapy. This survey contains much detailed tabulated information. The leukaemia incidence for the total population studied and for the various subgroups defined by type of therapy and by interval after radiation exposure were all compatible with general population rates. However, in the same year, Doll and Smith reported rather different results in a group of 2068 women treated for metropathia haemorrhagica and other benign gynaecological conditions by X-rays only. Five or more years after treatment the observed mortality was greater than expected for leukaemia (6 deaths against 1·0) and for cancers of other directly irradiated sites (31 deaths against 18·4). The average mean marrow-dose was calculated to be 136 rad. The authors state that 'previous observations on subjects who received more widespread irradiation suggest that this dose would produce about 6 cases of leukaemia in 10–20 years. The results are therefore consistent with the hypothesis that the risk of leukaemia induction is proportional to the total energy absorbed in the marrow irrespective of the volume irradiated.'

In 1817 patients treated with radium only and followed up for 10 years or till an earlier death, Dickson (1969) stated there was no excess of leukaemia but an excess of uterine cancer. However, when the detailed paper is read it does appear that five patients subsequently developed leukaemia. In 1971 Alderson and Jackson reviewed 2049 patients who had been treated by ovarian irradiation for menorrhagia. Of these, 14·5 per cent in the series had been treated with radium; the others received X-rays. Patients under 40 had received a total ovarian dose of 1250 rad orthovoltage or 1500 rad megavoltage. The average follow-up period was 14·6 years. There were 3 deaths from leukaemia in the patients treated with X-rays, compared with an expected figure of 0·37, the excess being significant at the 1 per cent level, and a slight excess of pelvic cancer. The authors consider that in some cases the advantages of radiation therapy outweigh the risks involved.

3.2.3.2. *Osteogenic sarcoma.* No mention is made in any of the four surveys of effects on osteogenic tissue, possibly because the follow-up period was relatively short. Apart from the occurrence of benign exostosis or osteochondroma occurring in the upper extremities of children irradiated for thymic enlargement, evidence that external irradiation might injure osteogenic tissue has come from isolated case reports or examination of small groups of patients. Among the 150 cases of osteosarcoma developing in bone during therapeutic irradiation, few appear to be associated with pelvic irradiation (Hatfield and Schulz 1970). Dahlin and Coventry (1967) mention three, two arising in the upper part of the femur and one in the ilium, but they give no details.

3.2.4. *Other localized skeletal irradiation*

3.2.4.1. *Osteogenic sarcoma.* Osteogenic sarcomas associated with external irradiation are reported as occurring both in normal bone involved in the irradiation of some non-bony lesion as for instance carcinoma of the breast, and secondly in bone irradiated for some supposedly benign bone lesion.

It should be noted that in the early cases treated, radiation was frequently given for tuberculosis of the knee joint and it has been suggested that the inflammation associated with the tubercle bacillus was in part responsible for the subsequent malignancy. As will be discussed (Chapter 6.5.2.3) much of the early experimental work in animals planned to study the effect of radiation on the skeleton combined artificial inflammation with irradiation (Lacassagne and Vinzent 1929), but it is now recognized that osteogenic sarcomas can be produced experimentally in perfectly healthy bone (Cater, Baserga, and Lisco 1959; Vaughan, Lamerton, and Lisco 1960; Baserga, Lisco, and Cater 1961). The first cases of sarcomas arising in previously normal bone included in the beam of roentgen rays given to adjacent tissue were described by Beck in 1922 and 1924.

The recorded cases occurring in both normal and abnormal bone now number at least 150 (Hatfield and Schulz 1970). Reviewing 552 cases of osteogenic sarcoma seen between 1925 and 1955, McKenna and his colleagues (McKenna, Schwinn, Soong, and Higinbotham 1966) classified 22 as irradiation induced on the basis of the following criteria: 'the sarcoma must be confirmed by microscopic examination; it must arise within a bone that has received at least 3000 rad five or more years prior to development of the sarcoma; the patient must have had an asymptomatic latent period following irradiation therapy; and the bone at the ultimate site of the sarcoma must have been roentgenographically normal or the site of a histologically proven benign condition.' In a review of 600 patients with osteogenic sarcomas Dahlin and Coventry (1967) recognized 16 that had had previous irradiation. Three developed after irradiation for carcinoma of the breast, all in the scapula, two after irradiation for squamous cancer of the cervix and one followed irradiation for adenocarcinoma of the endometrium. Two of these three involved the upper portion of a femur and one the ilium. They also mention briefly seven osteogenic sarcomas of the jaw bones, eighteen fibrosarcomas of various unspecified bones, one Ewing sarcoma, and one chondrosarcoma arising in bones that had been the site of prior irradiation. There are several other excellent reviews of cases recorded in the literature together with cases added from the authors' own experiences which include malignancies arising in both normal and abnormal bone (Cahan, Woodard, Higinbotham, Stewart, and Coley 1948; Jones 1953; Sabanas, Dahlin, Childs, and Ivins 1956; Cruz, Coley, and Stewart 1957; Cohen and D'Angio 1961; McKenna, Schwinn, Soong, and Higinbotham 1966; Dahlin and Coventry 1967; Hatfield and Schulz 1970).

Normal bone. It has been suggested that osteogenic tumours arising in normal bone included in a radiation field are not causally related to the radiation but examples of the so called 'double primary' phenomenon. Pendergrass (1968) recently published the case of a 63-year old female in whose right third rib an osteogenic sarcoma developed seven years after a right radical mastectomy. Since no radiotherapy had been given Pendergrass suggested that 'just because radiation is a cause of cancer does not, in my opinion, mean that the appearance of a second type of malignancy in an area which previously had received a large dose constitutes good evidence of radiation induction'. As it is true that the patient with a primary malignant tumour is statistically more at risk from a second unrelated neoplasm, Pendergrass certainly raises a valid point. However, the latent period for the double primary is usually shorter (Moertel, Dockerty, and Baggenstoss 1961; Pickren 1963) than that observed in the radiation cases (Hatfield and Schulz 1970). The possibility of multicentric origin of osteogenic sarcomas arose in 16 of 600 cases reviewed by Dahlin and Coventry (1967). Amstutz (1969) in his recent review of the recorded cases of possible double primaries in the case of osteogenic sarcoma is very sceptical and considers the second tumour to be a metastasis. It has also been suggested that the sarcomas observed after radiation may have really been spontaneous. Hatfield and Schulz (1970) however argue that the apparent frequency of post-irradiation sarcomas in patients who have received therapeutic irradiation and survived the primary lesion for ten years, at least in the case of carcinoma of the breast, is considerably greater than the frequency of spontaneous osteogenic sarcomas. They calculate that the incidence of bone sarcomas in the normal population is of the order of 0·50 per 100 000 and the incidence of post-irradiation sarcomas per 450 radiation-treated and cured breast cancers surviving 10 years is one. This latter figure is based on their own results and those of Phillips and Sheline (1963). Hatfield and Schulz (1970) themselves describe 4 osteogenic sarcomas, 1 chondrosarcoma, and 2 fibrosarcomas arising adjacent to bone following therapeutic irradiation involving normal bone.

Radiation dose. It is extremely difficult, if not impossible, to arrive at any measure of the radiation dose received by the early cases. It is probable that doses above 2000 R and certainly above 3000 R are carcinogenic. Jaffe (1958) mentions that in an occasional case the dose may have been as low as 1500 R.

The children with retinoblastoma nearly all received a very high radiation dose presumably because of the great malignancy of the original tumour. It is of interest to note that in the most recent analysis of skeletal changes following irradiation of childhood tumours only one post-irradiation sarcoma is included in a series of 31 cases (Katzman, Waugh, and Berdon 1969), probably because the radiation dose was moderated in the majority. The patient with osteosarcoma received 6800 R given for retinoblastoma while the maximum received by the others was 4248 R, the majority receiving less.

Forrest (1961) and Reese, Hyman, Merriam, and Forrest (1957) had earlier reported the development of sarcoma following treatment of retinoblastomas with radiation therapy in over 5 per cent of cases.

Hatfield and Schulz (1970) emphasize that megavoltage therapy was used in three of their cases of carcinoma of the breast who subsequently developed osteogenic sarcoma though it is usually supposed that orthovoltage therapy is more likely to induce sarcomatous degeneration. They consider that factors, such as the energy source and method of delivery have no effect on the latent period.

Latent period. In discussing latent period it is essential to remember that the latent period in an old person may well be different from the latent period in a child or young adult for the same type of malignancy. On the other hand relative growth rates for different tumours probably hold good throughout life (Stewart 1972*a,b*).

Soloway (1966) has studied the latent period particularly and has compared the latent period for the development of secondary tumours following radiotherapy for retinoblastoma in childhood with that for their appearance following radiotherapy in adulthood. In both groups the mode for mesenchymal tumours occurred between 6 and 10 years. However, examination of all the recorded cases suggests that the latent period tends to be longer in the older age-group. Here latent periods as long as 24 years (Wolfe and Platt 1949) and longer are recorded. Jones (1953) mentions 30 years in the case of an osteogenic sarcoma arising in the clavicle and sternum following irradiation (dose not given) for a naevus in the neck.

Type of tumour. The commonest tumour to arise in normal adult bone included in a radiation field is the osteosarcoma, but fibrosarcomas and chondrosarcomas do occur. Two chondrosarcomas are recorded in the early literature occurring in the femur irradiated for tuberculosis of the limb (Marsch 1922; Gruca 1934). Cohen and D'Angio (1961) note one case occurring in a rib in a child. Hatfield and Schulz (1970) mention four osteogenic sarcomas, one chondrosarcoma of clavicle, and two fibrosarcomas in their small series. Hatcher (1945) also records a chondrosarcoma of the rib. The chondrosarcoma he describes in a fibula as secondary to radiation occurred in a patient whose tibia had been irradiated four years before also for a chondrosarcoma. This is perhaps an odd story and difficult to interpret. The only chondrosarcoma reported in the skull occurred in a child of one year. There are also four fibrosarcomas and two osteosarcomas of the skull recorded (Skolnik, Fornatio, and Heydemann 1965; Berg, Landberg, and Lindgren 1966).

Some of the fibrosarcomas are noted as arising in relation to the periosteum and do not appear to invade (Hatfield and Schulz 1970).

Benign exostoses so characteristic of the response of young bone to irradiation are discussed in the next chapter.

Site of tumour. The site of the tumour is largely governed, especially in children, by the site which is likely to have been involved in heavy irradiation. Thus many tumours are recorded in the region of the orbit following irradiation of retinoblastoma. Soloway (1966) collected 17 from the literature and added three of his own. Others occur in the vertebrae, ilium, or ribs following irradiation for Wilm's tumour (Katzman, Waugh, and Berdon 1969). In the earlier literature describing radiation treatment for tuberculosis many tumours were recorded subsequently in the region of the knee joint.

In adults there is practically no bone which has not been the site of a radiation induced tumour though clearly they are more common in those sites which are likely to be heavily irradiated such as ribs, scapula, and clavicle (Hatfield and Schulz 1970). Seven cases of sarcoma of the skull following radiation are on record in adults not included in Soloway's series (Berg, Landberg, and Lindgren 1966).

Pathological bone. The majority of the recorded tumours in bone following external irradiation have occurred in bone irradiated for a condition that was considered to be benign. They are discussed in considerable detail in the recent review of McKenna, Schwinn, Soong, and Higinbotham (1966).

The fact that tumours do arise in normal bone suggests that the bone lesion in these cases is not necessarily an aetiological feature. The condition that gives most difficulty in interpretation is the *giant cell tumour* (Lichtenstein 1951, 1953; Williams, Dahlin, and Ghormley 1954) (Chapter 4.2.2.). Approximately 10 per cent of these tumours prove to be clinically malignant without radiation according to Murphy and Ackerman (1956); Sissons (1966) claims that increasing experience shows many of these tumours to be potentially malignant and that a significant proportion finally metastasise to the lung. Jaffe (1958) quotes three of his own cases which were carefully followed. He suggests that a differentiation can be made between malignancy caused by radiation of a benign tumour and malignancy arising from the tumour itself. The latter undergo malignant transformation progressively and fail to respond to treatment while those that respond to treatment and then some years later develop malignant change can be attributed to radiation.

Sabanas and his colleagues (1956) record nine giant cell tumours that became malignant following heavy radiation; with one exception not less than 3000 R had been given and they record one dose of 7200 R. The latent period varied from $3\frac{1}{2}$ to 13 years. Several giant cell tumours treated by radiation which became malignant are also described by Cruz, Coley, and Stewart (1957). Goldenberg, Campbell, and Bonfiglio (1970) have recently presented an analysis of 218 giant cell tumours. Three died of post irradiation osteosarcomas. They suggest that irradiation should only be employed for those tumours that cannot be treated surgically. Irradiation should not be combined with surgery. In another large series of 195 cases Bickel (1970) quotes Dahlin as concluding that radiation does not prevent recurrence. Seven patients (19 per cent) of the thirty-seven patients treated by radiation developed osteosarcomas but only one (three per cent) of the patients treated surgically.

Other benign bone lesions may become malignant after radiation therapy. Jaffe (1958) quotes the case of a boy who developed an osteosarcoma 5 years after the end of radiation therapy for a solitary bone cyst. Other cases of bone cysts in children becoming malignant have been described (Cahan, Woodard, Higinbotham, Stewart, and Coley 1948; Francisco, Pusitz, and Gerundo 1936; Cruz, Coley, and Stewart 1957). Several cases of fibrous dysplasia becoming malignant following radiation are recorded (Cahan, Woodard, Higinbotham, Stewart, and Coley 1948; Sabanas, Dahlin, Childs, and Ivins 1956). A benign chondroblastoma is also recorded as becoming malignant after curettage; 3600 R had been given to the site (Hatcher and Campbell 1951). The chondrosarcoma that later developed had no histological resemblance to the original lesion.

3.2.5. *Dysplasia*. Therapeutic irradiation of the skeleton, particularly in the young, may have serious results quite apart from inducing malignancy. The possible risks are better appreciated than they were in the early days of radiotherapy so that one may expect to see less skeletal damage in the future than is recorded in the literature.

Retardation of growth and abnormal but benign growth resulting in exostosis are the hazards of excessive local radiation of the skeleton in the young. In the elderly 'radiation necrosis', which may result in fracture, has in the past been common.

3.2.5.1. *Retardation of growth*. Perthes recognized as long ago as 1903 that therapeutic X-rays could retard the growth of the long bones. This has been confirmed by all subsequent observers (Katzman, Waugh, and Berdon 1969).

The most common lesions observed are changes in the vertebral bodies often resulting in scoliosis, hypoplasia of the ilium and rib cage, and epiphyseal injury resulting in discrepancy in length of the limbs. A detailed account of the vertebral body changes was first given in 1952 by Neuhauser and his colleagues. Similar lesions have since been well described by Rubin, Duthie, and Young (1962) and by Katzman, Waugh, and Berdon (1969). The latter report altered development of the vertebral bodies in 23 out of 28 patients (82 per cent). In the latter part of the first year following radiotherapy, subcortical lucent zones orientated parallel to the epiphyseal plates appeared and progressed to growth-arrest lines so giving the appearance of a bone within a bone. Subsequently the vertebral bodies appeared more bulbous than normal and later appeared to diminish in vertical height. Irregularities or scalloping of their superior and inferior surfaces often developed. The final picture may be like that seen in cretinism or Hurler's disease. Scoliosis has long been recognized as a complication of radiation therapy (Arkin, Pack, Ransohoff, and Simon 1950; Neuhauser, Wittenborg, Berman, and Cohen 1952; Whitehouse and Lampe 1953; Rubin, Duthie, and Young 1962; Vaeth, Levitt, Jones, and Holtfreter 1962;

Berdon, Baker, and Boyer 1965). The spinal curves that develop have their concavity to the side of the primary tumour provided there is no tumour invasion of bone. Katzman reports scoliosis in 71 per cent of his patients, twelve of whom had Wilm's tumour and eight a neuroblastoma.

Hypoplasia of the ilium and rib cage, though seldom mentioned by other observers, is thought by Katzman to be a frequent complication. He found it in fifteen of twenty-five patients treated through abdominal fields, and considers that patients with iliac hypoplasia had more severe scoliosis and vertebral body changes than those who showed no ilial changes. Only one of these patients had such hypoplasia without scoliosis and she had been followed for less than 4 years.

The most dramatic retardation of growth is probably that described by Franz (1950). A boy of 16 showed a shortening of $9\tfrac{3}{4}$ inches in one femur when examined. At the age of $2\tfrac{1}{2}$ weeks he had received, over a short period, 2992 R for treatment of a haemangioma overlying the distal femoral epiphysis. The shaft diameter and the thickness of the cortex were also diminished. Transverse dark lines were apparent in roentgenograms in the region of the metaphyseal bone and two exostoses cartilaginae were present in the field of irradiation. A similar radiological picture is described by Spiess (1969) in children treated with ^{224}Ra (see Chapter 7.4.1.2.) though he does not mention severe bone shortening. Benign exostoses are also mentioned in the region of the head of the humerus and in the scapula in the large series of children treated for enlargement of the thymus. A follow up of 2878 young people so treated was reported in 1967 (Hempelmann, Pifer, Burke, Terry, and Ames 1967); 10 had benign osteochondroma in the irradiated area and 5 were seen in adjacent bone. Only 4·5 would have been expected in a population of this size. The latent period was 10–16 years and the accumulated dose 150–680 R. Exostoses following radiation have also been reported by several other authors in ribs, ilium, humerus, scapula, femur, and tibia (Neuhauser, Wittenborg, Berman, and Cohen 1952; Franconi and Illig 1959; Kolar 1960; Murphy and Blount 1962; Cole and Darte 1963; Katzman, Waugh, and Berdon 1969). There is no evidence that they become malignant or differ in any way from the usual osteochondromas of idiopathic origin (Berden, Baker, and Boyer 1965). Two benign osteochondromas arising spontaneously have been reported as becoming sarcomatous (Dahlin and Coventry 1967).

Dose relationship. The severity of the changes observed appears to be related to the radiation dose received (Frantz 1950; Neuhauser, Wittenborg, Berman, and Cohen 1952; Whitehouse and Lampe 1953; Katzman, Waugh, and Berdon 1969). Certainly doses above 2000 R can be expected to produce skeletal injury in growing children. Doses as low as 150–600 R have produced benign exostosis (Hempelmann, Pifer, Burke, Terry, and Ames 1967).

Katzman and his colleagues, however, do not consider it is yet possible to establish a therapeutic dose at which the malignant disease under treatment will be cured and yet no bone damage result. They suggest that the extent of the disease at the time of treatment may well explain some differences in the results obtained. This conclusion is also reached by a recent review of 335 cases of Wilm's tumour (Ledlie, Mynors, Draper, and Gorbach 1970). Other factors are the inherent growth potential of the affected bone and the age of the child. Bones with the highest potential will be most affected. The ilium is more likely to be affected than the rib, and the femur than the fibula.

The younger the child the more severe are the lesions observed (Neuhauser, Wittenborg, Berman, and Cohen 1952; Whitehouse and Lampe 1953; Vaeth, Levitt, Jones, and Holtfreter 1962; Katzman, Waugh, and Berdon 1969). Katzman suggests however that this may be due in part to the fact that, for the neuroblastomas and Wilm's tumours, the earlier treatment is begun the better the prognosis and therefore the younger children survive to show later radiation dysplasia while the older children die before the dysplasia becomes apparent. The same may well be true of other conditions irradiated in children.

Pathology. The basic underlying pathological lesion, resulting in growth retardation following radiation, is interference with the normal pattern of epiphyseal growth. There is no reason to believe that what happens in man differs fundamentally from what occurs in the experimental animal, so the experimental results are next discussed.

There have been extensive experimental studies carried out to elucidate the mechanism of radiation injury to the cells of the epiphyseal plate. The effects of external radiation do not differ from those of internal radiation from, for instance, long-range beta particles such as those arising from the decay of $^{90}Sr^{90}Y$ (Woodard and Spiers 1953; Wilson 1956a,b; Kember 1960, 1965; Macpherson, Owen, and Vaughan 1962; Blackburn and Wells 1963; Macpherson 1961). It is clear from all such studies using both internal and external irradiation that there is damage to the cells, both of the cartilage and of the osteogenic tissue. This affects their enzymic content and capacity to handle ^{32}P and ^{45}Ca. There is also injury to the blood vessels, which will have secondary effects on the bone and cartilage cells. However careful analysis suggests that there are direct cellular effects independent of vascular damage. What is remarkable is the capacity of the epiphyseal plate and its blood supply to recover so that growth, though initially abnormal, is resumed (Kember 1967, 1969).

Woodard and her colleagues (Woodard and Spiers 1956, Woodard and Laughlin 1957) have studied the effect of X-rays of different qualities on the alkaline phosphatase content of the epiphyseal region of the knee in mice. There was an initial reduction following doses of X-rays generated at 100 kV, 185 kV, and 1000 kV, followed by a period of temporary recovery for doses not exceeding 1500 R succeeded by a later and permanent reduction in phosphatase activity. This is thought to be due to a decreased production rather than to destruction of preformed enzyme (Dziewiatkowski and Woodard 1959). There is also a depression in the uptake of ^{35}S sulphate

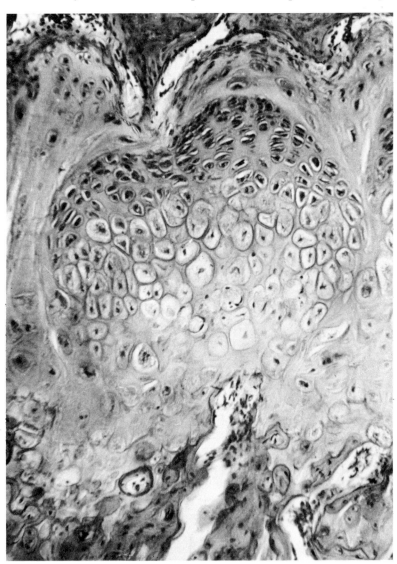

FIG. 5.17. Recovery clone in growth cartilage at 25 days after 1700 rad. Longitudinal section of rat tibia. Haematoxylin and eosin, × 160. (Kember 1967, by courtesy of author and publishers.)

and of ^{32}P phosphate (Wilson 1956a,b, 1958, 1959). In rat epiphyses a failure to utilize ^{45}Ca following radiation has been shown (Cohen and Gong 1953, Blackburn and Wells 1963). In association with these biochemical changes detailed studies of the histological changes have been made both in mouse and rat epiphyses (Blackburn and Wells 1963, Kember

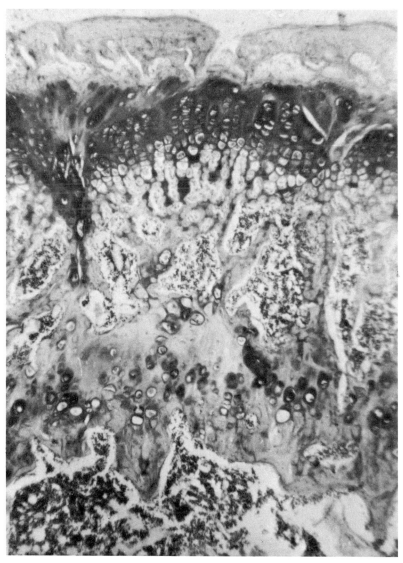

FIG. 5.18. The cartilage plate at 50 days after 1700 rad. A new active metaphysis separates the growth cartilage from the calcified cartilage 'bar'. Haematoxylin and eosin, × 70. (Kember 1967, by courtesy of author and publishers.)

1967, Kember and Coggins 1967, Kember 1969, Kember and Sadek 1970).

Beautiful experiments have been described by Kember (1965, 1967) showing the special pattern of recovery in the cartilage plate during which clones of active cells appear and in due course repopulate the plate. He is doubtful if recovery of the vascular supply of the plate plays a significant part. While 900 rad had little effect on the blood supply, 1800 rad reduced it and although a substantial number of vessels remained intact the cartilage cells showed severe damage. The clones of active cartilage cells appeared above the calcified cartilage bar of damaged cells which constitutes the characteristic dark band seen in radiographs of stunted bones. Figure 5.17 shows a recovery clone in growth cartilage at 25 days after 1700 rad in a rat's tibia and the cartilage plate at 50 days after 1700 rad. Here a new active metaphysis separates the growth cartilage from the calcified cartilage bar as shown in Fig. 5.18.

3.2.5.2. *Fracture.* In adults, fracture, particularly affecting the neck of the femur, is the most common result of radiation dysplasia following both external and internal radiation. It was first discussed by Baensch in 1927. That such fractures are much more common in women than in men is due to the fact that the neck of the femur is often exposed to heavy external irradiation given for gynaecological reasons. All such fractures cannot be attributed to senile or post menopausal osteoporosis. There is an early recorded case which occurred in a woman of 34 (Stampfli and Kerr 1947) and many are noted in women between the ages of 40 and 50 though the average age is about 65 (Kok 1953, Stephenson and Cohen 1956). Two cases are recorded in men who were irradiated for carcinoma of the penis (Batt and Hampton 1940, Hight 1941). Three other male cases are known (Bonfiglio 1953). Several cases of bilateral fracture not necessarily occurring at the same time have been recorded (Vaughan 1962). In several instances autopsy or biopsy examination showed no secondary deposits and in others the fractures healed well. Fractures in the ribs or clavicle following irradiation for breast cancer are far less common (Peck 1939, Slaughter 1942, Smithers and Rhys Lewis 1945). In a series of 369 such patients only 4 had fractures attributed to radionecrosis. Baudisch (1960) suggests however that such fractures may be unnoticed unless the chest is radiographed. Fractures have also been described in the pelvis (Peck 1939; Gratzek, Holmstrom, and Rigler 1945).

Pathology. There are numerous detailed accounts of the histological picture seen in radiation dysplasia in man (Ewing 1926; Baensch 1927; Dahl 1934, 1936; Dalby, Jacox, and Miller 1936; Hight 1941; Barden 1943; Stampfli and Kerr 1947; Bonfiglio 1953; Kok 1953; Stephenson and Cohen 1956; Jaffe 1958; Goodman and Sherman 1963). It is still not clear whether the underlying cause of the dysplasia is vascular injury, destruction of bone cells, or direct injury to bone collagen. They probably all play their part. The fact that the femur is subjected to heavy weight bearing that may render it specially liable to injury must not be forgotten in considering the frequency with which this bone is affected.

Ewing in 1926 gave the name radiation osteitis to the changes he described in the neck of the femur. He emphasized the importance of vascular injury and also the part played by the associated marrow aplasia in reducing the nutrition of the affected bone, and speculated on the direct effect of radiation on the collagen, causing the bones to become brittle. The latter was originally emphasized by Regaud (1922) as one of the effects of radiation on bone.

Probably the earlier reports in the literature are descriptive of a condition more rarely seen today, when the importance of maintaining as low a radiation dose as possible is appreciated. As a result of a more recent study Goodman and Sherman (1963) conclude that at the present time post-irradiation fracture of the femoral neck occurs in viable bone which has a blood supply sufficient to support new-bone formation. They noted minimal anatomical changes but suggest that these do not reflect the degree to which the functions of the cells have been disturbed. Such functional disturbance of bone and cartilage cells exposed to radiation has been well documented in experimental animals. They conclude that the actual pathogenesis of fracture is probably stress concentration at the interface between normal and abnormal bone. Jaffe (1958) describes the histological picture of florid radiation osteitis as representing in various combinations aseptic bone necrosis, disintegration of necrotic bone, and the representative reparative reaction. To a greater or lesser degree the radiation injures and/or destroys not only the cellular elements of the bone marrow but also the cells involved in the formation, maintenance and reconstruction of the osseous tissue proper, i.e. the osteoblasts, osteocytes, and osteoclasts. Death of the osteocytes results in degenerative changes in the collagen. Jaffe considers there are direct radiation effects upon the bone cells upon which may be superimposed on the effects of vascular injury.

3.2.5.3. *Osteomyelitis.* In 1926 Ewing drew attention to the fact that bone devitalized by irradiation is peculiarly liable to infection resulting in both periostitis and osteomyelitis. This is particularly important in the case of the jaw. Daland went so far in 1949 as to state that teeth and irradiation are not compatible. In a series of 1819 patients with intra-oral cancer treated with radiation, 235 developed osteonecrosis. If the lip cases were included the figure was considerably higher.

4. Experimental animals

The effects of experimental external irradiation of the skeleton in animals has largely taken the form of whole-body irradiation. No attempt is made here to review an extensive literature. Certain points only are made.

It is important in looking at radiation experiments on animals to appreciate the great species and strain differences in response to radiation (Lamerton 1961, 1966, 1968). For instance, Twentyman and Blackett (1970) have recently shown that the red cell system in C57Bl mice is much more easily affected by continuous gamma irradiation than the red cell system in F_1 hybrid rats of the inbred 'August' and Marshall strain, while Carworth rats respond differently from this hybrid strain. For this reason different species are discussed separately. A distinction is also made between whole-body and local irradiation.

4.1. *Whole body irradiation*

4.1.1. *Mice*. Finkel and Biskis (1968) have reported a low incidence of osteosarcomas in CF 1/Anl mice given external irradiation, e.g. 1·9–10·7 per cent. As already discussed, studies of tumour induction in this strain of mouse may be difficult to interpret (Chapter 1.1.3).

4.1.2. *Rats*. Two groups of parabiont rats have received whole body irradiation (Koletsky and Gustafson 1955; Warren and Chute 1963). Both developed some osteosarcomas but no leukaemia. The tumours were said to arise endosteally. The dose in rad was calculated to be 1000–1500 rad and the latent period was as long as 500 days. Lamerton (1966) gave 12 rats whole body radiation of 50 rad day^{-1} for 200 days from a ^{137}caesium source—no tumours developed. Among 80 rats given 1056 rad at 176 rad day^{-1} for 6 days there were 2 osteosarcomas, and 1 rat given 400 rad developed a similar tumour.

4.1.3. *Rabbits*. Hulse (1969) exposed adult rabbits to a brief period of whole body irradiation with gamma rays or fission neutrons. The mean energy of the neutrons was about 0·7 MeV and the dose rate about 2500 rad per hour. The mean energy of the gamma rays was about 2·5 MeV and the dose rate about 1950 R per hour. The results are shown in Table 5.9. Osteosarcomas, fibrosarcomas, and basal cell carcinomas appeared in animals receiving gamma rays or fission neutrons but the latent period was shorter following the latter. Hulse suggests that the RBE for the production of all these types of tumour is about 3. It is of some interest that nearly all the osteosarcomas occurred in the jaw as in adult rabbits given ^{90}Sr (see Chapter 8 Section 1.3.1.3) presumably because this is a site where there is still active proliferation of osteogenic tissue (Rushton, Owen, Holgate, and Vaughan 1961). The lowest dose to give tumours in the case of fast neutrons was 400–450 rad and in the case of gamma rays 1000–1300 R. This dose is considerably lower than that observed in the ^{90}Sr rabbits, 10 000–20 000 rad, but this was measured only at death and presumably contained much 'wasted radiation'.

TABLE 5.9

Adult rabbits given total body irradiation
(from Hulse 1969, by courtesy of author and publishers)

Type of radiation	Dose range: neutrons in rad gamma-rays in R	Number irradiated at least 1 yr ago	Number of rabbits with tumours			Total
			Osteo-sarcoma*	Fibro-sarcoma	Basal cell carcinomas	
Unirradiated	—	12	0	0	0	0
Neutrons	500–600	5	2(2)	1	0	3†
	400–450	11	2(2)	4	1	7
	200	9	0	0	0	0
Gamma rays	1400–1600	5	3(4)	1	0	3
	1000–1300	12	1(3)	0	4‡	5
	500	4	0	0	0	0

* Number of individual osteosarcomas given in parentheses.
† One rabbit had two osteosarcomas and a fibrosarcoma.
‡ Two rabbits still alive not confirmed histologically.

4.1.4. *Dogs.* Fritz and his colleagues (Fritz, Norris, Rehfeld, and Poole 1970; Norris, Fritz, Rehfeld, and Poole 1968) have described two cases of widespread myeloproliferative disorder in beagle dogs exposed to protracted whole-body irradiation from a ^{60}Co source. They calculate that the dogs received a total of 1275 rad at about 17 R day^{-1}. The authors describe the condition as anaplastic myelogenous leukaemia. The appearance of the tissues was similar to that seen in some of the swine and the dogs given ^{90}Sr, i.e. extensive invasion of many organs including the choroid plexus by neoplastic cells. Dogs receiving 72, 50, and 35 R day^{-1} and one receiving 24·5 R day^{-1} died with mild anaemia, profound leukopenia, and septicaemia during the 100 day irradiation period. No osteosarcomas were reported. It would be important to know if these would occur with a lower daily dose.

4.2. Partial body irradiation

4.2.1. *Mice.* Finkel and Biskis (1968) report experiments on partial exposure of CF 1/Anl mice to X-rays. 2000–5000 R caused gross tissue damage but when the hind legs as well as the pelvis and spine between the third sacral and ninth caudal vertebrae were exposed to 1000, 625, and 250 R no tissue damage occurred although osteosarcomas were induced as shown in Table 5.10. No tumours occurred in 420 control mice.

TABLE 5.10

Results 650 days after partial body exposure to X-irradiation
(adapted from Finkel and Biskis 1968)

Total Exposure dose (rad) (given 58–9 rad min^{-1})	No. mice	Osteosarcomas per 100 mice	Mortality %
1000	300	10·7	58·3
625	750	6·1	59·9
250	750	1·9	54·7

4.2.2. *Rats.* An extremely well documented study was made of the effect of the local application of gamma rays from an iridium source to the knee region of 60 day old Wistar rats by Cater and his colleagues (Cater, Baserga, and Lisco 1959, 1960; Baserga, Lisco, and Cater 1961). The dose given was 3000 R. Thirty-four separate osteosarcomas were found in 34 out of 116 irradiated rats between 300 and 800 days after irradiation. Only well developed tumours were counted, small nests of atypical bone formation frequently found in the metaphysis being excluded although they may well have been presarcomatous. Fifty rats developed fibrosarcomas. All the osteosarcomas arose within the bone. The fibrosarcomas could be divided into 3 groups: (1) those originating from the soft tissues of the leg in which the tumour cells did not even come into contact with the bone; (2) those in which the tumour cells did come into contact with the bone but did not invade, some of these being obviously of periosteal origin; (3) those that had replaced large portions of bone and in which an origin from within the bone could not be ruled out. No tumours arose from the patella even though radiation dysplasia there was as great as in the long bones. This is of particular interest as it is a bone which had ceased to grow at the time of the irradiation. The distribution of tumours correlated with the growth rate of the affected bones, the incidence being higher in the faster growing bones. Further, the incidence of tumours was higher in rats treated with growth hormone and the induction period was shorter in rats treated with thyroxine. The first bone tumours following external irradiation from X-rays were reported in 1910 by Marie, Clunet, and Raulot-Lapointe. After a latent period of 14 months a polymorphous cell sarcoma developed.

4.2.3. *Rabbits.* In 1929 Lacassagne and Vinzent induced osteogenic sarcomas in rabbits with an X-ray dose of the order of 1000 R. The tumours developed within 6–14 weeks. At this time it was thought that an inflammatory process was essential to tumour production and this was induced experimentally. It is now recognized that such tissue damage is not essential, though how far it may accelerate the effects of radiation is still not known.

4.2.4. *Monkeys.* A fibrosarcoma developed in the left mandible of a Macaca rhesus monkey 25 months after it received locally 7500 R of external, fractionated X-irradiation. The animal was the only one of fourteen monkeys irradiated to develop a malignancy. The monkey was about 28 months old at the time of irradiation (Gowgiel 1965).

6. Internal Irradiation from the Bone-Seeking Radionuclides

It has been customary, since Hamilton first published his classical paper in 1947, to describe those radionuclides that are concentrated and retained in the skeleton as 'bone seekers' and this peculiar characteristic attributed to them of actively seeking bone was further emphasized by Marshall (1969a) when he divided them into 'volume seekers' and 'surface seekers'. It should be emphasized that as understanding of the physiology of bone advances there are an increasing number of excellent biochemical reasons as to why the radionuclides behave as they do in the skeleton. The effects of internal radiation from these bone-seeking radionuclides depend on the physical, chemical, and biological characteristics of each radionuclide.

1. Physical characteristics

Some of the physical characteristics of the more important bone-seeking radionuclides as they affect the skeleton have already been shown in a summarized form in Table 4.1. They are shown in more detail in Table 6.1. These radionuclides vary greatly in the energy and range of their radiation, their half-life, and their decay products. Those that decay with an alpha emission like plutonium, americium, radium, and thorium might be expected to have a higher toxicity than those that decay with a beta emission like ^{45}Ca, because of the high energy of the alpha particle.

The injury they may do to cells depends not only on the physical character of the initially deposited element but also upon the physical character of the radioactive daughter products. For instance the danger of ^{90}Sr lies in the fact that it has a long half-life, 28 years, and that its decay product ^{90}Y has a high energy, 2·27 MeV, and therefore a long range in tissue. The decay chain of the thorium series which includes mesothorium and thorium X (^{228}Ra and ThX or ^{224}Ra) is shown in Table 6.2. The decay chain of the radium series including ^{226}Ra and the noble gas radon is shown in Table 6.3. It is evident that the radioactive daughter-products of some of these alpha-emitting bone-seeking isotopes must contribute to the dangers involved in their skeletal deposition. Those that are important are discussed in later sections and chapters (7.1.1, 7.3, 7.4, 11.1, 11.2).

IRRADIATION FROM BONE-SEEKING RADIONUCLIDES

TABLE 6.1

Physical characteristics of the more common bone-seeking radionuclides

	Half-life	Maximum particle energies MeV	Gamma ray energies MeV	Daughter products
^{45}Ca	165 days	β^- 0·254	—	Stable scandium
^{89}Sr	51 days	β^- 1.46	—	Stable yttsium
^{90}Sr	28 years	β^- 0·54	—	^{90}Y β
^{90}Y	64·2 hours	β^- 2·27	—	Stable zirconium
^{224}Ra (Thorium X)	3·64 days	α 5·68 95%	—	^{220}Rn
		α 5·44 4·9%	0·24	$\alpha + \beta$ to stable ^{208}Pb
^{226}Ra (Radium)	1620 years	α 4·78 94·3%	—	^{222}Rn
		α 4·59 5·7%	0·19	half-life 3·8 days $\alpha + \beta$ to stable ^{206}Pb
^{228}Ra (Mesothorium I)	5·7 years	β^- 0·04	—	^{228}Actinium (Mesothorium II) 6·13 hours $\alpha + \beta$ to stable ^{208}Pb
^{228}Th (Radiothorium)	1·91 years	α 5·42 71%	0·084	^{224}Ra
		α 5·34 28%		
^{239}Pu	2·4 × 10^4 years	α 5·15 72%	0·013 0·038	^{235}U
		α 5·13 17%		7 × 10^8 years
		α 5·10 11%	0·151	
^{241}Am	458 years	α 5·48 85%	0·06 36%	^{237}Np
		α 5·44 13%	0·03 2·5%	2·2 × 10^6 years
^{32}P	14·3 days	β^- 1·71	—	Stable suphur

In addition to the radionuclides listed in Table 6.1 there are a large number of others which are selectively concentrated in the skeleton but about which much less is known since hitherto they have only been available in minute amounts.

2. Chemical characteristics

The chemical characteristics of the bone-seeking radionuclides, to some extent account for their behaviour in the organism to which they gain access, and condition their biological characteristics. For instance calcium, strontium, and radium are alkaline earths. Their final pattern of distribution in the skeleton is the same though it may be achieved more slowly by radium and barium than by strontium and calcium, but their detailed pattern of retention and excretion is different (see Section 3.1.). Plutonium behaves

TABLE 6.2
Radioisotopes in the thorium series
(*Spiers 1968, by courtesy of author, and publishers*)

Radioisotope (historical name)	Element	Half-life	Particle energies* (MeV)	γ-Ray energies (MeV)
Thorium	^{232}Th	1.4×10^{10} years	α, 4·01 (76%) α, 3·95 (24%)	0·059 (24%)
Mesothorium 1	^{228}Ra	5·7 years	β^-, 0·02	
Mesothorium 2	^{228}Ac	6·13 hours	β^-, 0·45–2·18	0·057–1·64
Radiothorium	^{228}Th	1·91 years	α, 5·42 (71%) α, 5·34 (28%)	0·084–0·21
Thorium X	^{224}Ra	3·64 days	α, 5·68 (95%) α, 5·44 (4·9%)	0·241 (4·9%)
Thoron	^{220}Rn	55 s	α, 6·28 (99·7%) α, 5·75 (0·3%)	0·54 (0·3%)
Thorium A	^{216}Po	0·16 s	α, 6·78	
Thorium B	^{212}Pb	10·6 hours	β^-, 0·58 (12%) β^-, 0·34 (84%)	0·12–0·41 0·24 (84%)
Thorium C	^{212}Bi	60·5 min	β^-, 0·08–2·27 (64%) α, 6·09, 6·05 (36%)	0·72–2·2 0·04–0·47
Th C′ (64%)	^{212}Po	0·30 μs	α, 8·78	
Th C″ (36%)	^{208}Tl	3·1 min	β^-, 1·0–2·38	0·04–2·61
Thorium C	^{208}Pb	Stable		

* Where the β- or α-spectra contain many lines, only ranges of energy without abundances are given.

very differently from americium though they are adjacent in the periodic table. Plutonium exhibits all valency states in the range 2–7 though it is thought generally to revert to the tetravalent state once it reaches the bloodstream. Americium is trivalent (Stover, Atherton, Keller, and Buster 1960). The plutonium ion is readily hydrated giving rise to large polymeric particles and for the same reason forms complexes readily. These complexes are relatively stable compared to those formed by americium. Plutonium binds selectively to quite different proteins, as will be discussed, from those that bind americium. Some of the chemical characteristics of the important bone-seeking radionuclides are discussed in greater detail in sections dealing with their individual metabolism and skeletal distribution.

3. Biological characteristics

3.1. *Volume Seekers*

The volume seekers, calcium, strontium, radium, and barium, are alkaline earths. Characteristically they are all taken up in high concentration in

TABLE 6.3

Radioisotopes in the radium series
(*Spiers 1968, by courtesy of author and publishers*)

Radioisotope (historical name)	Element	Half-life	Particle energies* (MeV)	γ-Ray energies (MeV)
Radium	^{226}Ra	1620 years	α, 4·78 (94·3%)	0·187 (5·7%)
			α, 4·59 (5·7%)	
Radon	^{222}Rn	3·82 days	α, 5·49 (99+%)	0·51 (0·07%)
			α, 4·98 (<0·1%)	
Ra A	^{218}Po	3·05 min	α, 6·00 (99+%)	
Ra B	^{214}Pb	26·8 min	β^-, 0·67–1·03	0·053–0·352
Ra C	^{214}Bi	19·7 min	β^-, 0·4–3·18 (99+%)	0·61–2·43
			α, 5·51, 5·44 (0·04%)	
Ra C' (99+%)	^{214}Po	160 μsec	α, 7·68	
Ra C" (0·04%)	^{210}Tl	1·32 min	β^-, 1·96	0·30–2·36
Ra D	^{210}Pb	21·4 years	β^-, 0·017 (85%)	0·047 (85%)
			β^-, 0·064 (15%)	
Ra E	^{210}Bi	5·0 days	β^-, 1·16 (99+%)	
Ra F (polonium)	^{210}Po	138·4 days	α, 5·30 (99+%)	
Ra G	^{206}Pb	Stable		

* Where the β- or α-spectra contain many lines, only ranges of energy without abundances are given.

areas of bone where active mineralization is taking place at the time the blood level is high and also diffusely in low concentration throughout the bone mineral. It was originally thought that because calcium is the main constituent of bone mineral, i.e. calcium hydroxyapatite $Ca_{10}(PO_4)_6(OH)_2$, the other alkaline earths might replace calcium in the apatite structure and show an identical distribution in the skeleton. This may indeed prove to be true after a period of time but there are important differences in the way they behave in the skeleton. It is now known that the metabolic behaviour of the four elements is different (Vaughan 1970a) and further that their skeletal pattern of distribution may, at least at short times after reaching the blood stream, be different (Ellsasser, Farnham, and Marshall 1969). Rowland concluded from experimental work in 1966 that immediately following injection of ^{45}Ca to dogs calcium was taken up in concentration on all bone surfaces and that within a matter of days it became concentrated in sites of active accretion and diffusely distributed throughout bone mineral. More recently Ellsasser and his colleagues (Ellsasser, Farnham, and Marshall 1969) have compared the pattern of uptake of ^{45}Ca and ^{133}Ba. Up to six days after injection ^{133}Ba is retained largely on bone surfaces

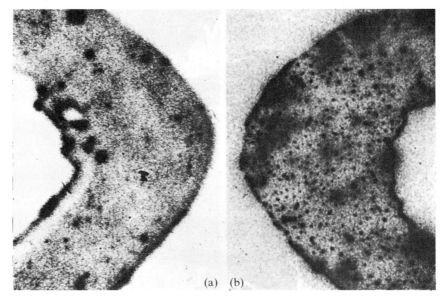

Fig. 6.1. The barium autoradiograph (b) from the mid-diaphysis of the tibia shows localization of the barium on Haversian canal walls in contrast to the calcium autoradiograph (a). The dogs were killed 6 days after injection. (Ellsasser et al. 1969, by courtesy of authors and publishers.)

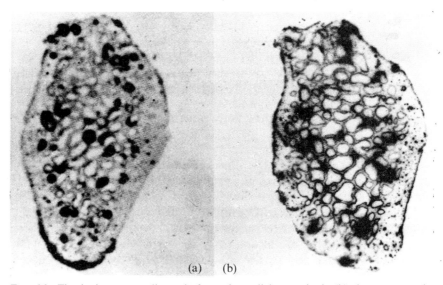

Fig. 6.2. The barium autoradiograph from the radial metaphysis (b) demonstrates the trabecular surface concentration of isotope which is present to a lesser degree in the calcium picture (a). The dogs were killed 6 days after injection. (Ellsasser et al. 1969, by courtesy of authors and publishers.)

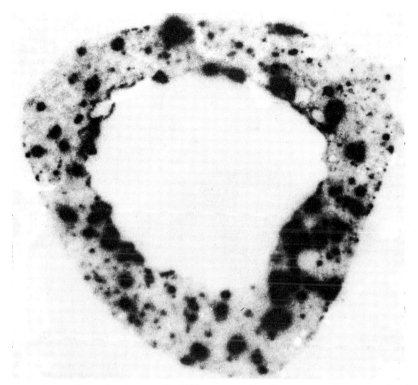

FIG. 6.3. Autoradiograph of a complete tibia cross-section from a case of radium poisoning (case 118) given ^{226}Ra therapeutically for 1 year at age 46, estimated body burden 10 μCi at death 36 years later. Note hot spots and diffuse distribution throughout bone. (McLean and Rowland 1963, by courtesy of authors and publishers.)

particularly round Haversian canals, in contrast to ^{45}Ca. This is well illustrated in Figs 6.1 and 6.2. Evidence from experimental work on rabbits (Kidman, Tutt, and Vaughan 1951a,b; Kidman, Rayner, Tutt, and Vaughan 1952; Tutt, Kidman, Rayner, and Vaughan 1952) suggests that ^{90}Sr follows the ^{45}Ca pattern. No evidence is yet available about ^{226}Ra in the early period after reaching the blood stream but its metabolic behaviour is so like that of barium that it may also remain for some time on bone surfaces.

The classical distribution in cortical bone of a 'volume seeker' some long time after reaching the blood stream is shown in Fig. 6.3, an autoradiograph of a cross-section of a tibia from a case of radium poisoning 36 years after a therapeutic injection of ^{226}Ra. Both the hot spots, where radium was initially taken up in areas of new bone formation, and the diffuse reaction throughout the bone can be seen. A high power view is shown in Fig. 6.4. On the left is an osteone heavily labelled with alpha tracks, on

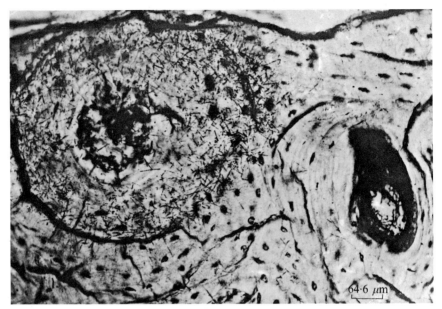

Fig. 6.4. Autoradiograph left on section of cortical bone in case of radium poisoning. Note one Haversian system labelled with alpha tracks while adjacent system contains none (× 175). (Vaughan 1962a, by courtesy of author and publishers.)

the right an unlabelled osteone. The labelled osteone was in process of laying down calcium at the time the blood level was high in radium, the unlabelled osteone was either fully calcified at that time or had been formed subsequently.

If an alkaline earth is fed continuously instead of being given by injection on either one or several occasions the distribution throughout the bone becomes more even depending on the age at which continuous feeding began as indicated in Fig. 8.3.

3.2. *Surface seekers*

The important radionuclides that concentrate on bone surfaces and which are not distributed throughout the mineral are plutonium, americium, thorium, cerium, curium, californium and yttrium. It must however be emphasized that though they are all surface-seekers the pattern of their distribution on surfaces is not the same. Plutonium and americium have been studied most intensively because they appear the greatest practical hazard at present. In Fig. 9.2 are seen autoradiographs from longitudinal sections of the femora of young rabbits. In the rabbit which was given $^{45}CaCl_2$ there is the characteristic concentration beneath the epiphyseal plate, on the endosteal young bone of the metaphysis, and the periosteal young bone of the diaphysis with a diffuse distribution in bone elsewhere. In Fig. 9.2 is shown a similar autoradiograph from a rabbit given ^{241}americium. Here again there is heavy uptake beneath the epiphyseal plate and on the periosteal surface, with a less significant concentration on the endosteal surface, and a patchy distribution around Haversian systems in the cortex. In the case of plutonium the concentration is noticeably greatest on the endosteal surface.

More detailed consideration of the pattern of distribution of different radionuclides in bone is given in the sections dealing with their radiation effects on the skeleton but it must immediately be obvious that the variations in distribution must have important implications for the measurement of the radiation dose they will deliver to sensitive tissues.

4. Relative toxicity

Certain experiments on the relative toxicity of some of the bone-seeking radionuclides may be discussed here before going on to give an account of the effect of the individual internal emitters. The most complete data comes at present from the experiment initiated in 1952 at Utah when young adult beagle dogs about 1·4 years of age were given a series of single intravenous injections of ^{228}Th, ^{239}Pu, ^{228}Ra, ^{226}Ra, and ^{90}Sr in citrate solution. ^{241}Am, ^{249}Cf, and ^{252}Cf have since been added to the programme.

The amount of each radionuclide injected at each dose level was so adjusted that the desired 'retained' μCi kg^{-1} was the same for all the radionuclides except ^{90}Sr in which case they were greater by a factor of 10. Dose level 1 was the basis of the scheme and it is 10 times the maximum permissible ^{226}Ra activity per kg in man.

Level 1 was $\quad 10 \times \dfrac{0 \cdot 1 \,\mu\text{Ci} \; ^{226}\text{Ra}}{70 \text{ kg man}} = 0 \cdot 0143 \,\mu\text{Ci kg}^{-1}$.

All other dose levels were simple multiples of level 1. The resulting injection dose levels are shown in Table 6.4. The actual injected doses were 4 times the desired 'retained' doses of ^{226}Ra, ^{228}Ra, and ^{90}Sr and 1·11 times the desired 'retained' doses of ^{239}Pu and ^{228}Th. Since radioactive decay and excretion occur continuously the term 'retained' dose was obviously meaningless unless the time after injection was specified. Measurements reported in 1962 (Dougherty 1962; Stover, Atherton, and Mays 1962) indicated that:

average ^{226}Ra retention = 0·25 after 330 days,
average ^{239}Pu retention = 0·90 after 6 days,
average ^{228}Ra retention = 0·25 after 235 days,
average ^{228}Th retention = 0·90 after 6 days,
average ^{90}Sr retention = 0·25 after 150 days.

TABLE 6.4
Osteosarcomas in Utah beagles (1 March 1969)
(from Dougherty and Mays 1969, by courtesy of authors and publishers)

Nuclide	Injection level	Injected μCi kg⁻¹	(Deaths) Total	(Deaths) Osteosar.	(Osteosarcoma dog averages) Years from inj. to death	(Osteosarcoma dog averages) rad 1 yr. before death
^{239}Pu	5	2·88	9*	7	4·05	4930
	4	0·909	12*	12	3·61	1310
	3	0·296	12*	12	4·52	602
	2	0·0951	12*	10	7·15	313
	1·7	0·0477	12	8	8·14	183
	1	0·0157	12	4	9·92	78
	0·5	0·00553				
	0·2	0·00189				
	0·1	0·00064				
	0	0·00000	11			
^{228}Th	5	2·70	2*	0	—	—
	4	0·858	4*	2	2·02	2870
	3	0·290	12*	11	2·39	1160
	2	0·0919	13*	12	3·26	516
	1·5	0·0302	9	7	6·38	243
	1	0·0152	5	2	8·75	130
	0·5	0·00518	5			
	0·2	0·00171	1			
	0	0·00000	5			
^{228}Ra (MsTh)†	5	8·49	4*	1	2·17	2830
	4	2·62	5*	4	3·08	2930
	3	0·973	9*	9	4·13	1650
	2	0·309	7	6	6·24	895
	1·7	0·148	6	5	7·99	501
	1	0·0505	3			
	0·5	0·0177				
	0	0·0000	2			
^{226}Ra	5	10·4	10*	9	3·04	10900
	4	3·21	13*	12	4·36	4530
	3	1·07	12*	11	6·28	1940
	2	0·339	13*	5	10·28	837
	1·7	0·166	9	1	11·25	458
	1	0·0621	12			
	0·5	0·0220				
	0·2	0·0074				
	0	0·0000	12			
^{90}Sr	5	97·9	14*	6	4·06	9480
	4·5	63·6	2	2	2·77	3890
	4	32·7	6			
	3	10·8	4			
	2	3·46	6			
	1·7	1·72	5			

TABLE 6.4 (contd)

	1	0·571	2
	0	0·000	4
Ageing controls‡		30	
Totals		326	158

* All injected dogs at these levels have died.
† Excludes MsTh dogs injected with over 1% ^{228}Th contamination.
‡ Excludes controls sacrificed for special studies.

The average dose-rates in 1-level beagle skeletons up to 20 years after injection are shown in Fig. 6.5 and the cumulative doses in 1-level beagle skeletons in Fig. 6.6. These figures illustrate the extremely different dose patterns given by the different radionuclides.

Radiation induced bone cancers were the chief cause of death except (a) at some of the highest levels, where animals died of other generalized radiation effects, and (b) at the lower levels where the natural diseases of old age proved to be the greater hazard. The published results up to March

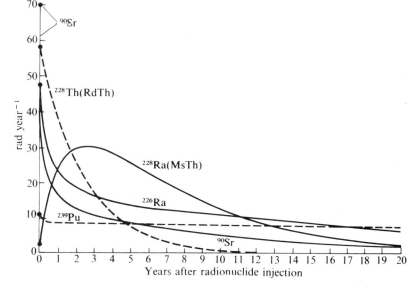

FIG. 6.5. Dose-rates in 1-level Beagle skeletons. The 1-level injections were: ^{239}Pu, 0·0157 µCi kg^{-1}; ^{228}Th; 0·0152 µCi kg^{-1}; ^{228}Ra (MsTh), 0·0505 µCi kg^{-1}; ^{226}Ra, 0·0621 µCi kg^{-1}; ^{90}Sr, 0·571 µCi kg^{-1}. The dose-rate for ^{239}Pu remains fairly constant; that for ^{228}Th decreases, primarily because of radioactive decay; that for ^{228}Ra (MsTh) increases initially due to the build-up of its α-emitting daughters—but it subsequently decreases because of radioactive decay and biological excretion; that for ^{226}Ra decreases because of biological excretion; while that for ^{90}Sr decreases because of both biological excretion and radioactive decay. (Mays et al. 1969, by courtesy of authors and publishers.)

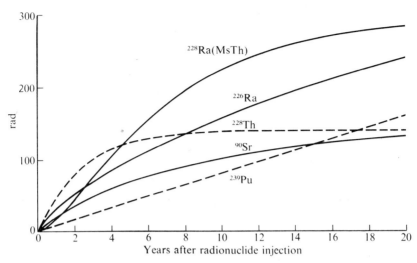

FIG. 6.6. Cumulative doses in 1-level Beagle skeletons. (Mays *et al.* 1969, by courtesy of authors and publishers.)

1969 are shown in Table 6.4* (Dougherty and Mays 1969). Skeletal dose was calculated at 1 year before death and expressed as the average dose in rad. The period of 1 year before death was chosen on the assumption that the osteosarcoma that killed the dog must have taken about 1 year to grow (Mays, Dougherty, Taylor, Lloyd, Stover, Jee, Christensen, Dougherty, and Atherton 1969). The expression 'the average dose in rad calculated from body burden' can be objected to on the basis that the distribution of different radionuclides in the skeleton differs, and therefore (as will be further discussed) radiation dose to sensitive tissues differs. It has however proved a useful unit for comparison of the toxicological effects of different radionuclides when given as a single injection. In Fig. 6.7 average survival times are plotted against dose for each group of osteosarcoma dogs except for the 4·5-level ^{90}Sr dogs. These were injected only 3 years beforehand and if any of the 10 survivors develop bone cancer the average survival time and the average dose will both increase. It is only possible to say at present that the dose level shown is one at which bone cancer is induced by ^{90}Sr. The further information about ^{90}Sr that can be gathered from the present table is discussed in Chapter 8.1.3.1.1.

It is clear from Fig. 6.7 that ^{228}Th, ^{239}Pu, and ^{228}Ra are all more carcinogenic than ^{226}Ra. The biological reasons for these differences will be discussed later.

The data in Table 6·4 were also used to calculate the relative biological

* Newer, updated results are available in Research in Radiobiology COO 119, 242, 1970.

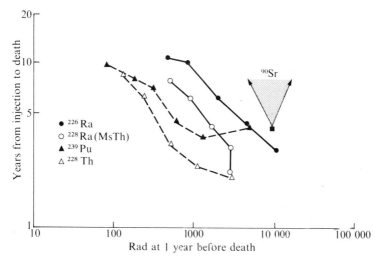

FIG. 6.7. Survival time for osteosarcoma beagles (March 1969). Each point is the average for the osteosarcoma dogs at a given injection level (see reference). In general the lower the dose the longer the survival time. (Dougherty and Mays 1969, by courtesy of authors and publishers.)

efficiency (RBE) of the different radionuclides in relation to radium. The RBE is defined as the ratio of the dose from a reference radiation to that from a tested radiation when each produces an equal biological effect (Chapter 1.2). In this case the biological effect was death from osteosarcoma. ^{226}Ra was chosen as the reference radiation because the permissible levels from bone seekers are based on the observed toxicity of ^{226}Ra in man (Mays, Dougherty, Taylor, Lloyd, Stover, Jee, Christensen, Dougherty, and Atherton 1969; Evans, Keane, Kolenkow, Neal, and Shanahan 1969; Evans, Keane, and Shanahan 1969). At the longest time (8 years) for which RBE could be compared among the five radionuclides, values relative to ^{226}Ra = 1 are ^{239}Pu = 6, ^{228}Th = 8, ^{228}Ra = 2·5, and ^{90}Sr = 0·07 to 0·24. Certain general dosimetric considerations arise from these results. Table 6.4 and Fig. 6.5 suggest that there is a tendency for the appearance time of bone cancers to lengthen as the dose decreases. This suggests that the true threshold below which no tumours will occur when the radionuclide is given as a single injection may be set by life span.

From these RBE values calculated in dogs and the observed toxicity for ^{226}Ra in man, Dougherty and Mays (1969) have attempted to calculate the life-time skeletal dose above which the induction of bone cancer may frequently occur in man. They use data from the human cases of radium poisoning which will later be discussed in detail (Evans, Keane, Kolenkow, Neal, and Shanahan 1969; Evans, Keane, and Shanahan 1969; Finkel, Miller, and Hasterlik 1969a,b). This suggested that hitherto osteosarcomas and carcinomas of overlying epithelium have only occurred when the average skeletal dose is above 1200 rad. They suggest that corresponding doses for the other emitters should then be 1200 rad divided by the appropriate RBE. The results are shown in Table 6.5 together with the calculated skeletal dose during a 50-year occupational exposure to constant permissible body-burdens calculated from ICRP assumptions (International Commission on Radiological Protection 1960) which may however need some revision (Dougherty and Mays 1969).

TABLE 6.5

Predicted doses causing bone cancer in humans
(from Dougherty and Mays (1969), by courtesy of authors and publishers)

	^{239}Pu	^{228}Th	^{228}Ra	^{226}Ra	^{90}Sr
Predicted rad for induction	200	150	480	1200	5000 to 17000
Permissible rad during 50 years	25	37	160	150	260
Predicted/permissible	8	4	3	8	19 to 65

The results of the Utah experiment on relative toxicities following a single injection are interesting in view of what is known about the biological behaviour of the different radionuclides and the character of the cells at risk discussed in Chapter 3. 'Volume-seekers' which are deposited throughout bone mineral will have far less opportunity to irradiate the proliferating cells found on bone surfaces than the 'surface-seekers' which are concentrated actually in or closely adjacent to the cells at risk. Furthermore the two 'surface-seekers' ^{239}Pu and ^{228}Th are also alpha emitters, ^{228}Th having a somewhat higher energy.

The fact that ^{228}Ra, a beta emitter, and a volume-seeker, is more toxic than ^{226}Ra, also a volume-seeker, but largely an alpha emitter,* can be explained by the following facts. ^{228}Ra is a weak beta-emitter but the average range of its alpha-emitting daughter-products is greater than the range of the alpha-emitting daughter-products of ^{226}Ra. Thus in the ^{228}Ra dogs at a given point in time a larger fraction of the alpha radiation from the daughter products is able to escape beyond the bone mineral, and furthermore the escaping fraction can irradiate cells which are more distant than in the case of ^{226}Ra. Secondly the small fraction of the daughter ^{228}Th which is released from the bone volume site of its parent will deposit on bone surfaces, while some of the escaping fraction of the next daughter ^{224}Ra should continuously redeposit in the critical sites of new bone formation where there is cellular activity (Mays, Dougherty, Taylor, Lloyd,

* The total contribution of the beta-particle dose to marrow is only about 2 per cent and to osteogenic tissue about 0·5 per cent of the respective alpha-particle dose (see Chapter 7.1.1.).

Stover, Jee, Christensen, Dougherty, and Atherton 1969). The particular problems associated with ^{224}Ra are discussed elsewhere (Chapter 7.1.3, 7.4.).

A second series of experiments, designed to give information about the relative toxicity of radionuclides given by a single injection, has been carried out by Finkel and her colleagues (Finkel and Biskis 1968; Finkel, Biskis, and Jinkins 1969). Reference will be made to the results in relation to individual radionuclides. Unfortunately the strain of mouse used, the CF 1/Anl mouse has recently been shown to carry an indigenous RNA virus of the C-type Gross serotype, and to show a high natural incidence of lymphomas, osteosarcomas, fibrosarcomas, and a myeloproliferative and osteosclerotic bone dyscrasia (Chapter 1.1.3) (Lisco, Rosenthal, and Vaughan 1971). This degree of pathology in the normal controls makes interpretation of experimental results difficult. The relative toxicity of the radionuclides was however of the same order as that in the Utah experiment.

The results of the experiment on the Utah dogs and Finkel mice give some measure of the relative biological efficiency of certain radionuclides under conditions which are unlikely to occur in real life, i.e. single intravenous administration. The end-point in this case was osteosarcoma. It will however be shown in subsequent chapters that continuous feeding of ^{90}Sr results, in dogs, in leukaemia as well as osteosarcoma, and that in pigs osteosarcoma is a rare event following continuous feeding. This at once raises difficult questions about the assessment of the radiobiological efficiency of radionuclides. How far can the efficiency of a radionuclide given in a single injection be compared with the efficiency of a radionuclide given continuously? How far can the radiobiological efficiency of a radionuclide that induces leukaemia be compared with the radiobiological efficiency of a radionuclide that induces osteosarcoma? How far can the radiobiological efficiency of a radionuclide in a young dog be used to estimate radiobiological efficiency in an adult man when as already discussed (2.1.1., 2.1.2) the pattern of trabeculation and rate of bone turnover varies with both species and age? As already discussed in Chapter 1.2, the Report of the RBE Committee to the International Commissions on Radiobiological Protection and on Radiobiological Units and Measurements (1963) laid down very stringent conditions for determination of RBE, namely the dose rate, fractionation of the dose, the dose distribution in the tissues, oxygenation in the tissue, and the nature of the biological effect should all be the same.

Further, it may be suggested that, with the information now coming from Spiers and his colleagues on experimental measurement of radiation dose to sensitive tissues, i.e. osteogenic tissue on bone surface and marrow tissue in trabecular cavities, the value of measurements of RBE based on absorbed dose *in bone* must be questioned. Such measurements have served a useful purpose but perhaps more precise measurements of actual dose to sensitive tissues may lead to rather different valuations of relative biological efficiency in the future. These precise measurements in relation to radium and strontium are discussed later (Chapter 7.3.6.2. and 8.1.3.4.) (Spiers 1968a,b, 1969; Spiers, Zanelli, Darley, Whitwell, and Goldman 1972; Spiers, Whitwell, and Darley 1971).

5. ^{32}P

^{32}P requires separate consideration. It is listed by the International Commission on Radiological Protection (1959) as a predominantly bone-seeking radionuclide, which is often forgotten in considering the problems of internal irradiation. It is taken up throughout bone, being incorporated in the hydroxyapatite crystal. In addition it is incorporated into the nucleoprotein fraction of all cells (Marshak 1949), particularly those of lymph nodes, spleen, and bone marrow. The concentration is greater in the nucleus, particularly in DNA, than in the cytoplasm. It has a relatively short half-life of 14·3 days. At decay, a beta particle, of energy up to 1·46 MeV, is emitted and a new element ^{32}S is formed which recoils with energy up to 70 eV. Therefore, particularly when the ^{32}P atom is incorporated into DNA, it is possible that there may be some further contribution to the biological effect as a result of the transmutation which takes place.

In discussion of the relative toxicities of the bone seeking radionuclides ^{32}P is not considered. As discussed in Chapter 12, it can produce osteosarcomas in experimental animals, and may be responsible for inducing myeloproliferative disorders in man given ^{32}P for therapeutic purposes.

7. Radium Isotopes

THE decay pattern of the radionuclides in the radium series is shown in Table 6.3. ^{228}Ra may be found in Table 6.2.

There is information available on the clinical effects on man of three isotopes of radium, ^{224}Ra, ^{226}Ra, and ^{228}Ra. ^{224}Ra is an alpha emitter with a short half-life, ^{226}Ra is an alpha emitter with a long half-life, and ^{228}Ra (mesothorium) is a beta emitter with a half-life of 5·7 years (see Tables 6.2, 6.3). An analysis of the behaviour of these three isotopes which have the same biological and chemical characteristics, but different physical characteristics is therefore of particular interest. It is valuable to be able to compare the biological effects of a short-lived radioisotope with the effects of a long-lived radioisotope which has the same chemical characteristics.

1. Physical characteristics

1.1. ^{226}Radium

^{226}Ra decays first to the noble gas radon, ^{222}Rn, which has a half-life of 3·82 days. The extent to which this gas escapes from any tissue in which ^{226}Ra is deposited profoundly affects the radiation dose. Rowland and Marshall (1959) calculated a 30 per cent radon retention in bone after 25 years. More recently Mays has given the Rn/Ra activity ratio in man 20 years after radium deposition as 30–40 per cent. In the beagle 3 years after deposition it is 25 per cent. In both man and dog the fraction of radon retained increases with time (Mays 1969). Radon and some of its daughter products (see Table 6.3) also emit alpha particles. An average of 2·1 alpha particles is given off for each ^{226}Ra atom fixed in the body. The number of alpha particles emitted per radium disintegration after 25 years in bone, assuming 30 per cent radon retention, is 1·03 (3+0·55) (Mays, Van Dilla, Floyd, and Arnold 1958; Lloyd 1961a). As will be discussed, some of the biological effects of ^{226}Ra are probably largely due to ^{222}Rn. Spiers and his colleagues (Spiers, Whitwell, and Darley 1971) have recently given alpha and beta particle energy graphs for the retained ^{226}Ra chain, which are shown in Tables 7.1 and 7.2. The total contribution of the beta-particle dose to marrow dose would be about 2 per cent and to the endosteal dose about 0·5 per cent of the respective alpha-particle doses. It is therefore usual to think of ^{226}Ra as essentially an alpha emitter.

TABLE 7.1

α-particle energy graphs for retained ^{226}Ra chain
(*Spiers, Whitwell, and Darley 1971, by courtesy of authors*)

Element	Particle energy MeV	Fractional numbers of particles
^{226}Ra	4·782	0·430
	4·599	0·024
^{222}Rn	5·490	0·182
^{218}Po (RaA)	6·000	0·182
^{214}Po (RaC¹)	7·680	0·182

N.B. RaD (half-life 21·4 years) and its daughters are ignored.

TABLE 7.2

β-particle energy graphs for retained ^{226}Ra chain
(*Spiers, Whitwell, and Darley 1971, by courtesy of authors*)

Element	Particle energy MeV	Fractional numbers of particles
^{214}Pb (RaB)	0·678	0·245
	0·735	0·225
	1·030	0·030
^{214}Bi (RaC)	0·400	0·045
	1·000	0·115
	1·510	0·200
	1·880	0·045
	3·260	0·095

N.B. RaD (half-life 21·4 years) and its daughters are ignored.

1.2. ^{228}*Radium* (*Mesothorium*)

^{228}Ra, though itself a beta emitter, decays to radiothorium, ^{228}Th, a powerful alpha emitter with a different distribution in bone, discussed Chapter 11.2.1.2. From the point of view of radiation hazard this is important since the dial painters, whose contamination was dependent on oral intake, were in particular contaminated with a mixture of ^{228}Ra and ^{226}Ra. The mesothorium component of the luminous dial-paint had in

RADIUM ISOTOPES

some cases been allowed to reach equilibrium with its grand-daughter ^{228}Th and was at one time therefore thought to be extremely toxic (Aub, Evans, Hempelmann, and Martland 1952). However, it is now known that radiothorium is poorly absorbed from the intestinal tract and therefore the ^{228}Th component of the paint is not significant (Maletskos, Keane, Telles, and Evans 1969) (see Section 3.1).

1.3. ^{224}Radium

This radionuclide has a short half-life of 3·64 days and therefore, if Rowland (1966) is correct about the initial uptake of alkaline earths on all bone surfaces and if radium behaves like barium and has a maintained retention on such surfaces (Ellsasser, Farnham, and Marshall 1969), the greater part of the ^{224}Ra will decay while actually on the surface, as discussed later (Section 2.3.). The first of the daughter products, ^{220}Rn, is a rare gas which has a short half-life of only 55 seconds. This is followed by ^{216}Po with a very short half-life, 0·16 seconds, and then by ^{212}Pb which has a half-life of 10·6 hours. ^{212}Pb, in mice, is excreted more rapidly than ^{224}Ra as shown in Fig. 7.1 (Hug, Gössner, Müller, Luz, and Hindringer 1969), which indicates that there is a preferential retention of ^{224}Ra. The excretion rates for ^{224}Ra and ^{212}Pb decrease rapidly and reach a level as low as 1 per cent of the initial value on the fifth day after injection. The amount of ^{224}Ra

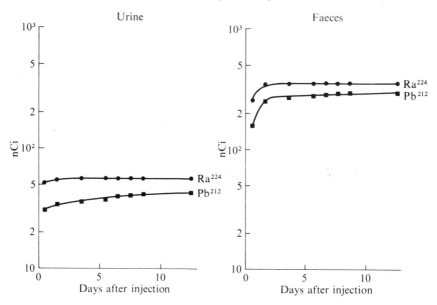

FIG. 7.1. Total excretion of ^{224}Ra and ^{212}Pb in urine and faeces of mice. (Hug *et al.* 1969, by courtesy of authors and publishers.)

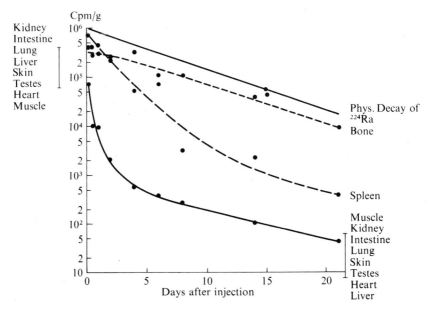

Fig. 7.2. Amount of ^{224}Ra in the different mouse organs. (Hug *et al.* 1969, by courtesy of authors and publishers.)

retained in different organs is shown in Fig. 7.2. Similar metabolic results are recorded in man (Grillmaier, Muth, and Oberhausen 1968).

The skeletal dose from the injected daughters of ^{224}Ra is insignificant but when the ^{224}Ra decays in bone the daughters so produced remain there and decay there (Mays, Haldin, and Van Dilla 1958; Stover, Atherton, Buster, and Bruenger 1965). Consequently the total alpha-particle energy of 26·5 MeV for the ^{224}Ra decay series results from each ^{224}Ra atom which decays in the skeleton (Spiess and Mays 1970). The dosimetry of ^{224}Ra in man is further discussed in Section 4.1.3. As in the case of ^{226}Ra the beta particle energy is small and its effects are usually ignored.

2. Biological characteristics

Radium is an alkaline earth. It is therefore not surprising to find it is concentrated in the skeleton. ^{226}Ra and ^{228}Ra can be described as 'volume-seekers' since they are found throughout the bone mineral and not particularly on surfaces. The half-life of ^{224}Ra is however so short (3·64 days) that it decays largely on surfaces and is not found diffusely throughout bone mineral. Because of these differences in their half-lives the metabolism and skeletal distribution of the three isotopes are discussed separately.

RADIUM ISOTOPES

2.1. $^{226}Radium$

In broad terms the metabolism of ^{226}Ra in man is like that of calcium but there are important metabolic differences which are recognized but not understood. They have been discussed in Chapter 6.3.1. Figure 7.3, prepared by Maletskos and his colleagues (Maletskos, Keane, Telles, and Evans 1969), illustrates most of the known information on ^{226}Ra retention by man as determined by different laboratories. There is some evidence, reviewed by Maletskos, Keane, Telles, and Evans (1969), that possibly immediately following injection much of the radium is retained in soft tissues and that the skeletal content is relatively low, about 10 per cent. This is indicated by the broad arrow R_3. Further data on the early retention of radium, both in the skeleton and in soft tissue, is urgently required. A modified power function model together with a final exponential appears to provide an accurate description of the retention of radium in man from minutes to many years after intake (Marshall, Rundo, and Harrison 1969). It must also be remembered that the pattern of excretion and retention varies in different species, as shown in Fig. 7.4 (Stara, Nelson, Della Rosa, and Bustad 1971).

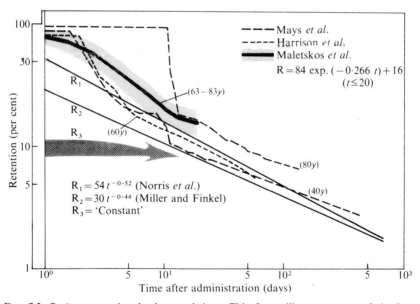

FIG. 7.3. Radium retention by human beings. This figure illustrates most of the known information on radium retention by man as determined by various laboratories. Data for the power function equations start at $\pm > 30$ days, and these equations may not represent correctly either whole-body or skeletal retention prior to this time. The broad arrow representing R_3 is intended to indicate another possibility of early skeletal retention at about 10 per cent of the injected Ra. The word 'constant' is used to suggest a relatively stable or slowly changing function during this short period. (Maletskos et al. 1969, by courtesy of authors and publishers.)

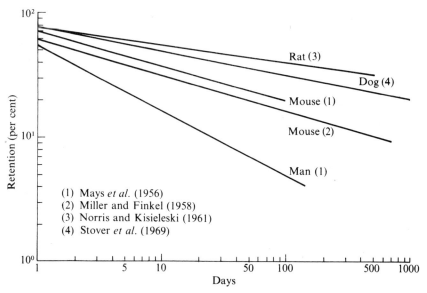

FIG. 7.4. Retention of ^{226}Ra in four species (IV). (Stara et al. 1971, by courtesy of authors and publishers.)

Evans and his colleagues have measured the distribution of ^{226}Ra and ^{228}Ra in 26 complete human skeletons, each divided for measurement into different bone groups as shown in Fig. 7.5 (Evans 1966; Evans, Keane, Kolenkow, Neal, and Shanahan 1969; Evans, Keane, and Shanahan 1969). On the average the skull has a radium concentration per gram of bone which is essentially equal to the skeletal average while the vertebrae tend to be about 1·2 times and the tibia about 0·8 times the skeletal average.*
In some individuals the macroscopic distribution is greatly different from the mean values for all individuals studied. Great differences have also been found both between and within individual bones (Lloyd 1961a; Lucas, Rowland, Miller, Holtzman, Hasterlik, and Finkel 1963). The characteristic distribution of the alkaline earths in bone takes two forms. There is heavy concentration in areas of active new bone formation and mineralization present at the time the blood level was high, and a diffuse distribution throughout the bone mineral. This characteristic distribution does not however occur at once. Rowland showed, in 1966, that immediately following injection of ^{45}Ca into young adult dogs the radionuclide was taken up in concentration on all bone surfaces and only within

---

* Recent observations suggest that the size of bone trabeculae and marrow cavities in the parietal bone are very different to those elsewhere and therefore radiation dose in this bone is different (Section 5.6.2.).

RADIUM ISOTOPES

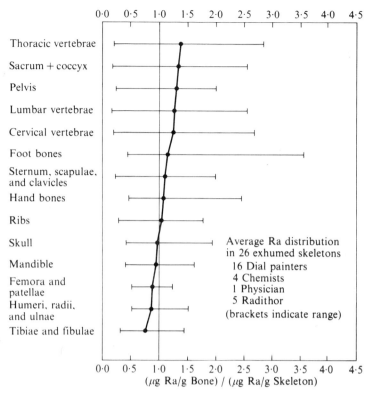

FIG. 7.5. Average radium distribution in 26 exhumed skeletons. (Evans 1966, by courtesy of author and publishers.)

a matter of one or two days was it found in concentration in areas of active mineralization and even later diffusely distributed throughout the bone mineral. More recently Ellsasser and his colleagues (1969) have shown in autoradiographic studies that ^{133}Barium remains on bone surfaces for at least six days after injection, i.e. longer than calcium. No study of how long radium may remain on bone surfaces after reaching the blood stream is available, but remembering its similar metabolic behaviour to barium it is not improbable that it may have a longer stay on the surface than has hitherto been recognized. The remarkable differences between the behaviour of calcium and barium have already been shown in Figs 6.1 and 6.2, where autoradiographs of the cortical bone of two adult dogs following the injection of ^{45}Ca and ^{133}Ba are illustrated. Six days after injection the calcium autoradiograph shows the classical picture of a diffuse distribution throughout the mineral and some areas of high concentration. The barium picture also shows a diffuse distribution but there is high concentration on many surfaces particularly around Haversian systems. In the case of the trabecular bone shown in Fig. 6.2 this surface distribution of barium is even more striking. If radium isotopes too are retained on surfaces in the same way for at least some days it will become particularly important in considering the dosimetry of ^{224}Ra.

The pattern of distribution of ^{226}Ra in the skeleton, many years after concentration of the radionuclide was high in the circulating blood, has already been illustrated in Fig. 6.3 as that of a typical volume-seeker. Mention has been made however of the fact that in certain bones, notably those of the skull, turnover rate is extremely slow. Evans and his colleagues have recorded the case of a man who developed a carcinoma of a frontal sinus 52 years after acquiring his radium. The radiation dose at 10 μ from the bone tissue interface was 82 000 rad. To this dose, due only to alpha rays emitted from the frontal bone, must be added an additional alpha ray dose due to ^{222}Rn diffusing out of the bone into the frontal sinus. The air withdrawn from the uninvolved right frontal sinus had a radon concentration at least 30 times that in the exhaled breath.

2.2. ^{228}Radium

2.2.1. *Distribution in the skeleton.* The distribution of ^{228}Ra itself in the skeleton is similar to that of ^{226}Ra but the behaviour of its daughter products may well be different. ^{228}Th, the second daughter, is a surface-seeker and may migrate therefore to surfaces. Although ^{224}Ra, the next daughter, resulting from decay of ^{228}Th, has a short half-life (3·64 days) it is known to be highly carcinogenic (see Section 4.1.1) and will initially be present on the surface together with its parent ^{228}Th. The half-life of thoron (^{220}Rn) is extremely short, a mere 55 seconds. Evans and his colleagues suggest this may account for the fact that carcinomas of the sinuses have not been found in those radium cases with a known high skeletal burden of ^{228}Ra. The gas may not have accumulated sufficiently in the sinuses to build up a high radiation dose as radon does when the body burden is largely ^{226}Ra.

2.3. ^{224}Radium

The case of ^{224}Ra is different again. It has an extremely short half life of 3·65 days. Therefore if initially like barium it is concentrated on bone surfaces it will almost entirely decay on such surfaces. The sensitive osteogenic cells will therefore receive an extremely heavy radiation dose as is discussed in Section 4.1.3.

3. Clinical effects of ^{226}Ra and ^{228}Ra

Information on the effects of ^{226}radium on man comes largely from two groups, one in Chicago and the other observed at the Massachussets Institute of Technology (M.I.T.), Boston.

Some of these radium cases came to attention as patients seeking help because of disabilities induced by their radium burdens, others have been actively sought for as individuals whom it was important to study. Many of them were girls who had painted watch-dials in the First World War and been taught to point their brushes with their lips. Others were radium chemists, and, in the Chicago area particularly, there were a large group of people who had been treated with radium in the thirties for a variety of medical conditions.

Human contamination has frequently involved a mixture of ^{226}Ra and ^{228}Ra, and while it is difficult to discriminate between their individual effects in man, ^{226}Ra has, however, usually predominated for reasons discussed below.

3.1. *The dial painters*

Two groups of dial painters have been extensively investigated. The dial painters on the east coast of America (the Boston M.I.T. Group) used a paint containing a variable mixture of ^{226}Ra, ^{228}Ra, ^{228}Th, and their decay products, so they might have been contaminated with several radionuclides which they swallowed as a result of pointing their brushes. The Chicago group used paint containing little if any ^{228}Ra or ^{228}Th. Recently a figure of about 20 per cent was obtained for the absorption of radium sulphate from the gut in normal though elderly human subjects. The absorption of ^{228}Th was practically negligible, being only 0.02 per cent (Maletskos, Keane, Tellers, and Evans 1969). ^{228}Th is therefore not likely to be present in the skeleton of the Boston M.I.T. dial painters except as a decay product of retained ^{228}Ra.

3.2. *Radium chemists*

The radium chemists in both the Boston and Chicago groups were probably contaminated with both radionuclides but ^{226}Ra tended to predominate over ^{228}Ra. They inhaled or swallowed the radium.

3.3. *Therapeutic radium (iatrogenic cases)*

^{226}Ra was used therapeutically for a wide variety of conditions especially in the Chicago area before the Second World War. It was sometimes taken by mouth but more often given by intravenous injection. Its use was removed from the American Medical Association's new and non-official remedies about 1932 but it continued to be used for some time. One clinic alone, during a five-year period about 1920, gave more than 14 000 intravenous injections of radium, usually 10 μCi each, and more than 22 000 oral administrations. Several radium-water nostrums were also in use, particularly 'Radiothor', which was a mixture of 1 μCi of radium and 1 μCi of mesothorium in $\frac{1}{2}$ oz of water. This was often kept on the family breakfast table for general use. Some 22 000 vials of 'Radiothor' were sold in 1925 alone (Evans 1966). From the point of view of scientific medicine it is fortunate that a large group of well documented patients in a mental hospital have been most carefully followed over a number of years. The initial injection schedule was recorded and repeated estimations of body burden have been made. Some of the ^{226}Ra given was also contaminated with ^{228}Ra but on the whole patients in the Chicago series received less ^{228}Ra than patients in the M.I.T. series.

Robley Evans and his colleagues have however made extensive measurements and calculations which have enabled them to convert the radiation from ^{228}Ra to terms of radiation dose from ^{226}Ra to the skeleton, which they call the pure radium equivalent (PRE). PRE is defined as 'that amount of ^{226}Ra which will deliver the same radiation dose as a mixture of ^{226}Ra and ^{228}Ra in the body' (Maletskos, Keane, Tellers, and Evans 1969). Since the distribution pattern of ^{228}Ra is the same as that of ^{226}Ra, using Evans' calculations for the purposes of dosimetry, ^{226}Ra and ^{228}Ra are usually considered together.

The literature covering the clinical, physiological, and dosimetric aspects of poisoning from ^{226}Ra and ^{228}Ra is immense. No attempt is made to cover it here. Many of the references quoted have further useful bibliographies. The first report on radium poisoning is usually attributed to a dentist who, in discussing osteomyelitis of the jaw drew attention to the condition of 'radium jaw' he had seen in luminizers, young women who painted dials with luminous paint. In 1925 Martland and his colleagues published their classical account of radium poisoning in American dial painters. As the first dial painters continued to die over the succeeding years Martland followed them, recording both their clinical histories and autopsy findings (Martland, Conlon, and Knef 1925; Martland 1926; Martland and Humphries 1929; Martland 1931; Aub, Evans, Hempelmann, and Martland 1952). A remarkable account of a young woman, first seen by Martland and published as case 3 by Martland, Conlon, and Knef in 1925, has recently been given by Sharpe (1971). She was one of the first dial painters and had great difficulty in obtaining any compensation for her crippling disease. Following Hamilton's account in 1947 of the number of fission products that concentrate in the skeleton this population of people with a known body burden of an internal emitter became of extreme scientific importance. Groups of dial painters, radium chemists, and persons given radium as a therapeutic measure have been sought out, followed up and studied since 1947. This has hitherto been done in three main centres, one at the Massachusetts Institute of Technology (Evans 1966; Evans, Keane, Kolenkow, Neal, and Shanahan 1969; Evans, Keane, and Shanahan 1969) one in Chicago (Finkel, Miller, and Hasterlik 1969 a,b; Miller, Hasterlik, and Finkel 1969), and one in New Jersey—the two former being the best documented. The records have recently all been gathered together at the Centre for Human Radiobiology at the Argonne National Laboratory, Chicago (Rowland, Failla, Keane, and Slehney 1969–70, 1970–71). Inevitably some patients appear in both the Chicago series and the M.I.T. series. Certain discrepancies still appear in the final figures of the most recent papers discussing these important people but the general picture becomes increasingly clear. It must be remembered in looking at the latest documented reports of people with a known radium burden that many of them are still alive and though they may, at time of reporting, be free of malignancies, such malignancies may subsequently develop.

The late effects of radium poisoning differ significantly from the relatively acute effects first described by Martland and his colleagues in 1925. This may be due in part to the fact that many of the dial painters who died in the early years were relatively young and therefore the pattern of radium distribution in the skeleton might differ somewhat from that seen in older individuals. They also, if Martland's estimations of their radium body burden are even approximately correct, had much heavier burdens than many of the individuals at present under investigation.

3.4. *Acute effects*

Martland (1931) himself classifies 13 cases as acute and states 'These cases showed during life a clinical picture quite different from that in cases which developed later. They were characterized by the presence of jaw necroses and the development of anaemias. Most of these cases occurred four and six years after the girls had left their employment as dial painters'. He attributes the jaw necrosis to bacterial infection from the teeth attacking the bone and marrow already grossly damaged by radiation. His account of the anaemias and the marrow picture is of considerable interest and importance. The anaemia was severe and associated with leukopenia. It proved rapidly fatal. At autopsy

... the marrow of the femurs was dark red throughout and the lesion more pronounced than that seen in the most characteristic cases of pernicious anaemia. Histologically the marrow showed an astonishing picture ... The general architectural structure and landmarks were entirely obscured by the extreme hyperplasia with a packing of immature and primitive cells ... Some 60 per cent of the cells in well packed areas were very large, 12–20 microns in diameter, with large vesicular nuclei containing 1 or more nucleoli. There was a distinct nuclear limiting membrane with condensation of nuclear chromatin at its edges. The cytoplasm was a dull bluish grey (haematoxylin eosin), smooth, and glassy and contained no granules. The cells contained no haemoglobin. Mixed with these predominating primitive cells were many megaloblasts. Many normoblasts in mitosis of all varieties were present. The only cells of the granulocytic series were innumerable eosinophil myelocytes ... Lymphocytes were not present ... Megakaryocytes were usually abundant.

Later Martland says that this 'irritative marrow', as he calls it, starts to heal or subside in a patchy manner. 'This is essentially a replacement fibrosis ... in the beginning the reaction is a very cellular one ... It is important to note that all stages of this radiation osteitis may be seen in a single bone.' Loutit (1970) suggests that these early cases should be classified as examples of malignant transformation whether one calls them leukaemia or malignant myelosclerosis. They are reminiscent of the lesions produced by ^{90}Sr in pigs and dogs which are regarded as malignant (Chapter 8.1.3.5.). These early cases had no osteosarcoma, but this is not surprising when it is appreciated that the latent period for osteosarcoma is so much longer than that for myeloproliferative malignancies as is discussed elsewhere (Chapter 5.3.1.1). In the more recent literature there is at least one further case that might fall into this group, namely Case 1 described by Ardran and Kemp (1958) and recently reviewed by Loutit (1970). He was a man in chronic ill health who died 2 years after treatment with 'German Radium Salt' with severe anaemia and his long bones full of red marrow.

3.5. *Chronic effects*

The chronic effects of radium poisoning are very different from the acute effects just described. Unfortunately there is little data available on the detailed histopathology. Interest has concentrated on roentgenographic changes and dosimetric calculations related to the appearance of osteogenic sarcoma and cranial carcinomas. The outstanding clinical lesions are indeed osteogenic sarcomas and carcinomas of the cranial sinuses, though it is becoming apparent that tumours of the central nervous system occur more frequently than would be expected in a population of the size of the radium patients (Hasterlik 1968; Rowland, Failla, Keane, and Slehney 1969–70; Loutit 1970). Hasterlik in 1968 prepared a table of the known malignant tumour experience of dial painters who worked with radium before 1930. The unexpectedly high number of central nervous system tumours should be noted: 4 when 0·2 would be expected (1 astrocytoma, 1 acoustic neuroma, 1 glioblastoma, 1 glioma) (Hasterlik 1968).

3.5.1. *Neoplasia.* The most recent analysis of 777 radium cases from the combined M.I.T. and Chicago lists, for whom the total dose received has been calculated (Section 3.6.1), considers 51 osteogenic sarcomas and 20 carcinomas to be radiation induced (Rowland, Failla, Keane, and Slehney 1969–70). The excess of osteogenic sarcoma over what might be expected in a population of this size is about 50:1 and the excess of carcinomas about 100:1 (International Commission on Radiological Protection 1968). There was a remarkably consistent distribution of malignancies within the two groups as shown in Table 7.3. Four cases developed two independent

TABLE 7.3

Distribution of malignancies within two independent studies In patients with a known radium burden

(*after Rowland et al. 1969–70, by courtesy of authors and publishers*)

Location studied	Cases	Sarcomas	Carcinomas
MIT	474	33	10
Chicago	316	20	11
Total*	777	51	20

* There are some cases that have been studied and reported on by both groups so that the total number of cases and malignancies is less than the sums from the independent studies.

tumours, in each instance a carcinoma and a sarcoma. One patient had an osteogenic sarcoma 19 years after radium injections and subsequently developed a carcinoma 42 years after injection. Possible haemopoietic neoplasma were not included in the analysis.

3.5.1.1. *Osteogenic sarcoma.* The osteogenic sarcomas tend to occur earlier than the carcinomas (Evans, Keane, and Shanahan 1969; Rowland, Failla, Keane, and Slehney 1969–70). No really detailed study of their histopathology has been published. From a review of the available histology I was privileged to make through the courtesy of Professor Robley Evans and Dr. Hasterlik in 1967 I would see no reason to suggest that the tumours differed in character from those occurring spontaneously though, (like all osteogenic sarcoma) as emphasized by Sissons (1966), the histological picture within the same tumour could be extremely variable. In both series, typical osteosarcomas with abnormal bone were found but fibrosarcomas predominated. Finkel, Miller, and Hasterlik (1969) list 7 fibrosarcomas in the Chicago cases and six osteogenic sarcomas, a very different proportion from that occurring normally. In 1748 malignant bone tumours at the Mayo Clinic, Dahlin listed 469 osteosarcoma and only 58 fibrosarcoma (Dahlin 1957). The radium cases therefore appear to have an unusually high incidence of fibrosarcoma. In neither the M.I.T. nor the Chicago cases were haemangioendothelioma or angiosarcoma noted. Only two chondrosarcomas were diagnosed in the series at M.I.T. and none in the Chicago group. Lisco in 1956 described one case which exhibited gross dysplasia throughout the skeleton as well as a tumour of the ischium. This was a well differentiated, very cellular fibrosarcoma composed of interlacing bundles of very small cells showing a fair degree of pleomorphism of nuclei and cytoplasm and very little intercellular substance. There was no bone formation. Lisco considers it is likely that the tumour originated from the cells of the old peritrabecular fibrous tissue that was so abundant in the ischium. This may indeed be the case in some of the malignancies in the radium patients. On the other hand there is good evidence that bone tumours do not necessarily arise from bone showing injury and subsequent repair (Vaughan, Lamerton, and Lisco 1960). The majority of tumours appear to have arisen endosteally when the origin can be determined. Aub, Evans, Hempelmann, and Martland (1952) record one periosteal fibrosarcoma. The bone tumours are widely distributed throughout the skeleton and have been recorded in all bones. The most frequent site appears to be the upper end of the femur. Serial radiographs taken at regular intervals over a period of years in patients with a known radium burden have enabled the development of both malignant change and increasing dysplasia in the skeleton to be followed (Figure 7.6). The malignant lesion

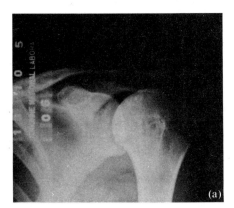 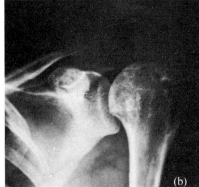

FIG. 7.6. (a) Left shoulder in 1951. Body-burden ^{226}Ra greater than 0·765 µg. (b) Left shoulder in 1959. Bone resorption and sclerosis in head of humerus now prominent. Body-burden ^{226}Ra = 0·590 µg. (Hasterlik *et al.* 1969, by courtesy of authors and publishers.)

may appear in what radiologically a year previously was apparently normal bone while gross dysplasia was present elsewhere in the skeleton (Hasterlik 1967, personal communication; Hasterlik, Miller, and Finkel 1969).

3.5.1.2. *Carcinomas of epithelium applied to bone.* Aub and his colleagues in 1952 described two cases of carcinomas arising in the nasal accessory sinuses or near the temporal bone. There are now in the latest analysis of the combined M.I.T. and Chicago patients twenty such cases (Rowland, Failla, Keane, and Slehney 1969–70). Again, no careful and detailed study of the histopathology of these tumours has been published. I had the privilege of seeing sections from many of these patients in 1967 through the courtesy of Dr. Robert Hasterlik and Professor Robley Evans and would tentatively suggest that they might be classified (as Hadfield has classified the carcinomas of the paranasal sinuses in the woodworkers in the furniture industry (Hadfield 1970)) as squamous carcinomas, adenocarcinomas, and a group of anaplastic unclassified carcinomas. Hasterlik and his colleagues have made careful and repeated radiological studies of the mastoid area in a series of the Chicago cases of radium poisoning and have been able to follow the development of mastoid involvement, so that early operative treatment at least on the mastoid tumours should now be possible.

The latent period for development of the carcinomas appears to be considerably longer than that for development of the sarcomas (Finkel, Miller, and Hasterlik 1969; Evans, Keane, and Shanahan 1969; Rowland, Failla, Keane, and Slehney 1969–70). It is of interest that the latent period in the woodworkers was also long-ranging, from 28 to 45 years, the mean being 38 years (Hadfield 1970).

3.5.1.3. *Haemopoietic neoplasms.* Unfortunately no careful study has been made of marrow histology or the peripheral blood picture in the population of radium patients at M.I.T. and Chicago. It is abundantly clear from reading the many reports that have been published about these patients that marrow dyscrasias were not an outstanding feature (Looney 1955, 1956*a,b*; Looney, Hasterlik, Brues, and Skirmont 1955; Hasterlik, Finkel, and Miller 1964; Finkel, Miller, and Hasterlik 1969*a*). Finkel and his colleagues in their assessment of the Chicago cases in 1969 list one case of chronic lymphatic leukaemia, which can be ignored since it is not a condition thought to be induced by skeletal radiation (Finch, Hoshino, Iloga, Ichimaru, and Ingram 1969), one case of myeloid leukaemia, one of splenic leukaemia, and two of aplastic anaemia. These were all relatively early cases and little further information is available.

3.5.2. *Dysplasia*

3.5.2.1. *Bone.* Severe dysplasia may occur in radium poisoning in man and animals. This may result clinically in fracture (particularly of the long bones), vertebral collapse, and severe bone pain. Figure 7.7 shows the common sites of fracture found in the M.I.T. series (Evans 1966). Apart from fracture, the bone lesions which may be radiologically severe are often

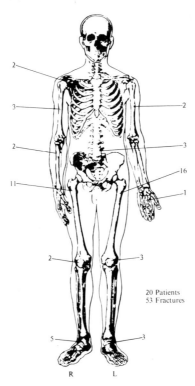

Fig. 7.7. Skeletal distribution of spontaneous fractures. (Evans 1966, by courtesy of author and publishers.)

symptomless and only found on routine radiological examination. Hasterlik and his colleagues (Hasterlik, Finkel, and Miller 1964) list the radiographic lesions as (i) coarsening of the trabecular pattern, (ii) localized areas of bone resorption, (iii) patchy sclerosis, (iv) small and large bone infarcts, and (v) aseptic necrosis.

The most typical lesions occur in the skull as shown in Fig. 7.8. There are clear-cut, pinched-out, round areas associated with other areas of increased sclerosis, a picture which is sometimes confused with that seen in myelomatosis. In the long and flat bones there are again areas of increased density and areas of rarefaction, giving the bone a moth-eaten appearance as shown in the pelvic radiograph in Fig. 7.9 and the long bone radiograph in Fig. 7.10. Sometimes the picture may be confused with the localized bone infarcts seen in Caisson disease (Looney, Hasterlik, Brues, and Skirmont 1955; Looney 1955, 1956a,b). On repeated restudy of a large series of patients it has been noted that an increase in the number and severity of the radiographically demonstrable lesions occurs concomitantly with a decrease in the body burden of radium. With the passage of time the

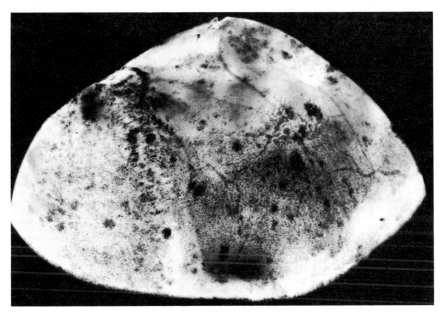

FIG. 7.8. Characteristic punched-out areas in the skull of a patient who had ingested radium, alternating with areas of increased density.

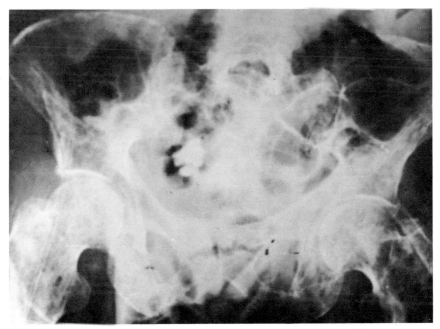

FIG. 7.9. Characteristic moth-eaten appearance of the pelvis and heads of the femora in a patient who had a radium burden.

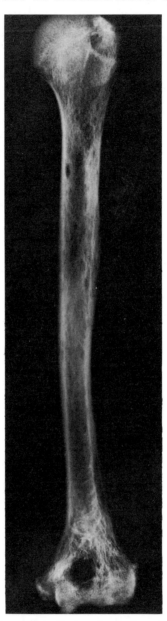

FIG. 7.10. Characteristic appearance in the long bone of a patient with a radium burden—areas of increased density alternate with areas of rarefaction.

FIG. 7.11. Autoradiograph of a 200 μ thick tooth-section showing intense line of radium in the dentine; adjacent to this line is a broad area of rather uniform deposition. (Rowland 1959b, by courtesy of author and publishers.)

relationship of the severity of the radiographically demonstrable skeletal lesions to current body-burdens of radium becomes less and less meaningful, as already seen in Fig. 7.6.

Hasterlik and his colleagues (Hasterlik, Finkel, and Miller 1964) have concluded that significant radiographic lesions distinguishable from usual ageing phenomena are not seen at body burdens below 0·1 μg ^{226}Ra.

Occasionally attention has been drawn to the possibility of diagnosing radium poisoning by the dentist noting the so-called 'pink teeth' which may be seen in patients who ingested radium when young (Looney, Hasterlik, Brues, and Skirmont 1955). One such patient, seen by a dentist

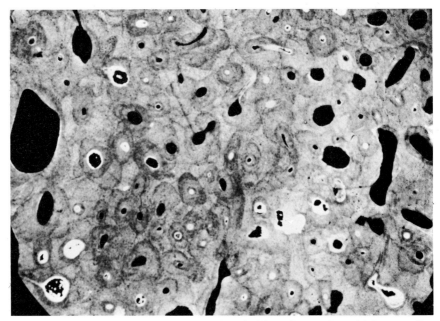

Fig. 7.12. Microradiograph of cross-section of cortical bone from patient with long standing radium burden. Note (i) loss of Haversian pattern, (ii) irregular patches of high calcification, (iii) high calcification round Haversian canals and plugged canals, and (iv) calcified lacunae. (Hindmarsh *et al.* 1959, by courtesy of authors and publishers.)

at the age of 32, showed a resorption of the teeth leading to the fragmentation and loss of the crown of a central incisor. Radiographs showed a bizarre widening of the pulp cavity and destruction of the dentine arising within the teeth. On questioning it was found that between the ages of 8 and 10 he had taken 'radium water' daily as a tonic.

Rowland (1959b) has investigated the problem of radium retention in the teeth in a dial painter (code no. 03473) who at the age of 55 had an infected third molar removed. His body burden was then 1·2 μCi ^{226}Ra and he had painted between the age of 19 and 21. The tooth had a radium content equal to about 0·2 per cent of the total body radium content. An autoradiograph of a 200 μ thick tooth-section is shown in Fig. 7.11. An intense line of radium is visible in the dentine and adjacent to this is a broad area of rather uniform deposition; a low diffuse level of activity exists in the dentine between this intense line and the enamel. The uptake in the bone of the jaw is negligible. The dentine hot line contained 21·1 pCi/mg compared with 3·3 pCi/mg in a bone hot spot. The ratio of dentine hot spot to dentine diffuse activity was 176 to 18. This was much higher than the ratio for hot spot to diffuse activity in the bone both of the same individual and of all other dial painters measured. Rowland suggests that the bone had lost radium by a process of long-term exchange but the tooth had not, since in the case of teeth, neither resorption nor exchange takes place.

These gross radiological changes are better understood when the characteristics microradiographic lesions are studied. The characteristic microradiographic appearance of a cross-section of cortical bone from a long standing human case of radium poisoning is shown in Fig. 7.12. The same appearance, though in less exaggerated form, is found in the bones of dogs treated with ^{226}Ra and killed some years later (Jee and Arnold 1960). The most characteristic finding is the plugging of Haversian canals with densely calcified material; the plugging may also affect osteocyte lacunae (Hindmarsh and Vaughan 1957; Hindmarsh, Owen, and Vaughan 1959; Rowland and Marshall 1959; Rowland, Marshall, and Jowsey 1959; Rowland 1960a,b,c,d; Lloyd 1961a). Rowland found such plugged canals in 14 out of 15 cases examined. The maximum proportion of canals found to be plugged in cross-sections of bone was 29 per cent, over 1000 different canals being counted in each individual. A three-dimensional analysis however indicated that although some canals were plugged for as much as 3 mm or more most of them were plugged for only about 0·5 mm and in some parts the plug was not complete. This observation implies that the fraction of osteones which contain a plug somewhere along their length is quite large and certainly larger than the percentage of Haversian canals showing no plugs. Patients with a relatively high terminal body-burden (over 1 μg) are more likely to show severe plugging than those with lower burdens (Rowland 1960b). In addition to complete plugging, canals showing a greater number of highly calcified inner lamellae are found (Hindmarsh and Vaughan 1957). Large and bizarre resorption cavities are also characteristic (Jee and Arnold 1960). Some of these are clearly dependent on abnormal resorption but others may be associated with neoplastic growth. Unless ordinary histological preparations as well as microradiographs are examined the presence of tumour as well as bone dysplasia may be missed (Vaughan 1961). Although extremely characteristic of vascular damage induced by radionuclides, the presence of plugged canals is also seen in conditions when vascular damage due to other causes is present. Rowland (1962) found an exactly similar picture in the bones of patients when amputations had been performed for gangrene.

3.5.2.2. *Marrow.* Dysplasia of the marrow is not characteristic of human radium poisoning. Haematologically the findings in these patients are neither striking nor diagnostic, as far as the peripheral blood or the marrow is concerned (Looney, Hasterlik, Brues, and Skirmont 1955).

3.6. *Dosimetric considerations*

The problems involved in the calculations of radiation dose to the skeleton from a deposited radionuclide are immense. They have been discussed and analysed in great detail in relation to those people known to have a body burden of ^{226}Ra and ^{228}Ra, since this is the only human data available. (Evans 1966; Evans, Keane, Kolenkow, Neal, and Shanahan 1969; Evans, Keane, and Shanahan 1969; Finkel, Miller, and Hasterlik 1969a,b; Rowland, Failla, Keane, and Slehney 1969–70, 1970–71). Calculation of the relative biological efficiency of different radionuclides, as already discussed in Chapter 5, is often related to that of radium for this reason. The techniques employed and the results of calculating radiation dose to the human skeleton from ^{226}Ra may be discussed under two headings, macroscopic and microscopic, though it appears probably that in future a combination of the two may prove extremely valuable since it will enable an estimate to be made of dose to sensitive tissues rather than of a mean average dose to the whole bone mass.

Macroscopic methods are based on calculations of whole-body burdens based on excretion data, whole body counting data, and measurements of exhaled radium. Microscopic data are based on a variety of techniques which involve examinstion of small sections of bone.

3.6.1. *Macroscopic*. In the case of human radium burdens of known value in μCi/kg, various methods of expressing dose in rad have been employed (Finkel, Miller, and Hasterlik 1969a,b; Evans 1966; Evans, Keane, Kolenkow, Neal, and Shanahan 1969; Rowland, Failla, Keane, and Slehney 1969–70, 1970–71; Spiess and Mays 1970). The most generally used calculation is of the total number of rad accumulated by the skeleton from acquisition of radioactive material to death or to diagnosis of a malignancy—the cumulative rads or CR of Evans (Evans 1966). This dose is the average skeletal dose to the whole skeleton in rad and is not related to sensitive tissues as emphasized both by Evans (Evans 1966) and Rowland (Rowland, Failla, Keane, and Slehney 1969–70, 1970–71). Evans himself prefers the CRY, cumulative rad years, which takes into account the element of time as well as of dose. Thus 1 rad delivered 20 years ago or 2 rad delivered 10 years ago or 5 rad delivered 4 years ago would all be equivalent to 20 rad years (Evans 1966).

In Fig. 7.13 (Evans 1966) is shown the relationship between body burden, expressed as PRE (μCi Ra), and years since first exposure to 1967, if living, or to death. Figure 7.14 (Evans 1966) shows the relationship between the X-rays lesions expressed as the classical X-ray score* and CR, i.e. average skeletal dose. Figure 7.15 (Evans 1966) shows the relationship between the X-ray lesions and CRY.

* The classical X-ray score is a scale of lesions seen radiographically in the skeleton which was agreed many years ago by the observers both at M.I.T. and Chicago (Evans 1966). Observations therefore made on both groups of patients are comparable and can be related to radiation dose.

RADIUM ISOTOPES

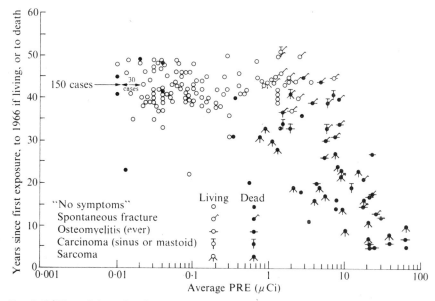

FIG. 7.13. Elapsed time after first exposure, related to body burden PRE (μCi). (Evans 1966, by courtesy of author and publishers.)

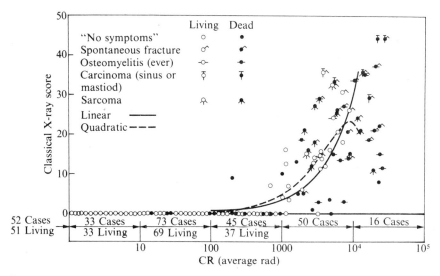

FIG. 7.14. Correlation between classical X-ray score and CR (cumulative rad) for about 270 dial painters, chemists, and iatrogenic cases (Evans 1966, by courtesy of author and publishers.)

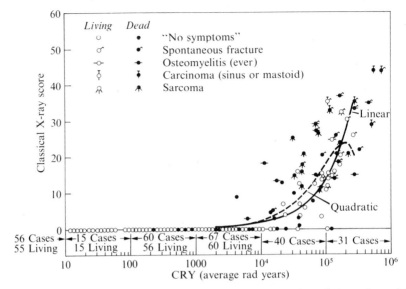

Fig. 7.15. Correlation between classical X-ray score and CRY (cumulative rad years) for about 270 dial painters, chemists, and iatrogenic cases. (Evans 1966, by courtesy of author and publishers.)

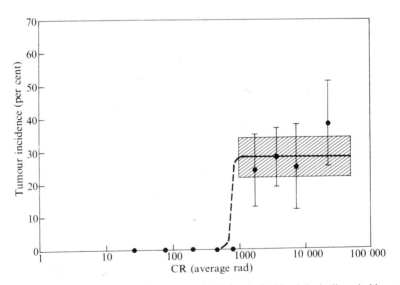

Fig. 7.16. The observed radiogenic tumour incidence in 'epidemiologically suitable cases'. (Evans, Keane, and Shanahan 1964, by courtesy of authors and publishers.)

Evans and his colleagues have used such data to show that the incidence of tumour is related to accumulated rad (CR) in a markedly non linear way (Fig. 7.16) i.e. that there is an effective threshold below which no tumours have occurred and which therefore justifies the acceptance of a permissible body burden of 0·1 μg ^{226}Ra.

Before accepting the proposal of a threshold it would appear wise to wait till all the patients at present under observation with known low body-burdens have lived their full life-span. A population of the order of 700 is a small one by epidemiological standards.

Finkel, Miller, and Hasterlik in 1969a published a curve which suggested linearity of dose response for the Chicago series. However, this curve is no longer acceptable since the point at 0·17 μCi is no longer correct. The maxillary tumour removed surgically from the patient with this body burden (no. 03–685), which has now been studied by a large number of oral pathologists, is not now regarded as malignant (Finkel, Miller, and Hasterlik 1969b). If this one case is excluded the data are still reconcilable with the threshold hypothesis of Evans *et al.*, the threshold being about 0·5 μg PRE.

The lowest average mean dose to the whole skeleton at which a tumour has been observed in man is 1200 rad, calculated by Evans *et al.* (1969); this is well above the permissible rad dose during 50 years based on the recommendations of the International Commission on Radiological Protection (1960). Dougherty and Mays (1969) have used this figure to calculate corresponding doses for the other alpha emitters, dividing 1200 by the appropriate RBE obtained as discussed in Chapter 1 and Chapter 6 in the Utah dog experiments. The predicted doses causing bone cancer in humans on this basis have already been shown in Table 6.5. How far such calculations are justified is questionable since the number of dogs on which the determinations of RBE are based are small. Further it is perhaps rash to assume that RBE values are the same in dog and man. In fact Dougherty and Mays (1969) themselves question whether human RBEs correspond exactly to those in the dog. They raise particularly the question of ^{228}Ra (mesothorium). The predicted dose that would give rise to a bone cancer in humans from this radionuclide as shown in Table 6.5 is 480 rad while the lowest dose actually observed to do so in man is about 1500 rad (Evans 1966).

Rowland and his colleagues (1969–70, 1970–71) have recently made a further analysis of dose–response relationships for tumour incidence in the combined M.I.T. and Chicago series, with particular reference to the low dose cases in which malignancies have not as yet developed. They state that they have attempted to fit response functions to the observed data, not to specify safety criteria which seek a linear relationship such that the observed incidence at any dose is less than predicted. The unit of dose employed is the CR of Evans. The calculation requires the evaluation of the expression

$$D = \int_0^T R(t)\,\delta t$$

where $R(t)$ is the average skeletal dose rate at any time and D is the dose accumulated in time T.

The osteosarcomas were well fitted by an empirical dose-response function for tumour incidence I, of the form

$$I = KD^2 e^{-D/D_0}$$

where D_0 and K are chosen to give the best fit to the data. The carcinoma cases, which are few in number, and occur after a much longer time interval than the sarcomas, cannot be fitted to any continuous function. At doses below 1000 rad, where only one tumour has been observed in 777 cases, Rowland et al. consider that there is only a 1 per cent probability of the incidence being greater than that given by the equation

$$I = 7 \times 10^{-5} D.$$

This analysis is valuable, but it does not take into account the shorter time interval between exposure and endpoint at larger accumulated doses; the time interval increases as the dose *rate* decreases, and when the time interval is so long that natural death occurs before tumour incidence, the observed incidence must depend on the dose rate as well as the accumulated dose.

3.6.2. *Microscopic.* No microscopic dosimetry in man in direct relation to an osteosarcoma induced by radium isotopes is available. There is however extremely interesting microdosimetry in relation to carcinomas of the cranial sinuses. In one case quoted by Evans and his colleagues (Evans, Keane, and Shanahan 1969) the cumulative soft-tissue dose, at a distance 10 μm from the bone tissue interface, was 82 000 rad when a carcinoma of the left frontal sinus appeared at a burden time of 52 years. To this dose, due only to alpha rays emitted from the frontal bone, there must be added an additional alpha-ray dose due to radon diffusing out of the bone through the sinus epithelial tissue and into the frontal sinus. In air withdrawn from the uninvolved right frontal sinus the radon concentration was at least 30 times that in the exhaled breath. The epithelial cells at risk had therefore been bombarded with alpha particles both from behind and in front. The importance of the additional alpha ray dose due to radon and its daughter products in the sinus is suggested by the fact that Evans has found no case of paranasal sinus or mastoid carcinoma in persons whose principal skeletal dose was due to ^{228}Ra (though this group has its full share of sarcomas) unless they carried a large burden of ^{226}Ra. The equivalent to radon in the ^{228}Ra series is thoron whose half-life is only 55 seconds, which may prevent

RADIUM ISOTOPES

the accumulation of a large concentration in the sinus, mastoid tissues, and air spaces, especially if the major supply mechanism is diffusion from the adjacent bone. The high dose of 82 000 rad at the bone surface is probably accounted for by a very slow turnover of radium in the bones of the skull. This has been found experimentally in the case of ^{90}Sr in the rabbit where similar carcinomas are found (Chapter 8.1.2) (Vaughan and Williamson 1969).

Several early attempts were made to use microscopic techniques on bone from patients with a radium burden to estimate total body burden and also to gain some idea of the degree of non-uniformity of deposition (Hindmarsh and Vaughan 1957; Hindmarsh, Owen, Vaughan, Lamerton, and Spiers 1958).

The most significant data on radiation dose to bone marrow cavities and to osteogenic tissue in trabecular surfaces from ^{226}Ra, using microscopic techniques, comes from the recent work of Spiers, Whitwell, and Darley (1971). These workers have measured mean path-lengths in trabecular bone and in marrow cavities in a number of human bones using techniques described elsewhere (Chapter 2.1.1). They have then calculated the mean dose from radium and its decay products to the marrow spaces in trabecular bone and the mean dose to the endosteum by averaging the dose throughout a tissue layer of thickness 10 μm adjacent to the trabecular surfaces. These dose calculations follow the method used to determine beta-particle dose in bone containing a uniform distribution of a beta-emitting radionuclide (Whitwell and Spiers 1971). The alpha and beta particle energies used in the radium calculations make use of the data already shown in Tables 7.1 and 7.2.

Such calculations indicate that the fraction of marrow irradiated in the parietal bone is about three times greater than that irradiated in the vertebra, ribs, iliac crest, or head of the femur. With the exception of the parietal bone, the mean path-lengths through the trabeculae are not markedly different and lie between extreme values of 220 and 320. The marrow spaces have mean path-lengths varying from 900 to 1700 μm—but in the case of the parietal bone the mean path length for the trabeculae is 515 μm and that for the marrow spaces only 390 μm. These small linear path-lengths in the parietal marrow spaces results in high mean-dose factors, which Spiers suggests may account for the typical punched-out areas seen so characteristically in the skulls of patients with a radium burden. 68 per cent of the tissues in the marrow cavities of the parietal bone are irradiated by deposited ^{226}Ra and its decay products, while in other bones the figure varies from 16 per cent to 28 per cent. It is therefore not surprising to find that osteosarcoma is the characteristic lesion in patients exposed to radium, while malignancies of marrow tissue are extremely rare. Spiers has indeed calculated that, on the assumption that a terminal radium burden of 10 μCi is evenly distributed through a skeletal mass of 7000 g (weight of bone without marrow), the accumulated mean marrow-dose would be 770 rad in 50 years. The endosteal tissue dose for a 10 μCi radium burden, he suggests, lies in the range 7900 to 9400 rad, and that for a burden of 0·5 μCi in the range 400–470 rad. Clearly, if by macroscopic methods the total body burden is known, it will be possible in future, especially if path lengths are also measured, to make a more realistic estimate of terminal radiation dose to sensitive tissues. Non-uniformity is of course not taken into account, nor is the fact that the radium burden has changed with time. Calculations based on other data might be used to correct for the time factor.

Careful estimations of the radiation dose that may be received from radium deposited both diffusely and in hot spots has been made by a number of workers (Hoecker and Roofe 1951; Hindmarsh and Vaughan 1957; Hindmarsh, Owen, Vaughan, Lamerton, and Spiers 1958; Hindmarsh, Owen, and Vaughan 1959; Rowland 1959a, 1960d; Rowland and

TABLE 7.4

Particulars of a series of radium patients with a known radium burden (Rowland 1960d, by courtesy of the author and publishers)

Patient	Terminal radium burden (μCi)	Sex	Age radium acquired	Age at death	Years radium carried	Method of acquisition
EE	50·?	F	21–4	32	11	Dial painter
KS	19·?	F	15–20	31	16	Dial painter
IL	8·?	F	16–20	30	14	Dial painter
FR	7·0	F	21–2	48	27	Dial painter
103	6·8	F	36	64	28	I.V. medically
R	3·6	F	32	56	24	I.V. medically
Q*	1·2	M	33	59	26	I.V. medically
BM	1·2	M	21–3?	51	28–30?	I.V. medically
112	1·2	F	25	48	23	Oral: medically
1b	0·8	F	28	53	26	I.V. medically
118*	10·5	F	48	84	36	Oral: medically
LS	8·0	F	17	57	41	Dial painter
302	3·8	F	15	47	32	Dial painter
HG	3·0	F	16–19	48	32	Dial painter
RB	2·7	F	20	53	33	Dial painter
313	1·3	F	18–47	48	30	Dial painter
IJ	1·2	F	18–21	—	31	Dial painter
03473*	1·2	F	19–21	—	36	Dial painter
ML	1·0	F	16–17	—	39	Dial painter

* Patient without skeletal tumours.

RADIUM ISOTOPES

TABLE 7.5

Measured hot spot and diffuse concentrations of radium in the same series of patients as shown in Table 7.4
(Rowland 1960d, by courtesy of author and publishers)

Patient	Terminal radium burden (μc)	Maximum hot spot ($\mu\mu c/mg$)	Average diffuse ($\mu\mu c/mg$)	Calculated uniform label ($\mu\mu c/mg$)	Ratio of hot spot to diffuse	Ratio of diffuse to uniform label
EE	50·?	19·0	1·2	—	16	—
KS	19·?	—	1·4	—	—	—
IL	8·?	9·0	0·28	—	32	—
FR	7·0	30·0	0·32	1·0	94	0·32
103	6·8	38·0	0·30	1·2	127	0·25
R	3·6	16·0	0·15	0·66	107	0·23
Q*	1·2	12·0	0·080	0·17	150	0·47
BM	1·2	6·0	0·043	0·17	140	0·25
112	1·2	1·4	0·075	0·20	19	0·38
I*	0·8	13·0	0·060	0·15	218	0·40
118*	10·5	41·0	0·50	2·0	82	0·25
LS	8·0	16·0	0·54	1·3	30	0·42
302	3·8	20·0	0·54	0·67	37	0·81
HG	3·0	9·3	0·39	0·50	24	0·78
RB	2·7	15·0	0·38	0·49	40	0·78
313	1·3	2·8	0·15	0·30	19	0·50
IJ	1·2	4·7	0·046	0·20	102	0·23
03473*	1·2	3·3	0·18	0·20	18	0·90
ML	1·0	6·0	0·10	0·17	60	0·59

* Patient without skeletal tumors.

Marshall 1959; Lloyd 1961a). The results of some of these measurements are shown in Tables 7.4, 7.5, and 7.6. In Table 7.4 are certain facts about each patient i.e. the terminal radium burden in μCi, the age at which the radium was acquired, the age at death, and the method of acquisition. In Table 7.5 are the measured hot spot and diffuse concentrations in pCi, and in Table 7.6 the terminal dose rate to a 10 μm lacuna from both the hot spot and diffuse distribution, and the minimum life-time dose to a 10 μm lacuna. It can be seen at once that the doses received are extremely high in some instances. The figures for radiation dose are, however, too low since about 90 per cent of the radiation originally present in the hot spots will have been removed by long-term exchange process over a 30 year period from both hot spots and diffuse reaction (Rowland and Marshall 1959). If, instead of the dose received by an osteocyte, that to a 10 μm thick layer of endosteum or periosteum is considered the value quoted should be halved (Rowland 1960d).

TABLE 7.6

Terminal dose rates from radium in the Series of Patients shown in Tables 7.4 and 7.5

(Rowland 1960d, by courtesy of the author and publishers)

Patient	Terminal body burden (μCi)	Terminal dose-rate to a 10 μm lacuna		Minimum lifetime dose to a 10 μm lacuna	
		Hot spot (rad day^{-1})	Diffuse (rad day^{-1})	Hot spot (rad)	Diffuse (rad)
EE	50 ?	15	0·95	60 000	3800
KS	19 ?	—	1·1	—	6400
IL	8 ?	7·2	0·22	37 000	1100
FR	7	24	0·25	240 000	2500
103	6·8	30	0·24	310 000	2500
R	3·6	13	0·12	110 000	1100
Q*	1·2	9·5	0·064	90 000	610
BM	1·2	4·8	0·034	49 000	350
112	1·2	1·1	0·060	9 200	500
1[b]	0·8	10	0·048	95 000	460
118*	10·5	33	0·40	430 000	5200
LS	8	13	0·43	190 000	6400
302	3·8	16	0·43	190 000	5000
HG	3	7·4	0·31	86 000	3600
RB	2·7	12	0·30	150 000	3700
313	1·3	2·2	0·12	24 000	1300
IJ	1·2	3·7	0·037	42 000	420
03473*	1·2	2·6	0·14	34 000	1800
ML	1	4·8	0·080	68 000	1100

* Patient without skeletal tumors.

3.7. Effect on animals

The effect of radium poisoning in animals has been studied particularly in mice, rats, rabbits, and dogs (Cluzet 1910; Albee 1920; Phemster 1926; Sabin, Doan, and Forkner 1932; Brooks and Hillstrom 1933; Regen and Wilkins 1936; Rosenthal and Grace 1936; Ross 1936; Dunlap, Aub, Evans, and Harris 1944; Evans, Harris, and Bunker 1944; Finkel 1953, 1956, 1959a,b; Finkel and Biskis 1959, 1960; Jee and Arnold 1960; Marshall and Finkel 1960; Dougherty 1962; Finkel, Jinkins, and Biskis 1965; Dougherty and Mays 1969; Finkel, Biskis, and Jinkins 1969; Mays, Dougherty, Taylor, Lloyd, Stover, Jee, Christensen, Dougherty, and Atherton 1969; Goldman, Dungworth, Bulgin, Rosenblatt, Richards, and Bustad 1969).

The effect of radium on the skeleton in all species is in broad terms similar to that seen in man, though the degree of retention, as illustrated in Fig. 7.4, will differ (Stara, Nelson, Della Rosa, and Bustad 1971). There may be gross dysplasia of bone. Osteogenic sarcomas are the common cause of death but carcinomas of epithelial tissue closely applied to bone in the skull also occur though they are not noted by the earlier workers, who used large doses of radium, and the animals therefore probably died with osteogenic sarcomas or profound marrow aplasia before the carcinomas had an opportunity to develop. Haemopoietic malignancies are rare.

3.7.1. *Neoplasia.* Many of the earlier investigations induced osteosarcomas in small rodents exposed to ^{226}Ra. Such experiments have been reviewed elsewhere (Vaughan 1962*a*). Of particular interest is the observation of Dunlap and his colleagues (1944) that the tumours in rats were very rich in alkaline phosphatase, which suggests perhaps that they arise from osteoblastic precursors, since the osteoblast is rich in this enzyme (Vaughan 1970*a*). More recent investigations have been planned with a view to determining the relative toxicity of different internally deposited radionuclides and to defining a dose-response curve both for radium and for other radionuclides, since data is available on the effect of ^{226}Ra in man. This is particularly true of the work of Finkel and her group, and of the Utah group, i.e. Dougherty and Mays (1969). In attempting to assess the significance of the results obtained it should be pointed out that all experience goes to show that the metabolism of many radionuclides in the dog appears to be fairly comparable to that in man. Further the size of a dog bone is more closely related to that of man than the size of a mouse bone so that the geometric considerations involved in dosimetric calculations are more likely to be comparable between man and dog than between man and mouse. The beagle dog which has been used in the majority of the dog studies is further known to have a very low natural incidence of osteosarcoma (Mays and Taylor 1964). To find a single naturally occurring osteosarcoma would require the life-span observation of 1000–100 000 control beagles, so that any bone tumours that arise in experimental beagles can probably be accepted as due to the radionuclide.

7.1.1. *Mice.* Many strains of mice are known to have a high leukaemia incidence which is enhanced by any form of radiation (Kaplan 1966). The strain, CF 1/Anl, used by Finkel and her colleagues in the great majority of her studies has a natural osteosarcoma incidence of at least 1–2 per cent (Finkel and Biskis 1968). A further study made of the same strain of normal CF 1/Anl mice used as controls in a different experiment gave an osteosarcoma incidence of 4 per cent and indicated that this strain has a strong tendency to develop a skeletal dysplasia which may influence the carcinogenic response (Lisco, Rosenthal, and Vaughan 1971). In assessing the results of Finkel's mouse experiments on tumour response to all radionuclides, including radium, these facts must be remembered (Chapter 1.3).

Finkel has summarized her experience of the toxicity of radium in female mice (CF 1/Anl strain) (Finkel, Biskis, and Jinkins 1969). In all, 2685 mice were given injections of ^{226}radium varying from 120 μCi kg^{-1} to 0·05 μCi kg^{-1}. There were 525 control mice. The diagnosis of osteosarcoma was made on radiological examination—few lesions were thought to require histological examination. The authors comment that even the lowest level (0·05 μCi kg^{-1} ^{226}Ra) was responsible for some of the observed osteosarcoma, since in this group both the number of mice dying with osteosarcoma and the average number of tumours per mouse were four times higher than in the control group. As the dose level decreased from 120 to 2·5 μCi kg^{-1} there was a gradual increase in average latent period though Finkel herself doubts whether the figures for differences in the latent period were significant (see Table 7.7). In all there were 711 osteosarcomas* (diagnosed on radiographic examination), 49·5 per cent in the spine, 21·5 per cent in the legs, 16·5 per cent in girdles and ribs, and 12·5 per cent in the head. Radiographically detected lesions of the mandible and maxilla were frequent as shown in Table 7.8. Those tabulated were regarded as

TABLE 7.7

Summary of osteosarcoma data for mice injected with ^{226}Ra
(*Finkel et al. 1969, by courtesy of authors and publishers*)

Amount Injected (μCi kg^{-1})	Mice with osteosarcoma (%)	Average number per mouse	Average osteosarcoma expectancy†	Estimated time (days) to radiographic appearance	
				Range	Average
120·0	31·1	0·467	1·011	194–455	328·5
80·0	68·9	1·330	1·880	108–515	359·0
40·0	73·5	1·360	1·724	229–550	393·9
20·0	84·5	1·400	1·721	120–583	428·3
10·0	75·5	1·000	1·250	127–679	484·3
5·0	62·3	0·956	0·980	281–726	543·5
2·5	42·8	0·600	0·851	307–887	638·7
1·25	21·0	0·305	0·405	392–910	657·2
1·0	23·4	0·317	0·377	347–965	643·4
0·75	18·5	0·208	0·267	332–919	686·0
0·5	11·6	0·143	0·195	312–987	654·7
0·25	7·45	0·078	0·081	327–843	580·0
0·1	1·96	0·020	0·058	621–1059	852·6
0·05	4·32	0·043	0·077	390–852	709·7
0·0	1·14	0·011	0·014	633–875	730·5

* For definition see Section 3.2.1.1.

† Of the 711 osteosarcomas four were parosteal and 2 of these occurred in controls. It is perhaps of interest that the tumours induced in mice by FBJ virus are characteristically parosteal (Finkel and Biskis 1969) and further that Huebner and his colleagues have found FBJ virus in normal CF 1 mice (Huebner, Hartley, Lane, Turner, and Kelloff 1969).

TABLE 7.8

Non-malignant lesions of mandible and Maxilla in mice given ^{226}Ra
(*Finkel* et al. *1969, by courtesy of authors and publishers*)

Amount injected (μCi kg^{-1})	Mandibular lesion with hair (%)	Abnormal maxillary dental tissue (%)	Fibrous dysplasia in mandible and maxilla (%)
120·0	0	4·4	0
80·0	0	4·4	0
40·0	0	17·8	0
20·0	0	0	0
10·0	0	0	0
5·0	4·4	0	0
2·5	1·9	0	2·9
1·25	1·0	0	1·9
1·0	0·4	0	2·5
0·75	0·6	0	5·9
0·5	0·6	0	1·2
0·25	0	0	2·0
0·1	0	0	0·4
0·05	0	0	2·4
0·0	0	0	0·2

TABLE 7.9

Other skeletal and lymphoid lesions in mice injected with ^{226}Ra
(*Finkel* et al. *1969, by courtesy of authors and publishers*)

Amount injected (μCi kg^{-1})	Monostotic osteoblastic-resorptive lesion (%)	Osteoma (%)	Hemangio endothelioma of bone (%)	Reticular tissue tumor (%)
120·0	0	0	0	11·1
80·0	0	0	0	13·3
40·0	0	2·2	0	17·8
20·0	0	0	0	6·7
10·0	0	2·2	0	17·8
5·0	0	6·7	0	31·3
2·5	0	1·9	0	39·0
1·25	0	10·5	1·9	45·7
1·0	0·8	10·0	1·7	34·6
0·75	0·4	12·6	1·8	25·9
0·5	0·7	16·4	1·3	42·2
0·25	1·2	14·1	0·8	31·8
0·1	0·8	11·4	0·4	33·3
0·05	0·4	12·5	0·4	38·4
0·0	0·4	11·4	1·3	46·4

142   THE EFFECTS OF IRRADIATION ON THE SKELETON

non-malignant but in addition to the osteosarcomas there was one fibrosarcoma of the mandible and a giant cell tumour of the maxilla. Other skeletal and lymphoid lesions are shown in Table 7.9. It should be noted that osteomas, haemangioendotheliomas, and reticular tissue tumours were recorded in the controls. They were not examined microscopically. Finkel does not attribute those in the injected mice to radiation since their incidence was not affected by dose, but they make one uneasy about using a strain with such a high incidence of skeletal and reticular tissue abnormalities to test the toxicity of radionuclides.

Tumour expectancy is a figure much used by Finkel. She defines it as follows: 'Tumour expectancy, a statistic very similar to life expectancy and obtained by dividing the number of osteosarcomas still to appear by the number of animals still alive, is initially equivalent to the final cumulative

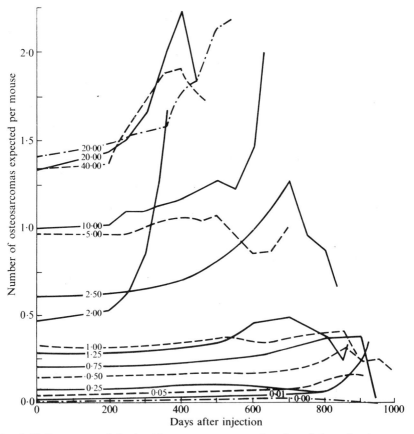

FIG. 7.17. Expectancy of death with osteosarcoma as a function of time after intravenous injection of radium—(number of osteosarcoma still to appear)/(number of mice still alive). (Finkel *et al.* 1969, by courtesy of author and publishers.)

RADIUM ISOTOPES

incidence. Subsequently as animals die without tumour, expectancy increases; as they die with tumour expectancy decreases.' (Finkel, Biskis, and Jinkins 1969). In Fig. 7.17 it is shown that tumour expectancy among mice that received 10 or more μCi kg^{-1} increased with time whereas expectancy among those that received the smallest amounts of radium remained more or less constant. She calculated average osteosarcoma expectancy for each group of mice shown in the tables from the expectancy values determined

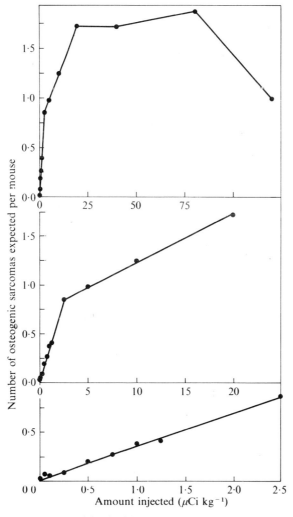

FIG. 7.18. Average expectancy of death with osteosarcoma as a function of amount of radium injected (expectancy values calculated at 10 day intervals and averaged from 250 days after injection until less than 6 mice remained alive. (Finkel et al. 1969, by courtesy of authors and publishers.)

at 10 day intervals from 250 days after injection until less than six animals remained alive, and plotted the results as a function of the amount of radium injected, to give a dose-response curve as shown in Fig. 7.18. She concludes 'The expectancy of death with osteosarcoma increased linearly with increasing amount of injected radium from the control value of 0·014 to 0·851, after the injection of 2·5 μCi kg.$^{-1}$. A linear response with different slope was present from 2·5 to 20 μCi kg^{-1}, where the expectancy was 1·721. There was no further increase at higher dose levels.'

The first change in the slope of the dose-response curve, she suggests, may be due to the existence of several mechanisms of oncogenesis with their relative importance changing at this level of irradiation. The second change, i.e. the flattening of the curve and subsequent fall, she considers is probably due to severe lethal damage to bone at high doses. She points out that this curve differs from those that she has obtained for other bone-seeking nuclides which have shown tumour expectancy increasing as the square or cube of the injected amount of radioactivity. How far tumour expectancy rather than straightforward tumour incidence is a valid basis for calculating dose–response curves is perhaps open to question. Certainly, as already seen, the human data at present available do not support linearity.

3.7.1.2. *Dogs.* Significant observations on the effects of bone-seeking radionuclides on the skeleton comes from the Utah experiment where ^{239}Pu, ^{228}Th, ^{228}Ra, ^{226}Ra, and ^{90}Sr have been given to young adult beagle dogs. These experiments have already been discussed in Chapter 6.4 in relation to the relative biological efficiency of different radionuclides. The Utah experiment was concerned with the effects of a single injection of ^{226}Ra. Another experiment where repeated injections of ^{226}Ra have been given to beagles is in progress at the University of California, Davis. A group of beagle dogs received eight intravenous injections of ^{226}Ra at fortnightly intervals at the age of 435 days (Goldman, Dungworth, Bulgin, Rosenblatt, Richards, and Bustad 1969). This will give a rather different dose distribution to that found in the Utah dogs. The injection levels were 0·000, 0·003, 0·008, 0·047, 0·14, 0·42, and 1·25 μg ^{226}Ra kg^{-1}. The workers themselves point out the difficulties of estimating the radiation dose to sensitive tissues where skeletal distribution of the radium is uneven. They show the curve illustrated in Fig. 7.19, which they describe as illustrating the average dose rate; but they do not make it clear what tissue is involved. This figure also shows an average dose rate from ^{90}Sr given to beagle dams and their offspring, which is discussed in Chapter 8.

The acute ^{226}Ra effect was an abrupt dose-related leukopaenia and a subsequent mild anaemia (Rosenblatt and Goldman 1967; Goldman, Dungworth, Bulgin, Rosenblatt, Richards, and Bustad 1969). Partial recovery, suggest to the workers, on haematological criteria, that tissue repair and cellular proliferation occurs to some extent and may be supported

RADIUM ISOTOPES

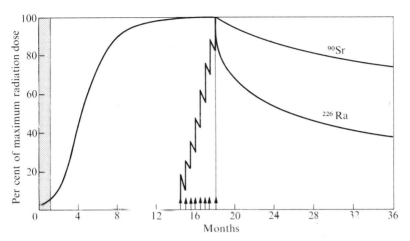

FIG. 7.19. Radiation dose rates to ^{90}Sr fed and ^{226}Ra injected beagles. Shaded area represents pre-weaning age. (Goldman *et al.* 1969, by courtesy of authors and publishers.)

by the initiation of extramedullary haemopoiesis. One of the dogs, receiving the highest dose (1·25 μg ^{226}Ra kg^{-1}), died with refractory anaemia 10 months after the injections ceased. Other dogs have developed osteosarcomas at dose levels as low as 0·14 μg kg^{-1} (given in eight successive doses). Radiographic features include trabecular coarsening seen in the distal femora, multiple areas of endosteal cortical sclerosis and thickening in the long bones, early fracture in the rib and spinous processes of the vertebrae, and fractures in the femora and occasionally in the humeri. Many of the older dogs, receiving the higher injection doses, showed a variety of dental lesions, including a peculiar type of caries, tooth root abscesses, and loss of teeth. There was a widespread distribution of osteosarcomas. All the tumours are reported to have some cells forming osteoid, but chondrosarcoma-like and fibrosarcoma-like areas were also found. In some animals secondary deposits were present in the lungs. The incidence of bone tumours appears to be dose-related. In Fig. 7.20 the osteosarcoma incidence is plotted together with that from the Utah experiment when the dogs received a single injection of ^{226}Ra. The three highest dose levels at Utah (10·4 μCi kg^{-1}, 3·21 μCi kg^{-1}, and 1·07 μCi kg^{-1}) compared reasonably well with the Davis high level dose. The shape of the curves obtained is very similar as shown in Fig. 7.20. In none of the dogs has a myeloproliferative disorder been noted. This experiment is still in progress, so effects are expected at lower dose levels.

3.7.2. *Dysplasia.* Injury, other than neoplasia, attributable to the presence of ^{226}Ra and ^{228}Ra in the bones of animals, has been described by many investigators (Dunlap, Aub, Evans, and Harris 1944; Heller 1948). The histopathology and clinical features of the bone lesions are similar to those

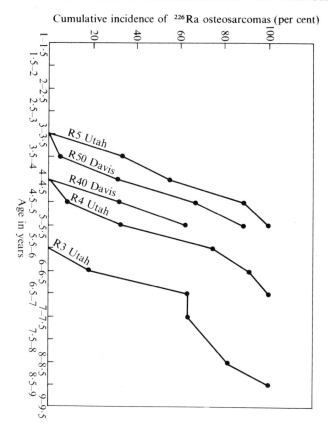

FIG. 7.20. Cumulative incidence rates of osteosarcoma in ^{226}Ra injected dogs in Davis study compared to those at Utah. (Goldman *et al.* 1969, by courtesy of authors and publishers.)

already described as occurring in man (Dunlap, Aub, Evans, and Harris 1944). There are patchy areas of sterile bone necrosis characterized by an absence of living osteocytes in the lacunae and also atypical new bone formation. Clarke (1962) who made a comparative study of the changes caused by ^{239}Pu, ^{226}Ra, and ^{90}Sr in pig bone concluded that ^{239}Pu produced more damage per rad than either ^{90}Sr or ^{226}Ra, irrespective of age. Jee and Arnold (1960) consider that the microradiographic picture illustrated in Fig. 7.21 is characteristic for ^{226}Ra compared with that resulting from ^{239}Pu or ^{228}Th. In the case of ^{226}Ra there are larger resorption cavities, hypocalcified osteones, and hypercalcified Haversian canal plugs. Frequent fractures were noted by the earlier investigators (Cluzet 1910, Albee 1920, Phemister 1926, Brooks and Hillstrom 1933, Regen and Wilkins 1936, Rosenthal and Grace 1936). Few fractures are recorded in the Utah beagles (COO 119, 242).

Fig. 7.21. Microradiograms of cross-section of the tibial diaphysis (× 16).
(a) Compacta of control dog. Note that osteones are slightly less mineralized than interstitial lamellae (from Jee and Arnold 1960, by courtesy of authors and publishers).
(b) Compacta of dog given 10 μCi kg^{-1} of radium. Note large resorption cavities, hypocalcified osteones, and hypercalcified Haversian canal plugs. High calcification appears white in the photograph.
(c) Compacta of dog given 0·9 μCi kg^{-1} of radiothorium. Note large area of periosteal resorption, ragged endosteal surface, few small cavities, and bland mineralization.
(d) Compacta of dog given 2·5 μCi kg^{-1} of plutonium. Note occasional large cavities, numerous hypocalcified osteones, and a few hypercalcified plugs.

Dunlap, Aub, Evans, and Harris (1944) describe focal atrophy, particularly in the marrow nearest to mineral surfaces with an excess of haemosiderin, throughout marrow and spleen. The latter in some cases showed focal eythropoiesis, but no proliferative hyperplasia like that occurring in the early Martland human cases of the marrow has been noted.

3.8. *Histopathology of* ^{226}Ra, ^{228}Ra *induced tumours*

Unfortunately there is no detailed description of the histopathology of either the bone tumours or the carcinomas in human radium poisoning. It is however clear from the somewhat inadequate data available that haemangiosarcomas or angiosarcomas do not occur, and that fibrosarcomas are at least as common as osteosarcomas. In animals it is also clear that haemangiosarcoma and angiosarcoma are rare. None are recorded among 38 bone tumours in the Utah dogs, which are all described as osteosarcomas, nor among the Davis dogs (Goldman, Dungworth, Bulgin, Rosenblatt, Richards, and Bustad 1969). As will be seen (Chapter 8.1.3.5) this picture is different from that presented by the neoplastic effects of ^{90}Sr, where many angiosarcomas or haemangiosarcomas are recorded. It is tempting to suggest that the short-range alpha particle of radium irradiates primarily only the osteogenic cells while the longer-range beta particle of $^{90}Sr + ^{90}Y$ also irradiates mesenchymal marrow cells.

Finkel and Biskis (1968) have described the histopathology of 830 radium-induced osteosarcomas in mice. Metastases, present in 20 per cent of the mice, were most commonly found in lungs; liver, spleen, and kidney were less often involved. Most of the osteosarcomas consisted of well-differentiated bone forming osteoblastic tissue arising on the endosteal surfaces and sometimes breaking through to the periosteal surface to give the sunburst appearance of rapid exogenous growth. Other morphological types appeared, some with fibroblastic, myxomatosis, chondroblastic, or telangiectatic features. They emphasized that several different components occurred in a single tumour, which is a feature also of tumours induced by ^{90}Sr in rabbits (Downie, Macpherson, Ramsden, Sissons, and Vaughan 1959). She also describes tumours of an unusual type occurring in low-dose animals that survived longer than 600 days. 'In these mice there was first an extensive spotty resorption of apparently normal cortical bone which was replaced by well vascularized tissue containing spindle cells and numerous giant cells. The spindle cells proliferated and formed small bundles of osteoid and it was within this tissue that the atypical tumours arose.' No mention is made in any of her papers of any lesions of the marrow or carcinomas.

3.9. *Conclusion as to the hazard of* ^{226}Ra

It appears justifiable to conclude from human experience, and from

RADIUM ISOTOPES
149

experiments on dogs given both single and repeated injections, that though there may be some degree of anaemia, malignancy of haemopoietic or other marrow elements is an unlikely result of a skeletal burden of ^{226}Ra. The characteristic malignancy is either an osteosarcoma, a filiosarcoma or sinus carcinoma.

4. ^{224}Radium (Thorium X)

The physical and biological characteristics of ^{224}Ra have been discussed earlier, in Sections 1.3. and 2.3 of this Chapter.

4.1. *Effects on man*

4.1.1. *Neoplasia*. Repeated injections of ^{224}Ra were given to about 2000 German patients during the years 1944–51 for a variety of conditions, particularly tuberculosis of bone or soft tissue and ankylosing spondylitis. The duration of the injections varied from a week to a few years. Patients received total amounts of 40–5000 μCi ^{224}Ra in repeated injections each of 8–70 μCi. The preparation used had the trade name 'Peteosthor'. The injection into children was stopped in 1951 when it became obvious that the radionuclide was ineffective in the treatment of tuberculosis. It is still given in certain German clinics, but at lower doses, for ankylosing spondylitis in adults. Spiess first drew attention in 1950 to some of the clinical effects of this therapy in children and has since published important follow-up studies (Spiess 1950, 1956, 1957, 1969; Spiess and Mays 1970). Exact information (including ^{224}Ra dosage) is available on 891 patients, 221 between the ages of 1–20 years when first injected and 670 patients 21 years and older (Spiess and Mays 1970)*. This group have 43 osteosarcomas and 6 chondrosarcomas. Twenty-eight of the osteosarcomas and 6 chondrosarcomas occurred in the younger age group. The osteosarcomas occurred most frequently in the metaphysis of the long bones, which is the common site for osteosarcomas, but they were also found in sites where naturally occurring osteosarcomas are rare, i.e. in the ribs, in the diaphysis of the long bones, and in the metacarpal and tarsal bones (Spiess 1969). In two patients a second osteosarcoma appeared 3 and 4 years after the first was diagnosed. The absence of pulmonary metastases at the time of the second sarcoma suggested to Spiess that they were of new origin rather than regrowths of the original tumour. The younger the age at injection the higher the incidence of bone sarcoma.

Twenty-seven soft-tissue tumours, mostly cancers of the lung and digestive tract, were observed in the older age-group but Spiess does not relate

* These are the lastest figures taken from Spiess and Mays 1970. They differ slightly from those given in Spiess 1969.

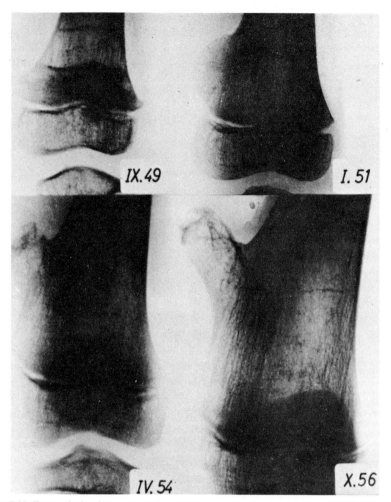

Fig. 7.22. Exostosis development on the distal femur metaphysis 13 months after the injection of 536 μCi ^{224}Ra. The injections started at 4 years of age and continued for 16 weeks. (Spiess 1969, by courtesy of author and publishers.)

these to the radiation. Only three leukaemias are recorded, a number that cannot be considered significant (Mole 1966). Benign bone tumours are discussed under dysplasia (Section 4.1.2).

4.1.2. *Dysplasia.* Dysplasia of bone occurs with ^{224}Ra as with all bone seeking radionuclides. The effects have been well documented by Spiess (1969). Growth retardation has occurred in 42 per cent of 99 injected youths whose heights could be measured. Benign osteochondromas were identified in 36 per cent of 50 children who had a radiological follow up. Such an osteochondroma is illustrated in Fig. 7.22. It developed on the distal femoral metaphysis, 15 months sfter the injection of 536 μCi ^{224}Ra, which was started at the age of 4 years and given over a 16-week period. Exostoses are a common result also of external irradiation of the metaphysis of a young person's long bones.

Acute haematological effects were also noted. After only a few weekly injections a brief rise in the leucocyte count was reported followed by the appearance of leucopenia. The count returned to normal when treatment was discontinued. Haemoglobin and red-cell counts were only reduced after long injection treatment and then returned to normal (Spiess 1969).

4.1.3. *Dosimetric considerations.* Spiess and Mays (1970) have published a comprehensive analysis of the radiation dose received by patients treated with ^{224}Ra, and compared it with the radiation dose in patients with a body burden of ^{226}Ra. At death with a bone sarcoma the lowest known average skeletal dose from ^{226}Ra or ^{228}Ra in a dial painter is 1200 rad for the M.I.T. series (Evans, Keane, Kilenkow, Neal, and Shanahan 1969) and 1290 for the Chicago series*, whereas in a ^{224}Ra patient it is only 90 rad. Thus at the lowest doses which have induced cancer the average skeletal dose from ^{224}Ra appears to be about 13 times (1200/90) more effective than that from the long-lived radium isotope. This difference is explained by the short half-life of ^{224}Ra. A significant proportion of the decay takes place while the ^{224}Ra is still on the bone surface (Sections 1.3 and 2.3) so that the dose to the endosteal surface is much higher than the average skeletal dose. Spiess and Mays (1970) estimate that it is 9 times higher, and that for ^{226}Ra the surface dose is only 0·63 times the average skeletal dose. They conclude that the smallest mean *local* doses to the endosteal lining of the bone surface which have thus far induced bone sarcomas appear comparable for the two isotopes: about 810 rad for ^{224}Ra and about 760 rad for ^{226}Ra. It is possible that further bone tumours will still appear at even lower doses in the ^{224}Ra patients by the time these patients reach the 40–50 years average exposure time which have been accumulated by the American dial painters.

4.2. *Effects on animals*

The toxicity of ^{224}Ra in rats and mice has recently been studied by Hug and his colleagues (Hug, Gössner, Müller, Luz, and Hindringer 1969). The results confirm the findings in man already discussed. The radionuclide is an excellent inducer of osteosarcomas, being slightly more effective, i.e. giving a higher tumour incidence, in rats than in mice. No leukaemias or epithelial cancers are recorded. The experiments initiated by Hug are still in progress. They involved 2506 mice given ^{224}Ra and 519 control mice

---

* This figure is computed by Spiess and Mays (1970) from data given by Finkel, Miller, and Hasterlik (1969a). It differs slightly from the figure (1160 rad) given by Rowland, Failla, Keane, and Slehney (1969–70).

between the ages of 3 and 4 weeks. Certain points of interest arise. The incidence of osteosarcomas was higher in female than in male mice. In some animals osteosarcomas appeared after only 5 μCi kg^{-1} of ^{224}Ra; increasing the dose to a moderate degree did not increase the number of tumours, but increasing the dose above 50 μCi kg^{-1} markedly reduced the number of tumours, due to a high acute mortality rate. The control mice, a strain used by Hug for many years, had a bone tumour incidence of 1·5 per cent. Osteosarcomas were also induced in 18·9 per cent of rats after incorporation of 5–40 μCi kg^{-1} of ^{224}Ra. The histopathology of the bone tumours is not yet available. Unfortunately it is impossible to compare these results in detail with those of Finkel, involving ^{226}Ra being given to mice, since the CF 1/Anl mice she used are known to have a basic bone dyscrasia, probably virus induced (Lisco, Rosenthal, and Vaughan 1971). Furthermore, her mice were injected when 70 days old while Hug's mice were aged only 20–30 days. Hug himself has attempted such a comparison and suggests that the available data suggest that for a comparison of the life-shortening effects of the two isotopes for equal μCi kg^{-1} injected. ^{226}Ra appears to be about 10 times more effective in the low and medium dose range than ^{224}Ra. A comparison of the medium doses of both isotopes, which produce approximately the same life-shortening effect, shows a striking difference in the oncogenic effects. Only 11·6 per cent of the animals developed tumours after administration of 50 μCi ^{224}Ra kg^{-1} whereas 5 and 10 μCi kg^{-1} ^{226}Ra produced osteosarcoma in 62·3 and 75·5 per cent of the animals. The same incidence of bone tumours as produced by 50 μCi kg^{-1} ^{224}Ra was produced by only 0·5 μCi kg^{-1} of ^{226}Ra.

Hug suggests that these differences may be due to the fact that the dose from ^{226}Ra is protracted over time while that from ^{224}Ra is short-term (Hug, Gössner, Müller, Luz, and Hindringer 1969; Hug, Kellerer, and Zuppinger 1966).

It is difficult to relate these comparative results with ^{224}Ra and ^{226}Ra in mice to comparative results with the two isotopes already discussed in man (Section 4.1.3), since trabecular sizes are so different in the two species, and therefore the radiation dose to sensitive tissue. In man, the average size of the trabeculae is between 220 and 515 μm (Spiers, Whitwell, and Darley 1971) which is long compared to the range of the alpha particles in bone. In mice, the only measurement available is given by Lloyd and Hodges (1971) as in the region of 80 μm for the trabeculae of a lumbar vertebra in an adult mouse. Much of the radium distributed through the bone volume will therefore still be able to irradiate the endosteal cells, and the ^{226}Ra, which emits most of the radiation when in the bone volume, will be more effective relative to ^{224}Ra than is the case is man, where a large proportion of the ^{226}Ra radiation does not reach the endosteal surface. It is impossible to make a direct numerical comparison between the results in man in

RADIUM ISOTOPES

Section 4.1.3, measured in terms of average skeletal dose, and those in mouse measured in terms of μCi kg^{-1} injected, but when the radiation characteristics of the two isotopes and the dimensions of the bones are considered, the results are not inconsistent.

Comparative investigations using ^{224}Ra and ^{226}Ra are in progress on beagle dogs at Utah. These results will be of great theoretical, if not practical, importance.

8. Strontium Isotopes

INITIALLY experimental work on the biological effect of bone-seeking radionuclides was largely concentrated on the strontium isotopes for two reasons. First, as alkaline earths, they were expected to behave in the skeleton like radium, a known powerful bone carcinogen, and secondly they were recognized as a fall-out product from A-bomb explosions to which the whole world was exposed. Much useful physiological and pathological information is therefore available, particularly about ^{90}Sr, though it has become apparent that strontium isotopes are less dangerous than other bone-seeking nuclides, such as plutonium. There are many strontium isotopes; among them ^{85}Sr, ^{87}Sr, ^{89}Sr, and ^{90}Sr have all been used in metabolic and toxicity studies.

1. Strontium 90

^{90}Sr is important as a bone-seeking radionuclide because of its long half-life, 28 years, and the high energy of its daughter product ^{90}Y. As a decay product ^{90}Y does not leave the skeleton nor does it appear to translocate significantly in bone, though ^{91}Y if given independently has a somewhat different skeletal distribution, behaving like the actinide elements (Arnold, Stover, and Van Dilla 1955; Jowsey, Sissons, and Vaughan 1956; Lloyd 1961b; Durbin 1962). The most likely source of human contamination from ^{90}Sr is from fall-out and subsequent ingestion through food chains, discussed in Chapter 6.1.2.1. Accidental injury to those handling radioactive strontium is also a possible risk. Experiments on the effect of ^{90}Sr inhalation will also be discussed.

The beta particle from ^{89}Sr also has a high energy, 1·463 MeV, but ^{89}Sr has a much shorter half-life, 53 days, than ^{90}Sr. It has been used in some interesting comparative studies with ^{45}Ca (Kuzma and Zander 1957a,b) which emits a beta particle of low energy (0·254 MeV) and has a half-life of 163 days.

1.1. *Metabolism*

It is often assumed that, as an alkaline earth, the metabolic behaviour of strontium is similar to that of calcium and that strontium can be used as a tracer for calcium. It is important to remember that this is not the case. There are differences in the retention, excretion, and distribution patterns of the four alkaline earths in man, as well as similarity (Vaughan 1970a). It must also be remembered that the pattern of retention of strontium in the skeleton varies in different species, as shown in Fig. 8.1 (Stara, Nelson,

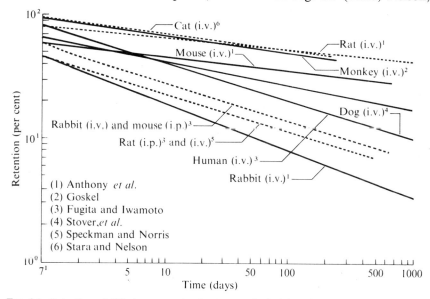

FIG. 8.1. Retention of ^{90}Sr in mammals after parental administration. (Stara *et al.* 1971, by courtesy of authors and publishers.)

TABLE 8.1

Numerical data for models of Ca and Sr metabolism
(from Dophin and Eve 1963, by courtesy of authors and publishers)

	Calcium	Strontium*
Daily absorption from gut	0.450 g/day	0.3 mg/day
Exchangeable pool	5 g	1.2 mg
Total in plasma	0.35 g	0.08 mg
Accretion rate in bone	0.5 g	0.12 mg
Skeletal content	1040 g	250 mg
Half-life in exchangeable pool	3.7 days	2.8 days
Half-life in bone	2770 days	2000 days
Fractional transfer		
Diet to exchangeable pool	0.38	0.19
Exchangeable pool to bone	0.53	0.29
Diet to bone	0.20	0.055

* Renal discrimination against Sr relative to Ca is taken to be 4.

Della Rosa, and Bustad 1971). The best numerical data for models of strontium and calcium metabolism in man are given in Table 8.1 (Dolphin and Eve 1963, as modified by Spiers 1968a). The differences observed are dependent on at least two known factors. The human body exerts two recognized discriminatory mechanisms against strontium. Less strontium than calcium is absorbed from the gut and more strontium is excreted by the kidney (Wasserman and Comar 1961, Walser 1969, Wasserman and Taylor 1969, Vaughan 1970a). Whether there is also discrimination by bone in man is still not known although in rabbits there is an overall factor of about 1·6 in favour of calcium in the transfer from blood to bone (Kshirsagar, Lloyd, and Vaughan 1966; Lloyd 1968). *In vitro* also the behaviour of strontium differs from that of calcium. Harrison and his colleagues (Harrison, Howells, and Pollard 1967) analysed the uptake and elution of ^{45}Ca, ^{85}Sr, ^{133}Ba, and ^{223}Ra in rat bone powder. There was an increased retention of barium and radium compared to strontium and calcium. This is the exact reverse of what happens *in vivo*. The only explanation the authors offer is that since it is known that radium can replace calcium in hydroxyapatite (Neuman, Hursh, Boyd, and Hodge 1955) it may be that the heavier nuclides become bound to protein matrix in the bone powder.

1.2. *Distribution*

In spite of differences in the metabolism of strontium and calcium the general pattern of distribution of the two isotopes is similar. Following a single injection of strontium there is within a few days a concentration of the radionuclide in sites of mineral accretion and a diffuse distribution throughout bone mineral. Immediately following injection, i.e. for a matter of hours, there is probably a similar concentration of strontium on all bone surfaces to that which Rowland has shown to occur with calcium (Rowland 1966, Marshall 1969b). This early surface concentration may have some importance dosimetrically. In animals fed ^{90}Sr continuously from birth or before, the whole skeleton is labelled and the characteristic hot spots are not apparent.

Figure 8.2 shows an autoradiograph of the tibia of a young rabbit given ^{90}Sr by a single intravenous injection and killed 10 minutes later. The ^{90}Sr distribution is extremely uneven. ^{90}Sr is concentrated beneath the epiphyseal plate, on the endosteal surface in the metaphysis, and on the periosteal surface of the diaphysis, i.e. in sites of active bone growth. In Fig. 8.3 is the autoradiograph of a six-month old rabbit fed ^{90}Sr from the age of 6 weeks. The whole bone formed after feeding started is evenly labelled. In an adult rabbit, given a single injection, the uptake is uneven, since active mineralization at this age is uneven, as shown in Fig. 8.4. This

STRONTIUM ISOTOPES 157

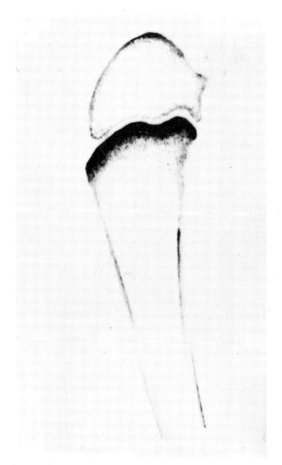

FIG. 8.2. Autoradiograph of longitudinal section of tibia of a young rabbit 10 minutes following a single injection of ^{90}Sr (600 µCi kg^{-1}). (Jowsey et al. 1953.)

difference in distribution depending on age and method of administration is important to remember when dosimetry is later considered (Chapters 8.1.3.4, 8.1.3.5). Furthermore, long-term retention of strontium is not necessarily the same throughout the skeleton since the pattern of 'bone turnover rate', the rate of movement of the alkaline earth ions in and out of the skeleton, differs in different parts even of the same bone, and certainly in different parts of the skeleton (Vaughan and Williamson 1967). This difference in 'turnover rates' is offered as an explanation of carcinomas of the sinuses occurring in rabbits and dogs following ^{90}Sr administration and of carcinomas of the sinuses occurring in radium poisoning (Mays, Dougherty, Taylor, Lloyd, Stover, Jee, Christensen, Dougherty, and Atherton 1969; Vaughan and Williamson 1969; Evans, Keane, Kolenkow,

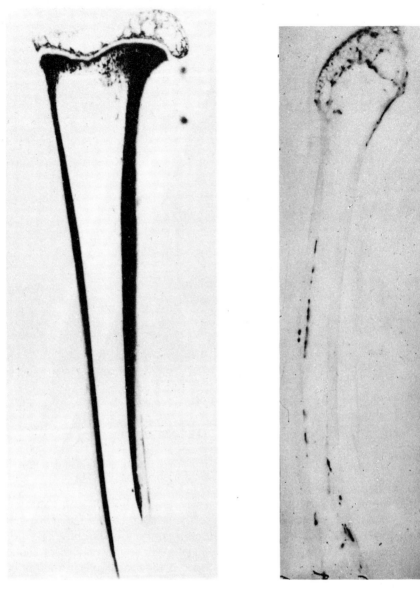

Fig. 8.3. Autoradiograph of longitudinal section of tibia of rabbit started on daily feeding of ^{90}Sr 10 μCi day^{-1} when 6 weeks old, killed 6 months later—note even distribution of isotope in bone formed after feeding started. (Vaughan 1962a, by courtesy of author and publishers.)

Fig. 8.4. Autoradiograph from thick longitudinal section of lower end of femur of rabbit 12 months old when injected with ^{90}Sr 80 μCi kg^{-1} killed 6 months later. Patchy distribution of concentrated reaction in epiphysis and cortical bone. (Jowsey, Owen, Tutt, and Vaughan 1955, by courtesy of authors and publishers.)

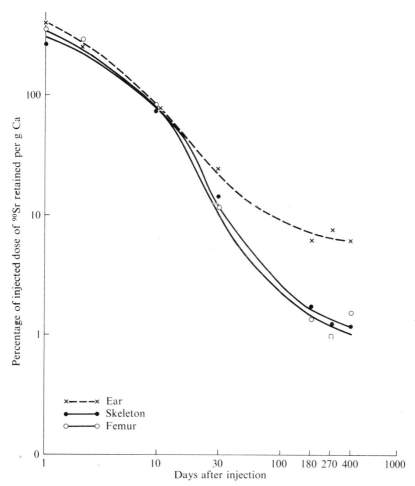

Fig. 8.5. Per cent injected dose ^{90}Sr per gram Ca in skeleton, femur, and 'ear bone' of rabbits killed at different intervals after intraperitoneal injection of 500 μCi kg^{-1} at 2 days old. (Vaughan and Williamson 1969, by courtesy of author and publishers.)

Neal, and Shanahan 1969). Figure 8.5 shows the per cent injected dose ^{90}Sr per gram Ca in the skeleton, femur, and 'ear bone' of rabbits killed at different time-intervals after an intravenous injection (100 μCi kg^{-1}) when 2 days old. It is at once apparent that ^{90}Sr is rapidly lost from the skeleton as a whole and from the femur but remains relatively high in the 'ear bone'. This difference in 'turnover rate' will result in differences in radiation dose. Figure 8.6 shows the accumulated radiation dose up to 1400 days after injection in the mid diaphysis and ear bone of rabbits injected with 100 μc/kg ^{90}Sr when six weeks old. The dose in the metaphysis, initially high, is measured for a shorter time but is falling behind, as it is known to do,

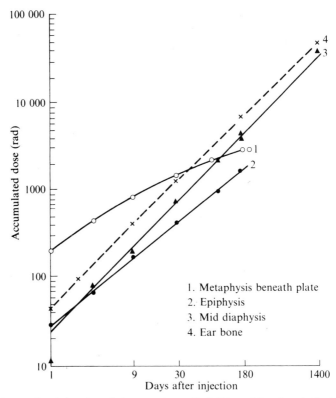

FIG. 8.6. Accumulated dose in rads in posterior wall of tibia and 'ear bone' of rabbits killed at different time intervals after an intravenous injection of ^{90}Sr 100 μCi kg^{-1} when 6 weeks old. (Vaughan and Williamson 1969, by courtesy of authors and publishers.)

because of remodelling, (Vaughan and Williamson 1969) while the dose in the epiphysis, also measured for a shorter time is always low. It is clear that though the intial uptake of a radionuclide may be low the final radiation dose to sensitive tissues may be high because the radionuclide is retained in a bone with a low turnover rate. On the other hand in some instances the initial uptake is high and subsequently may be considerably reduced by resorption and further growth. The changing pattern of dose rate that may result is illustrated in Fig. 8.7, (Owen and Vaughan 1959a). The difference in pattern of maximum dose when ^{90}Sr is given as a single injection or fed continuously to young animals is illustrated in Fig. 8.8, where maximum dose has been measured over the length of the tibia of the rabbit. There is a peak of high accumulated dose in the region of the epiphyseal plate at the time of injection in the rabbit given a single injection, while in the rabbit fed strontium there is a relatively even distribution down the whole length of the bone.

STRONTIUM ISOTOPES

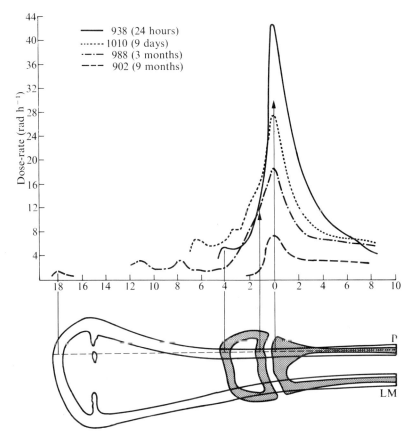

FIG. 8.7. Variation in dose rate at different times after injection, measured along middle of posterior wall of proximal end of tibia of rabbit given single injection of ^{90}Sr 600 µCi kg^{-1} at 5–8 weeks.

1.3. Effects of ^{90}Sr deposition

It must be evident from this description of ^{90}Sr distribution and consequent radiation dose delivered to sensitive tissues, that a distinction must be made between the effects of deposition of ^{90}Sr following a single administration by any route and those of continuous ingestion.

There is little information available about ^{90}Sr toxicity in humans other than some studies made on European dial painters who used a ^{90}Sr paint. These are discussed later (Section 1.3.2.1). Extensive experimental results are however available on mice, rats, rabbits, pigs, dogs, and a few monkeys. Some of these results are shown in Tables 8.2 and 8.3. These are discussed separately for different species because of the differences both of bone structure, particularly bone trabeculation, and metabolic patterns in different species.

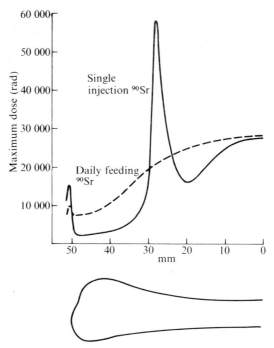

Fig. 8.8. Variation in the maximum accumulated dose received at different levels in the posterior wall of the upper half of the rabbit tibia 200 days after a single injection of ^{90}Sr 600 μCi kg^{-1} at the age of 5–8 weeks, and 250 days after daily feeding of 8·5 μCi ^{90}Sr from the age of 5–8 weeks. 0 is approximately in the middle of the bone shown below, the level of the epiphyseal plate at the age of 5–8 weeks is approximately 28.

1.3.1. *Single administration.* The effects of a single administration of ^{90}Sr are perhaps of theoretical rather than practical interest since the most likely hazard to man is from continuous ingestion of contaminated food as a result of fall-out.

1.3.1.1. *Dogs.* The Utah experiments in which young adult beagle dogs were given a single intravenous injection of ^{90}Sr have already been discussed (Chapter 6.4) (Dougherty and Mays 1969; Mays, Dougherty, Taylor, Lloyd, Stover, Jee, Christensen, Dougherty, and Atherton 1969; Mays, Dougherty, Taylor, Stover, Jee, Christensen, Dougherty, Stevens, Jr., and Nabors 1970; Mays and Lloyd (1972). Osteosarcoma was the commonest cause of death. Eight dogs died with osteosarcomas and two with haemangiosarcomas in bone*, 1 with a squamous carcinoma in the frontal sinus, and one with a squamous carcinoma in gingival tissue (Mays, Dougherty, Taylor, Lloyd, Stover, Jee, Christensen, Dougherty, and

* Dougherty, Taylor, and Mays (1971) have now mentioned 11 bone sarcomas. These are listed in Research in Radiobiology COO 119, 242.

STRONTIUM ISOTOPES

TABLE 8.2

Effects of a single injection ^{90}Sr or ^{89}Sr

Species	No.	Route	Dose administered $\mu Ci\ kg^{-2}$	Received, average rad dose	Leukaemia myeloid	Leukaemia lympho	Haemopoietic Dyscrasia	Osteosarcoma (sometimes multiple)	Fibrosarcoma	Chondrosarcoma	Osteochondrosarcoma	Angiosarcoma	Giant cell tumour	Cancer(a) head	Reference
Swine	2	intravenous	1869–6160					2						1	Howard et al. 1969
Dog		intravenous	97.9–63.6	6435			3	6			2			2	Dougherty and Mays 1969
Rabbits		intravenous	10–500		1	2		1	3		3			3	Finkel et al. 1972
		inhalation $^{90}SrCl_2$	?					5	3		2	7			McClellan et al. 1972
		intravenous													Vaughan and Williamson 1969
2 days old	8		500											7	
6–8 weeks	7		50–200					2						6	
6–8 weeks	13		500–1000					20						1	
old	22		200–1000					12							
Rats		intraperitoneal	5–500		+			+							Moskalev et al. 1969
Mice CF 1	810	intravenous	44–2200		+(c)			++	++	+		+			Finkel, Biskis, and Scribner 1959
(b) CBA	11	intravenous	20 μCi per mouse					6				5			Barnes et al. 1970
(b) CBA	219		$\mu Ci\ g^{-1}$ 1.6		+(c)			87.2	11.0			0.9	16		Nilsson 1970
	292		0.8		+(c)			68.8	29.8			2.1	3		
	90		0.4		+(c)			24.4	71.1			4.4			
	8		0.2					(50.0)	(37.5)			(12.5)			

expressed as percentage

(a) Tumours in soft tissues, i.e. sinuses of skull and mucous membranes of mouth, etc.
(b) No tumours in controls.
(c) The mouse leukaemias are not classified.

TABLE 8.3

Effects of multiple injections or ingestion of ^{90}Sr or ^{89}Sr

Animal	Route of administration	μCi kg⁻¹	Leukaemia myeloid	Leukaemia lympho	Myeloid hyperplasia	Osteosarcoma	Fibrosarcoma	Chondrosarcoma	Angiosarcoma	Reticulum cell tumour	Giant cell tumour	Cancer head*	Reference
Monkey	ingestion gavage 5–10 days	500–1000	1 'monocytic'			1		1					Casarett, Tuttle, and Baxter 1962
Dog	ingestion in utero + 540 days	150	14			17	2	1	1			4	Pool et al. 1972 } same population of dogs / Dungworth et al. 1969
	injection multiple from 6 months	150				38	1		4			2	Finkel et al. 1972
	multiple from birth	15	1			24	1	1	1	1			Finkel et al. 1972
	from 6 months repeated over 2–3 yr period	15	2	1		1	1	1	1				Finek et al. 1972
Swine	ingestion	100–200	+			+							Finkel et al. 1972
			23	17 29		7					3		Clarke et al. 1972
Rats	injection multiple at monthly intervals F₁ and F₂ generations	μCi g⁻¹ 4·4 total				+++	+	+	+		+		Skoryna and Kahn 1959 / Skoryna et al. 1958
	injection 10 consecutive daily or monthly	Total μCi 0·1–3·5				+	+		+	+		+	Kuzma and Zander 1957
	fed 10–30 days	330–790 total	+	++		+			+	+		+	Casarett, Tuttle, and Baxter 1962

* Tumours in soft tissues i.e. sinuses of skull and mucous membranes of mouth, etc.

Atherton 1969). Three dogs had a haemopoietic death. The lowest average dose received 1 year before death was 3840 rad (Mays, Dougherty, Taylor, Stover, Jee, Christensen, Dougherty, Stevens, and Nabors) the death being described by Dougherty, Taylor, and Mays (1972) as due not to a neoplasm, but to severe progressive anaemia, leucopaenia, and thrombocytopaenia. It is to be noted however that perivascular cuffing of central veins in the liver, which is characteristic of myeloid leukaemia and the myeloproliferative lesions described in swine, was present. Mention is also made of myelofibrosis in some cases. Detailed accounts of the effect of a single injection on the blood picture of the Utah dogs has been given by Dougherty and her colleagues (Dougherty J. H. 1962; Dougherty and Rosenblatt 1969; Dougherty, Taylor, and Mays 1972). They state in their latest report that no haematopoietic neoplasms have so far been observed. However their detailed reports suggest that, particularly at the higher dose levels, i.e. 64 or 98 μCi kg^{-1}, there is a profound disturbance of marrow function.

McClellan and his colleagues (McClellan, Boecker, Jones, Barnes, Chiffelle, Hobbs, and Redman 1972) have reported inhalation studies in dogs. Relatively soluble forms such as ^{90}SrCl$_2$ are rapidly absorbed from the respiratory and gastrointestinal tract and translocated to the skeleton so that the effects are similar to an intravenous injection, while insoluble forms are much more slowly translocated. The results in 19 dogs that died are shown in Table 8.4. There was one myeloid leukaemia, 3 fibrosarcomas, 5 osteosarcomas, 7 angiosarcomas, and one osteochondrosarcoma. As shown in Table 8.2, Finkel and her colleagues obtained a rather similar distribution of lesions in a small group of dogs given ^{90}Sr by the intravenous route (Finkel, Biskis, Greco, and Camden 1972).

1.3.1.2. *Miniature swine**. Fifteen miniature swine were given intravenous injections of ^{90}Sr in a dose of about 64 μCi kg^{-1}. The age at the time of injection was 1·5 to 53 months. Two animals aged 6 and 34 months at the time of injection developed bone tumours, one had a giant cell tumour of the mandible, and one an osteosarcoma of the skull and of the mandible. The tumours appeared to be periosteal in origin (Howard, Clarke, Karagianes, and Palmer 1969).

1.3.1.3. *Rabbits*

Bone. The effects of ^{90}Sr given to a relatively small number of rabbits of different ages have been studied by Vaughan and her colleagues. These are discussed in some detail as information is available on the radiation dose in relation to the lesions noted (Vaughan and Williamson 1969). Some of the results are shown in Table 8.5. Carcinomas of the external ear occurred

---

* The so called miniature swine is a very large pig when fully grown, weighing at least 60–70 kg. It has proved an extremely valuable experimental animal since its dietary requirements, intestinal tract, and bone mass are quite comparable to those of man. The life span is over 12 years.

TABLE 8.4

Radiation dose parameters and major findings associated with death for dogs that died or were euthanized at late time periods (>31 days) following inhalation of $^{90}SrCl_2$ (from McClellan et al. 1972, by courtesy of author and publishers)

Dog no.	Radiation Dose to Skeleton			Survival (Days after ^{90}Sr inhalation)	Major findings associated with death
	Dose rate (rad day^{-1})		Cumulative to death (rad)		
	Initial	Death			
12D	13	5·4	4000	585	Myelogenous leukaemia
164A	54	24	17 000	585	Epileptic seizure
157E	55	21	19 000	759	Fibrosarcoma
160B	44	17	17 000	864	Angiosarcoma and osteosarcoma
162F	47	20	22 000	886	Osteochondrofibrosarcoma
158E	54	15	17 000	927	Angiosarcoma
7B	31	14	15 000	928	Angiosarcoma
12F	26	8	10 000	1046	Angiosarcoma
23C	37	14	18 000	1099	Osteochondrosarcoma
160C	31	6·8	9900	1142	Angiosarcoma
10A	25	8·9	13 000	1168	Angiosarcoma
9D	22	9·0	9900	1316	Angiosarcoma
9C	17	5·4	10 000	1318	Osteosarcoma
13A	23	5·4	10 000	1361	Cerebellar hemorrhage
8A	23	8·4	15 000	1362	Osteosarcoma
9B	18	3·8	7500	1367	Fibrosarcoma
26A	19	5·3	10 000	1404	Fibrosarcoma
22F	16	5·1	10 000	1540	Osteosarcoma
23B	27	5·9	15 000	1787	Osteosarcoma

TABLE 8.5

Tumour sites in rabbits injected with ^{90}Sr (from Vaughan and Williamson 1969, by courtesy of author and publishers)

No. rabbits	Age at injection	μCi kg^{-1}	Survival (months)	Osteosarcoma				Carcinoma
				Meta-physis	Diaphysis	Jaw	Spine	Ear
8	2 days	500	6–22	0	0	0	0	7
7	6–8 weeks	50–200	16–69	0	2	0	0	6
13	6–8 weeks	500–1000	3–9	11	0	9	0	1
22	52–165 weeks	200–1000	6–73	0	0	10	2	0

within 22 months in two-day old rabbits given 500 μCi ^{90}Sr kg^{-1} no osteosarcomas were seen. Rabbits aged 6 weeks given a low dose of 50–200 μCi kg^{-1} also developed carcinomas of the ear but occasional osteosarcoma also occurred. Rabbits aged 6 weeks given 500 μCi kg^{-1} and killed at 6 months, developed osteosarcomas largely in the long bones. A mastoid carcinoma occurred in one rabbit that survived 9 months. In rabbits at least 1 year old when injected with high doses of ^{90}Sr, 200–1000 μCi kg^{-1}, only tumours of the jaw and spine were found. No animal in this age group however was allowed to survive longer than 38 months.

A detailed analysis was made of radiation dose rate, accumulated dose, and associated damage in these rabbits. These observations suggest, as stated in Section 1.2, that the pattern of turnover rate in different bones at different ages, affecting as it does the achievement of an appropriate accumulated dose to cells capable of proliferative activity in osteogenic or epithelial tissue, determines which type of tumour occurs most commonly at different ages. What is true of the rabbit skeleton is likely to be true for other species, though such detailed measurements are not available. Figure 8.9 shows autoradiographs of thick sections of the tibiae of rabbits injected with ^{90}Sr and killed 1 day, 9 days, 30 days, and six months later. Group A were given 500 μCi ^{90}Sr at 2 days of age. There is initially a particularly heavy uptake of the radionuclide beneath the plate and to a lesser extent throughout the diaphysis. The bone is growing rapidly at this age and 9 days later the ^{90}Sr has been lost by remodelling except in the mid diaphysis. This effect is even more marked 30 days later and by 6 months little ^{90}Sr is left except at two small areas in mid-diaphysis. In Group B are shown autoradiographs from bone of rabbits injected with a much lower dose (50–100 μCi kg^{-2}) when 6 weeks old. The immediate heavy uptake beneath the epiphysis is again characteristic but as the bone grows in length the metaphyseal trabeculae are resorbed and subsequently ^{90}Sr is restricted to the diaphysis. The figures under each autoradiograph give the time of exposure needed to obtain the autoradiograph and it should be noted that this is much longer for Group B than for the Group C animals of the same age group which were given a much higher dose (600 μCi kg^{-1}).

In this group very considerable damage was done to the bone as evidenced by microscopic examination and the result was failure of complete resorption of metaphyseal trabeculae. In the young animals given 600 μCi kg^{-1}, tumours were found to arise largely in the metaphyseal region, which is clearly the site of maximum uptake, but not in those areas where the dose was highest. Here the osteogenic tissue was destroyed and tumours arose in adjacent areas (Macpherson, Owen, and Vaughan 1962; Owen, Sissons, and Vaughan 1957; Owen and Vaughan 1959a,b). The high incidence of jaw tumours in both young and old animals was shown to be dependent on the fact that at the site where the tumours arose there is extremely actively

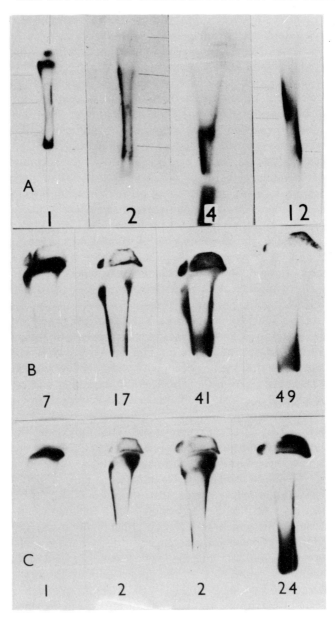

FIG. 8.9. Autoradiographs of thick sections of tibiae of rabbits injected with ^{90}Sr and killed 1 day, 9 days, 80 days and 6 months later.
    A. Group 1 (500 μCi ^{90}Sr kg^{-1} at 2 days of age).
    B. Group 2 (50–100 μCi ^{90}Sr kg^{-1} at 6 weeks of age).
    C. Group 3 (600 μCi ^{90}Sr kg^{-1} at 6 weeks of age). Figures refer to time of exposure in hours.
       (Vaughan and Williamson 1969 by courtesy of author and publishers.)

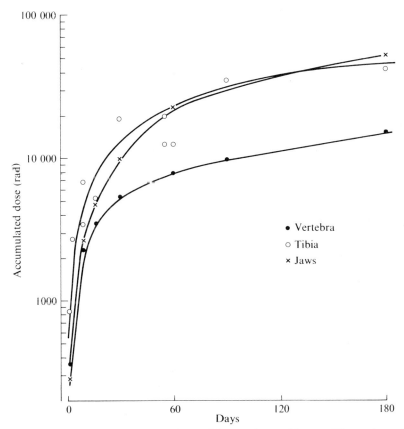

Fig. 8.10. Maximum accumulated doses in the tibia, vertebra, and jaws at different times after injection in rabbits aged 5–8 weeks given 600 μCi ^{90}Sr kg^{-1}. (Owen 1962, by courtesy of author and publishers.)

proliferating osteogenic tissue (Rushton, Owen, Holgate, and Vaughan 1961). This active bone growth is maintained even in old animals where there is very little growth and therefore no tumours in the long bones. Measurements of maximum accumulated radiation dose in vertebrae, long bones, and jaws of old animals at the site at which tumours arise in both young and old animals are shown in Figs. 8.10 and 8.11. This shows that in animals injected as weanlings and in which tumours occur both in tibia and jaws the accumulated dose is of the same order, while in old animals the dose to the jaw is very much greater than that to the tibia (Owen 1962).

Marrow. A careful analysis of the peripheral blood picture, the bone marrow picture, and precisely measured radiation dose in relation to marrow changes in young rabbits given ^{90}Sr (600 μCi kg^{-1}) made by Vaughan (1962) indicated that the relationships were extremely complex

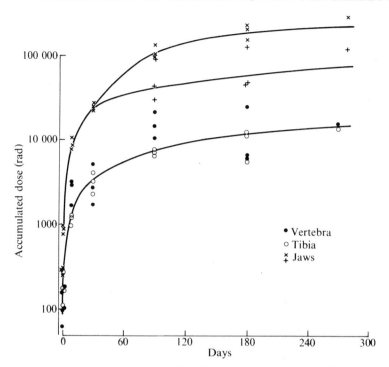

Fig. 8.11. Maximum accumulated doses in the tibia, vertebra, and jaws at different times after injection in adult rabbits given 600 μCi ^{90}Sr kg^{-1}. The maximum dose received in the heavy deposits in dentine and enamel is indicated by × whereas + indicates the maximum dose received in the soft tissue region between the alveolar bone and the teeth. (Owen 1962, by courtesy of author and publishers.)

even within one bone. In areas of marrow, receiving a high initial dose-rate, aplasia occurred within 24–48 hours, but as the dose rate fell, due to bone remodelling, there was partial or complete recovery either due to repopulation from undamaged marrow or to recovery of stem cells *in situ*. This initial aplasia was associated with anaemia and leucopaenia, affecting largely the myeloid elements. Death occasionally occurred about the 18th day but there was more usually recovery to low normal levels unless very high doses were given. The relation of radiation dose rate measured in rad to marrow cellularity in the marrow of the tibia beneath the plate at the time of injection is shown in Fig. 8.12, and that of the circulating haemoglobin in grams per 100 ml to marrow cellularity in the vertebrae and three sites in the tibia is shown in Fig. 8.13. When high doses were given (1000 μCi kg^{-1}) the rabbits died with severe anaemia and gelatinous degeneration of the marrow (Owen, Sissons, and Vaughan 1957).

1.3.1.4. *Rats*. Moskalev and his colleagues (Moskalev, Streltsova, and Buldahov 1969) have studied the effects of ^{90}Sr given in a single injection

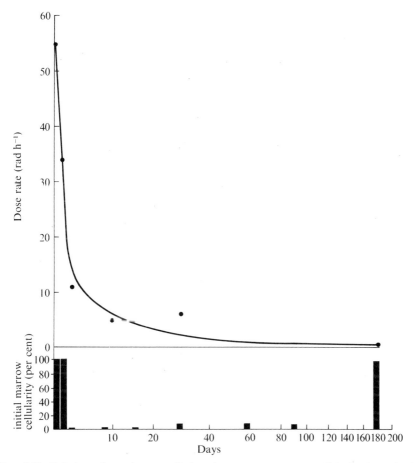

Fig. 8.12. Relation of maximum radiation dose-rate to marrow cellularity beneath the proximal tibial plate at time of injection 600 μCi ^{90}Sr (600 μCi kg^{-1}). (Vaughan 1962b, by courtesy of publishers.)

to rats. The lowest average bone dose when the animals died with osteosarcoma was 800 rad in rats given 5 μCi kg^{-1}. A leukaemia incidence of 6·1 per cent was noted in 82 rats injected with ^{90}Sr 10μCi kg^{-1} with a resulting marrow dose of 530 rad. Higher incidences were noted in some of the higher dose groups. In the 402 controls the incidence was 1·7 per cent. The leukaemia took the form of reticulosis, haemocytoblastosis, or myeloid leukaemia. Reticulosis was the most common form of leukaemia in exposed rats but osteosarcoma predominated when a high ^{90}Sr injection was given.

Kuzma and Zander (1957a,b) have carried out interesting experiments in which they gave ^{89}Sr or ^{45}Ca to rats, so comparing the effects of a high-

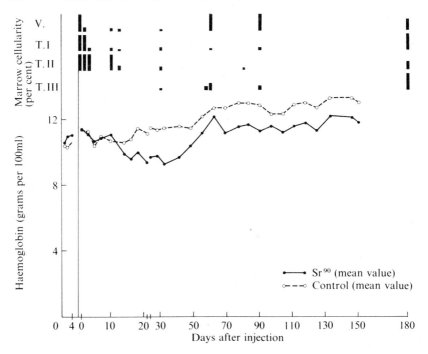

Fig. 8.13. Effect of intravenous injection of Sr^{90} (600 μCi kg^{-1}) on the haemoglobin of young rabbits. (Vaughan 1962b, by courtesy of publishers.)

energy and low-energy beta radiation. As might be expected ^{45}Ca was not as powerful a carcinogen as ^{89}Sr but produced many benign exostoses.

1.3.1.5. *Mice.* Extensive studies have been made of the effect of ^{90}Sr in mice. Mice have some advantages as an experimental animal for the study of radionuclide toxicity, i.e. they are small and large numbers of animals can be used. On the other hand, the geometry of their skeletons must, because of their small size, be different from man's and further, certain strains are known to have a natural high incidence of virus induced mesenchymal dyscrasias (Lisco, Rosenthal, and Vaughan 1971). As discussed in some detail in Chapter 1, radiation has long been known to potentiate the effectiveness of the leukaemia viruses and might therefore be expected to have the same influence on the osteogenic sarcoma viruses, as indeed it has been shown to have on the FBJ virus in the case of CF 1/Anl mice (Finkel and Biskis 1969). Finkel and Biskis (1968) note that CF 1/Anl mice which have a highly prevalent C-type RNA virus of the Gross serotype (Kelloff, Lane, Turner, and Huebner 1969) develop osteosarcomas after ^{90}Sr much earlier than CBA mice which are at present not known to have a natural osteogenic virus and which have an extremely low natural sarcoma incidence. As shown in Fig. 8.14, Finkel (1959b) noted a

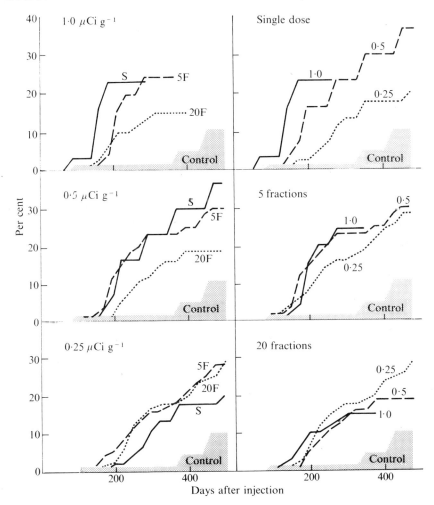

FIG. 8.14. Incidence of reticular tissue tumours to 475 days after the first injection of ^{90}Sr, as a function of dosage pattern. (Finkel 1959b, by courtesy of author and publishers.)

high incidence of reticular tumours in mice (CF 1) given ^{90}Sr, both in single and fractionated doses, compared with controls. She also noted a high incidence of osteogenic sarcomas, and osteochondrosarcomas. There were also an appreciable number of haemangioendotheliomas of bone marrow and epidermoid carcinomas of the oral cavity. The tumours were often multiple; as many as 13 tumours were detected in one mouse on gross examination. There was a pronounced association between the dose of ^{90}Sr and the incidence of both osteosarcoma and haemangioendotheoliomas (Finkel, Biskis, and Scribner 1959).

TABLE 8.6

Relationship between radiation dose and the incidence, latent period, and type of tumour in mice
(Nilsson 1969, by courtesy of author and publishers)

No. of mice	^{90}Sr dose (μCi/g body weight)	Carcinoma				Osteosarcomas				Leukaemia	
		Per cent mice with tumours		Total number of tumours	Latency time (days ± SE)	Per cent mice with tumours		Total number of tumours	Latency time (days ± SE)	Per cent	Latency time (days)
		Dead	Killed			Dead	Killed				
120	1·6	40·0	27·7	56	237·0 ± 11·8	84·0	32·3	219	267·6 ± 4·6	1·7	255
120	0·8	13·0	14·7	21	339·5 ± 11·7	89·1	30·7	292	320·7 ± 5·4	3·3	275
120	0·4	3·7	2·1	3	436·0	40·0	22·1	90	426·7 ± 9·8	10·0	255
120	0·2	0	0	—	—	5·9	4·9	8	485·1 ± 37·8	3·3	252
95	Controls	0	0	0	—	0	0	0	—	0	—

Nilsson (1968, 1969, 1970a) has analysed the dose-dependent carcinogenic response of CBA mice, a strain with a low incidence of both spontaneous osteosarcoma and leukaemia. He reports the absence of any malignancy in his controls. The results are shown in Table 8.6. At intervals of 7, 14, 21, and 30 days after injection of ^{90}Sr and then at monthly intervals five mice from each group were selected at random and sacrificed until all mice in each series were utilized. The highest percentage of skeletal tumours was found in mice receiving ^{90}Sr at 0·8 μCi g^{-1}. There was a tendency to multiple tumours except in the 0·2 μCi g^{-1} group and this tendency seemed to be a function of both dose and time. Latency time for tumour induction was clearly correlated to dose; with decreasing dose the first tumour was detected after 120, 210, 270, and 300 days. The dose also appeared to have an effect on the site of tumour development. In the 0·8 to 0·2 μCi g^{-1} groups more than 50 per cent of the tumours were located largely in the long bones, particularly in the femur. In the 1·6 μCi g^{-1} group on the other hand most tumours (50 per cent) were sited in the lumbar and

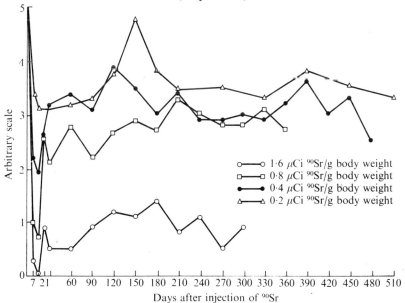

FIG. 8.15. Bone marrow cellularity in relation to dose and time after injection of ^{90}Sr according to an arbitrary scale. 0 equals aplasia, 1–4 varying degrees of hypoplasia, and 5 normal marrow cellularity; each point on the diagram represents the mean marrow cellularity from 155 estimations.

Five mice were investigated, in each of which the marrow cellularity was estimated in the distal, diaphyseal, and proximal parts of the femora, tibiae, and humeri respectively; in each of the two pelvic bones at the tuber ischii, tuber sacrale, the ilium, and the acetabular region; in the spine in one marrow cavity in each of the cervical, thoracic, lumbar, and sacral vertebrae; and in the base of the cranium in the head. (Nilsson 1970b, by courtesy of author and publisher.)

sacral spine and pelvis. A total of 80 invasive carcinomas, and carcinogenesis *in situ* of the mucous membranes of the head were found, the majority being in the highest dosage group. Leukaemia, which was considered with one exception to be lymphatic in origin, was found in 22 cases. There appeared to be no relation to dose. Nilsson (1969, 1970b) raises the question as to whether the site of origin was the bone marrow or the thymus.

Nilsson (1970b) has made a detailed study of changes in the haematopoietic system in male CBA mice following different doses of ^{90}Sr. The doses given were 1·6, 0·8, 0·4, and 0·2 μCi/g body weight and the results on marrow cellularity as shown in Fig. 8.15. The histological changes differed quantitatively but were of the same type in all the dose groups. Dilatation or hyperaemia of the medullary sinusoids became accentuated with increasing dose. Marrow necrosis was most marked in the femur in all the dose groups but appeared earlier at the higher doses. In many aplastic or markedly hypoplastic marrows where only reticular cells remained, they were increased in numbers or were in a state of proliferation. He considers that these cells initiate the development of predominantly fibroblastic osteosarcomas (Sundelin and Nilsson 1968). Early fibrosis occurred only

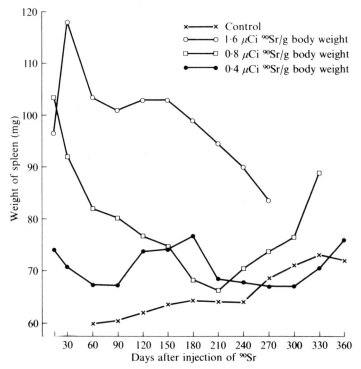

FIG. 8.16. Weight of spleen in relation to dose and time after injection of ^{90}Sr (Nilsson 1970b, by courtesy of author and publisher.)

STRONTIUM ISOTOPES

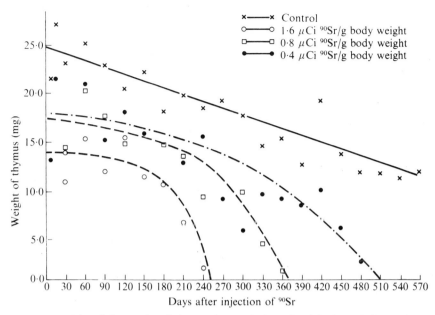

FIG. 8.17. Weight of thymus in relation to dose and time after injection of ^{90}Sr. (Nilsson 1970b, by courtesy of author and publisher.)

in the 1·6 μCi g^{-1} group. The erythrocyte series showed a tendency to predominate in regeneration during a short period (about 30 days). The granulocytic series started to dominate at 90 days in the 0·8 μCi g^{-1} group and after about 120–50 days in the others. Topographically, the onset and intensity of regeneration varied according to the bone in which the marrow was located. The femur was most affected—aplasia in this bone was to some extent compensated for by hyperplasia in the thoracic vertebrae. Changes in the weight of the spleen are illustrated in Fig. 8.16 (Nilsson 1970b). In spite of a heavy loss of lymphatic tissue in the higher dose groups and a moderate loss in the two lower dose groups a significant increase in weight in the spleen, compared with the controls, persisted over a long period. This was caused by hyperaemia and a heavy erythropoiesis magakaryocytopoiesis and granulocytopoiesis. The degree of lymphoid depletion depended upon the dose. In the 1·6 μCi group increased extra-medullary haemopoiesis, which was initially dominated by the erythroid series, was, after about 200 days, completely replaced by the granulocytic series. Megakaryocytopoiesis was much more prominent in the two lower dose groups. Changes in the weight of the thymus are illustrated in Fig. 8.17 (Nilsson 1970b). In discussing these changes Nilsson suggests that particularly in the two lower dose groups the granulocytosis, accompanied by a proliferation of myeloid elements, might be identical to the myeloproliferative disorders in beagles described by Goldman, Dungworth, Bulgin, Rosenblatt, Richards, and Bustad (1969). These dogs were fed ^{90}Sr continuously and are discussed in Section 1.3.2.3.

Van Putten and De Vries (1962) also carried out a careful histological study of the toxicity of ^{90}Sr in mice, using a strain in which no osteosarcomas or leukaemias occurred in the controls. They recorded only osteosarcoma in the injected mice.

1.3.2. Continuous administration

1.3.2.1. Man. Small groups of workers in the dial painting industry in Czechoslovakia, Switzerland, and Poland were contaminated at one time with paint containing varying amounts of ^{90}Sr and ^{226}Ra (Müller, David, Rejskova, and Brezikova 1961; Müller, Klener, Tuscany, Thomas, Brezikova, and Houskova 1966; Müller and Thomas 1969; Wenger and Cassinatis 1962; Wenger and Miller 1962; Wenger and Soucas 1963, 1965; Volf 1972). The level in the Polish workers was too low to justify prolonged study but important data on metabolism in man was obtained by Müller and his colleagues. The cumulative body burden from both radionuclides did not exceed the maximum permissible body burden (MPBB), and hitherto no untoward effects that can be definitely attributed to ^{90}Sr have been observed.

1.3.2.2. Monkey. Casarett and his colleagues (Casarett, Tuttle, and Baxter 1962) gave 7 monkeys 500–1000 μCi kg^{-1} of ^{90}Sr by gavage for 5–10 days; one monkey developed monocytic leukaemia, one a chondrosarcoma, one an osteosarcoma, and one pancytopaenia.

1.3.2.3. Dogs. An intensive study of the effect of continuous ingestion of ^{90}Sr in beagle dogs is being carried out at the University of California, Davis. The experiment was set up to obtain a skeleton uniformly labelled with ^{90}Sr. For this purpose the pregnant beagle was fed ^{90}Sr from mid-gestation and the puppy, when weaned, fed ^{90}Sr until 1·5 years of age, at which time the dogs are mature and adult skeletons have been developed (Bustad, Goldman, Rosenblatt, McKelvie, and Hertzendorf 1969; Goldman, Della Rosa, and McKelvie 1969). As shown in Table 8·3 the Davis dogs so fed developed some osteogenic sarcomas at the highest dose levels, 17 osteosarcomas, 2 fibrosarcomas, 1 haemangiosarcoma, 1 chondrosarcoma, and 4 squamous-celled carcinomas were recorded. At lower dose-levels 2 fibrosarcomas occurred (Pool, Williams, and Goldman 1972). The more striking effect was a myeloproliferative disorder varying from myeloid leukaemia to myelofibrosis with myeloid metaplasia. The picture differs from that in the pigs to be described in that no lymphoproliferation was seen. Fourteen cases are described in dogs fed 12 or 4 μCi daily from mid-gestation of their dam to 1·5 years of age (Dungworth, Goldman, Switzer, and McKelvie 1969). The age of the dogs at onset of the blood dyscrasia ranged from 14–72 months. The terminal leucocyte count varied

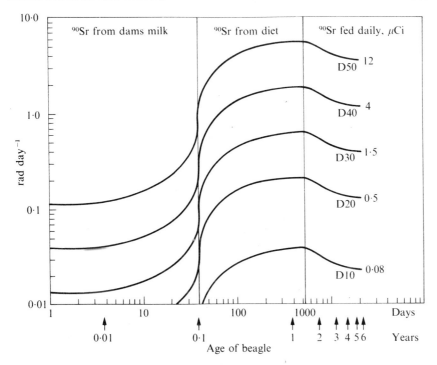

FIG. 8.18. Absorbed skeletal dose for ^{90}Sr fed beagles (Dungworth *et al.* 1969, by courtesy of authors and publishers.)

from about 3,000 to 40 000 cells mm^{-3} with degrees of shift to the left. Splenomegaly was usually present. Proliferation of granulocytic elements in marrow and spleen together with erythroid and megakaryocytic depletion were constant findings. The degree of involvement of liver, lymph nodes, and other organs was variable. In some instances the cells were all blast cells (Goldman, Dungworth, Bulgin, Rosenblatt, Richards, and Bustad 1969).

Finkel and her colleagues (Finkel, Biskis, Greco, and Camden 1971), using a relatively small number of dogs given repeated ^{90}Sr injections throughout one year, also record a few oral cancers, myeloid leukaemias, fibrosarcomas, haemangiosarcomas, and chondrosarcomas and a large number of osteosarcomas.

Dungworth and his colleagues (Dungworth, Goldman, Switzer and McKelvie 1969) have calculated the absorbed skeletal dose rates for beagles fed ^{90}Sr continuously. These are shown for the different feeding levels in Fig. 8.18. As has been said both myeloproliferative and osteogenic lesions occurred at the two higher levels. Bone tumours other than fibrosarcomas were seen after an accumulation of a mean skeletal dose between

8000 and 11 000 rad. The fibrosarcomas occurred with a dose of approximately 3700 rad.

1.3.2.4. *Miniature swine.* Since 1958 seven hundred and fifty miniature swine of the Pitman–Moore strain have been fed ^{90}Sr continuously (Clarke, Howard, and Hackett 1969; Howard and Clarke 1970; Clarke, Busch, Hackett, Howard, Frazier, McClanahan, Ragan, and Vogt 1972). The animals include parents and F_1 and F_2 generations. The parent generation was started on the experiment at 9 months of age and the females were mated after reaching radionuclide equilibrium. The daily dose varied from 1 to 3100 μCi of ^{90}Sr. The feeding level of offspring was gradually raised to the same level as that of the dam by 6 months of age. There were 225 litter-mate controls. Various animals from each group were killed periodically to evaluate radionuclide deposition and early effects of treatment. The animals in each treatment group that were considered at risk are shown in Table 8.7.

All animals receiving 3100 μCi day^{-1} died within 3–4 months with profound marrow aplasia. Fewer animals fed 625 μCi day^{-1} or less died of profound aplasia but in all exposure levels they developed haemopoietic disorders. The incidence of haemopoietic dysplasia was highest in swine fed 125 μCi day^{-2}: 21 myeloid, 8 lymphoid, and 3 stem-cell proliferative

TABLE 8.7

^{90}Sr-Fed pigs
(*Howard* et al. *1969, by courtesy of author and publisher*)

^{90}Sr diet (μCi day^{-1})	Pigs At risk	Pigs Dead	Bone tumour cases‡	Incidence tumours dead	Approximate age at death (months)	Rad 16 months before death
3100*	5	5	0	0	12	—
625	6	6	0	0	18	—
125	15	11	0	0	62	8900
25	20	12	0	0	88	3000
5	3	3	0	0	102	740
1	9	9	0	0	93	130
0	44	15	0	0	—	0
625†	21	21	0	0	3	—
125	35	35	5	14%	35	6300
25	57	35	0	0	72	3000
5	49	27	0	0	82	710
1	56	12	0	0	76	130
0	86	53	0	0	—	0

* Parent
† Offspring
‡ The number of bone tumours was 7.

Hematopoietic disorders occurring in swine ingesting ^{90}Sr
(Howard and Clarke 1970, by courtesy of authors and publisher)

^{90}Sr level (μCi day^{-1})	History					Pathologic diagnosis			
	Generation	Number of animals		Age at death (months)*	Average rad dose ($\times 10^3$) to skeleton*	Lymphoid neoplasms	Myeloid neoplasms	Stem-cell neoplasms	Myeloid metaplasia
		At risk	Now alive						
3100	Parent	5	0	12–13	7–10	—	—	—	2
625	Parent, F_1	6 27	0 0	17–19 3–34	6–9 2–10	— 1	— 1	— —	2 15
125	Parent, F_1, F_2	10 33	3 0	66–88 24–57	13–17 10–20	1 7	3 14	— 3	2 2
25	Parent, F_1, F_2	12 36	6 19	54–123 50–93	2–6 3–6	— 2	1 1	— —	1 3
5	Parent, F_1, F_2	3 26	0 20	— 82	— 1	— 1	— —	— —	— —
1	Parent, F_1, F_2	9 57	0 40	92–6 52–81	0.04 0.2	1 4	— —	— —	1 1
0	Parent, F_1, F_2	52 94	28 31	103–5 —	— —	— —	2 —	— —	— —
					Total	17	22	3	29

* For those animals with hæmoproliferative disorders.

disorders out of a group of 50 parents and offspring at risk. The neoplasms occurred in a shorter period and more frequently in the F_1 and F_2 generations than in the parents. It is important to note however that haemopoietic neoplasia did occur in the parent generation and not only in swine exposed *in utero*. It is apparent from Table 8.8 that at the higher radiation exposure levels there was a greater tendency for the development of myeloid metaplasia and neoplasia, while at lower feeding levels lymphoid neoplasia was more common.

The haemopoietic neoplasms in these swine covered a broad spectrum of morphological types and may well be simply described as a haemoproliferative disorder which varied in its invasive and morphological character. In view of the fact that the whole marrow was exposed to radiation it is not surprising that the resulting lesion took many forms. The marrow probably contains many different stem-cells (Loutit 1967) together perhaps with one extremely primitive pluripotent haemopoietic stem-cell so that generalized radiation may have resulted in abnormal cloning of one line rather than another in different pigs.

The induction period for the development of haemopoietic disorders was as short as 3 months for offspring exposed *in utero*. There appeared to be a general time-dose relationship for the induction period. Bone tumours were found at post mortem in only five of these swine (7 tumours), and were secondary to the presence of a myeloproliferative disorder, again illustrating the fact that the latent period for leukaemia appears to be shorter than that for osteosarcoma. Two of the control animals developed myeloid leukaemia, but none developed lymphoma though this is known to occur in normal pigs. It is of interest that one of the control myeloid leukaemias showed a chromosome abnormality similar to the Philadelphia chromosome seen in human myeloid leukaemia. The irradiated pigs however had no consistent chromosome aberrations (McClanahan, Hackett, and Beamer 1970). An adeno-virus has been isolated from three of the pigs with myelogenous leukaemia and C-type particles are constantly seen (Howard, Frazier, and Jannke 1970; Howard, Jannke, Frazier, and Adee 1970). The significance of this finding is discussed elsewhere (Chapter 1.1.3).

Radiation dose has been measured in swine fed ^{90}Sr continuously. The results of radioanalysis were used to calculate the average radiation dose rates to bones for the original dams and the adult F_1 and F_2 offspring at ^{90}Sr equilibrium and for the F_1 and F_2 foetuses *in utero*. These are shown in Table 8.9. Equilibrium reached approximately 10 times the daily intake of ^{90}Sr in the F_1 and F_2 generations, and 7·5 times the daily intake in the original dams. In studies with thermoluminescent dosemeters implanted in foetuses 55–110 days old no significant maternal contribution to the radiation dose to critical foetal tissues was noted. Essentially all foetal

TABLE 8.9

Radiation dose rates in bones of swine fed ^{90}Sr continuously (Palmer et al. 1970, by courtesy of authors and publishers)

^{90}Sr feeding level (μCi day^{-1})	Dose rates to bone (rad day^{-1})		
	Original dams*	F$_1$ and F$_2$ generations	
		in utero†	adults*
1	0·064	0·004	0·072
5	0·32	0·02	0·36
25	1·6	0·1	1·8
125	8	0·5	11
625	40	2·5	

* At ^{90}Sr skeletal equilibrium.
† Average to foetal skeleton during second half of gestation.

radiation is derived from the ^{90}Sr deposited in their own tissues (Palmer, Thomas, Watson, and Beamer 1970; Clarke, Busch, Hackett, Howard, Frazier, McClanahan, Ragan, and Vogt 1972).

1.3.2.5. *Rabbits.* A small series of rabbits fed ^{90}Sr continuously has been reported (Downie, Macpherson, Ramsden, Sissons, and Vaughan 1959) and the radiation dose analysed (Owen 1962). Osteosarcoma occurred in rabbits fed 8–11·8 μCi daily from the age of 5–8 weeks. No skeletal lesions were seen in rabbits fed when over 3 years old nor in sucklings born to rabbits fed ^{90}Sr. Varying degrees of anaemia, leucopaenia, thrombocytopaenia, and marrow aplasia occurred in the rabbits developing osteosarcoma but no haemoproliferative disorder was noted in any animals. No explanation of this difference in response from dogs and pigs is at present available except that in normal rabbits naturally occurring leukaemias are not reported.

1.3.2.6. *Rats.* An excellent histological study has been made of the effect of ^{90}Sr injected at 6-monthly intervals into rats, the first dose being given when the rats were 42 days old. Only osteogenic sarcomas, fibrosarcomas, chondrosarcomas, and what are described as telangiectactic sarcomas resulted; the latter may probably be classified as angiosarcomas. There was considerable marrow fibrosis but no note is made of haemopoietic neoplasms (Skoryna, Kahn, and Webster 1958; Skoryna and Kahn 1959).

Casarett and his colleagues have also given rats ^{90}Sr daily in their drinking water for 10 to 30 days. The dose received varied from 330 to 790 μCi in total. The rats developed a wide variety of tumours among them reticulum-cell tumours, angiosarcomas, and osteosarcomas. All of the latter contained some bone or osteoid but some also contained cartilage and much fibrous tissue. The leukaemias were generally lymphatic in type but there was one myeloid leukaemia.

1.3.3. *Consideration of the effects of ^{90}Sr administration.* Though it is not possible to obtain precise and comparative measurements of radiation dose in the experiments discussed above on the effects of single and continuous administration of ^{90}Sr, certain conclusions can be drawn from Tables 8.2 and 8.3. The most important is that though single injections of ^{90}Sr may result in osteosarcoma, and other tumours arising from marrow elements, continuous ingestion by mouth over both long and short periods will result in haemoproliferative dyscrasias in both dogs and swine as well as osteogenic tumours. Why in the case of the pig this may affect all haemopoietic cell lines and in the case of the dog only the granular cells is not at present clear. It is known however that proliferative lesions affecting the lymphoid elements are normally not uncommon in pigs. In dogs there appears also to be a relatively high incidence of osteosarcomas as well as of myeloid proliferation following continuous ingestion. In pigs the osteosarcoma incidence is very low. Why there is this difference must await more detailed dosimetry, preferably both by autoradiographic techniques (Owen and Vaughan 1959a) and by the methods used by Spiers and his colleagues (Spiers, Zanelli, Darley, Whitwell, and Goldman 1972). The pattern of feeding also appears to have been rather different in the two experiments so that exact comparisons are difficult.

The basic difference in the dosimetry of single and continuous administration can probably be explained by the results of measurements, made by autoradiographic techniques, of the radiation dose obtained in both marrow and bone in rabbits reported by Owen and Vaughan (1959a,b) and Vaughan (1962). It was shown that the maximum dose-rate in the length of a long bone increased over time in the animal ingesting ^{90}Sr in which the skeleton was evenly labelled, while in the animal receiving a single injection it falls, as shown in Fig. 8.19, dependent on uneven skeletal labelling. Furthermore, the accumulated dose measured on the endosteal surface became greater in the fed animals with time, while it fell in the injected animals as shown in Fig. 8.20. It was indicated in the section on the effect of ^{90}Sr on the bone marrow (in Section 1.3.1.3) that as the marrow dose fell the marrow activity recovered. Such recovery of a normal marrow would be unlikely to occur in the face of a rising rather than a falling marrow radiation dose and bone radiation dose. Haemopoietic tissue has been classified as of high sensitivity and bone (i.e. osteogenic tissue) as of low sensitivity (International Commission on Radiological Protection 1969). Therefore it is not surprising that if the whole marrow is irradiated by a long range emitter haemopoietic neoplasia is more likely to result than osteogenic neoplasia. Furthermore, the latent period of leukaemia is now recognized as being

STRONTIUM ISOTOPES 185

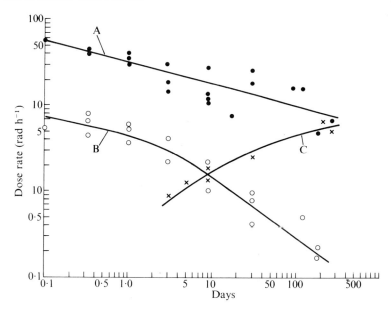

FIG. 8.19. Graph of maximum dose-rate at different times after injection or start of ^{90}Sr feeding. Rabbits injected with ^{90}Sr at 600 μCi kg^{-1} (curve A), 100 μCi kg^{-1} (curve B) at the age of 5–8 weeks; each figure has been corrected assuming the rabbits weigh 600 g when injected. Curve C is for rabbits fed 8·5 μCi of ^{90}Sr daily from the age of 5–8 weeks. (Owen and Vaughan 1959b, by courtesy of authors and publishers.)

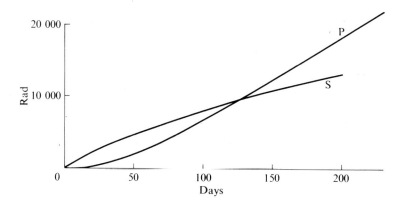

FIG. 8.20. Graph of the accumulated dose received by the endosteal surface at different times after a single injection of ^{90}Sr (600 μCi kg^{-1}, curve S) and after start of feeding 8·5 μCi of ^{90}Sr (curve P) at the age of 5–8 weeks. (Owen and Vaughan 1959b, by courtesy of authors and publishers.)

shorter than that of osteosarcoma (Vaughan 1970b) so that the animals or men may well die of leukaemia before they have time to develop an osteosarcoma. In this connection it is interesting to note that the few osteosarcoma that did occur in the swine fed ^{90}Sr were found only at post mortem on animals that had died of a myeloproliferative lesion (Howard, Clarke, Karagianes, and Palmer 1969).

It might be questioned as to why the dogs given repeated injections over a long period by Finkel and her colleagues developed predominantly osteosarcoma (Finkel, Biskis, Greco, and Camden 1972). There are no measurements of radiation dose in bone and marrow in such experimental conditions but it seems probably that intermittent ^{90}Sr administration will not produce the same radiation dose conditions as continuous ingestion.

Broadly speaking it may be said that a single injection of ^{90}Sr will result in an uneven distribution of the radionuclide in the skeleton and therefore an uneven distribution of radiation dose to both marrow and osteogenic tissue, while continuous ingestion will result in a relatively even distribution of the radionuclide throughout the bone mineral and therefore an even and continuous radiation dose distribution to the sensitive haemopoietic cells. The long range of the ^{90}Sr + ^{90}Y beta particle will result in irradiation of the whole marrow so resulting in a wide variety of tumours arising from marrow elements in addition to those tumours arising from osteogenic tissue.

1.3.4. *Dosimetric measurements.* Various methods have been used to estimate skeletal dose from deposited ^{90}Sr. Dougherty and his colleagues at Utah have based their analysis on an average skeletal dose calculated from the ^{90}Sr content of the whole skeleton (Mays, Dougherty, Taylor, Lloyd, Stover, Jee, Christensen, Dougherty, and Atherton 1969; Dougherty and Mays 1969). Osteosarcoma have occurred with an injection dose of 63·6 μCi kg^{-1}. The conclusions that are drawn about ^{90}Sr dosimetry by this group are discussed later (Section 1.3.5). Vaughan and her colleagues have made use of autoradiographic techniques (Owen and Vaughan 1959a, Owen and Vaughan 1959b, Vaughan and Williamson 1969). Some of their results have already been discussed. These authors emphasize that any measurement of dose rate or accumulated dose that may be obtained at the site of tumour induction will be too high, since the original malignant transformation will have occurred before the tumour was histologically recognized. The calculated lowest accumulated dose at the site of early malignancy was about 12 000 rads (Rushton, Owen, Holgate, and Vaughan 1961). Their observations emphasize the great differences in radiation dose to sensitive tissues that occur when ^{90}Sr is given by continuous administration compared with a single injection. In the latter case particularly, there are appreciable differences in different bones and in different parts of the same bone in radiation dose received.

It would appear, however, that the techniques developed by Spiers and his coworkers, already noted in connection with radium (Chapter 7.3.6.2), are likely to prove the method of choice in the future. These methods enable the following estimates to be made of dose to sensitive tissues: (a) the mean absorbed dose to haematopoietic marrow in trabecular bone,

TABLE 8.10

Trabecular contribution to the mean marrow dose—$\bar{D}_M/D_0(T)$
(*Spiers et al. 1972, by courtesy of authors and publishers*)

Bone \ Isotope	^{14}C	^{45}Ca	^{90}Sr	^{22}Na	^{18}F	^{32}P	^{90}Y	$^{90}Sr+{}^{90}Y$
Man								
Cranium	0·1300	0·2448	0·5002	0·5002	0·5259	0·5164	0·4595	0·4789
Mandible } Ribs/sternum }	0·0291	0·0547	0·1183	0·1178	0·1272	0·1527	0·1438	0·1431
Clavicle/hip	0·0501	0·0970	0·2047	0·2066	0·2203	0·2595	0·2572	0·2494
Cervical vert.	0·0559	0·1089	0·2365	0·2384	0·2563	0·3176	0·3199	0·3057
Thoracic vert.	0·0485	0·0935	0·2004	0·2019	0·2175	0·2740	0·2768	0·2638
Lumbar vert.	0·0412	0·0781	0·1642	0·1654	0·1786	0·2303	0·2336	0·2218
Femora/humeri	0·0380	0·0749	0·1663	0·1676	0·1803	0·2304	0·2339	0·2223
Beagle								
Tibia								0·393†

* Corrections introduced when the range of the particle is greater than the dimensions of the trabeculae (Spiers 1966).
† equilibrium factor not included.

TABLE 8.11

Trabecular contribution to the mean endosteal dose—$\bar{D}_s/D_0(T)$
(*Spiers et al. 1972, by courtesy of authors and publishers*)

Bone \ Isotope	^{14}C	^{45}Ca	^{90}Sr	^{22}Na	^{18}F	^{32}P	^{90}Y	$^{90}Sr+{}^{90}Y$
Cranium	0·4164	0·5184	0·6368	0·6383	0·6396	0·5418	0·4755	0·5114
Mandible } Ribs/sternum }	0·3459	0·3732	0·3074	0·3088	0·2922	0·2034	0·1770	0·1989
Clavicle/hip	0·3635	0·4028	0·3578	0·3626	0·3505	0·2968	0·2832	0·2964
Cervical vert.	0·3742	0·4294	0·4092	0·4153	0·4054	0·3613	0·3511	0·3613
Thoracic vert.	0·3687	0·4129	0·3755	0·3809	0·3699	0·3216	0·2762	0·3221
Lumbar vert. Sacrum	0·3631	0·3964	0·3418	0·3465	0·3344	0·2820	0·2012	0·2829
Femora/humeri	0·3566	0·3977	0·3443	0·3497	0·3354	0·2792	0·2688	0·2821

* Corrections introduced when the range of the particle is greater than the dimensions of the trabeculae (Spiers 1966).

(b) the mean absorbed dose averaged over a distance 0–10 μm from the trabecular surface, (c) the mean absorbed dose averaged over a distance 0–10 μm from the endosteal surface in cortical bone (Spiers 1966, 1968, 1969; Whitwell and Spiers 1971; Spiers, Zanelli, Darley, Whitwell, and Goldman 1971).

In Table 8.10 are shown the trabecular contributions to the mean dose for a variety of radionuclides calculated from a knowledge of path lengths in the bones of a man of 44 and of beta-particle energy spectra for the radionuclides listed. Table 8.11 shows the trabecular contribution to the mean endosteal dose. The results are expressed as ratios $\bar{D}_M/D_0$ and $\bar{D}_S/D_0$ (for mean marrow and mean endosteal absorbed dose rates respectively), where D_0 is the absorbed dose rate to a very small tissue-filled (Bragg–Gray) cavity surrounded by bone in all directions to a distance breater than the particle range (Spiers, Zanelli, Darley, Whitwell, and Goldman 1971). In order to obtain the total dose to the marrow and trabecular endosteum in a given bone the contribution from the radionuclide in the cortex must also be determined and added to that from the trabeculae. This has been done in Table 8.12 using methods described by Spiers (1966).

TABLE 8.12

Cortical contribution to the mean marrow dose and mean endosteal dose
(Spiers et al. 1972, by courtesy of authors and publishers)

Bone	^{14}C	^{45}Ca	^{90}Sr	^{22}Na	^{18}F	^{22}P	^{90}Y	$^{90}Sr+^{90}Y$
Cranium	0·0059	0·0108	0·0408	0·0384	0·0498	0·1829	0·2345	0·2028
Mandible } Ribs/sternum }	0·0046	0·0084	0·0317	0·0299	0·0383	0·1095	0·1294	0·1231
Clavicle/hip	0·0016	0·0029	0·0111	0·0104	0·0135	0·0474	0·0600	0·0528
Cervical vert.	—	—	—	—	—	—	—	—
Thoracic vert.	—	—	—	—	—	—	—	—
Lumbar vert.	—	—	—	—	—	—	—	—
Femora/humeri	0·0006	0·0011	0·0037	0·0042	0·0050	0·0197	0·0287	0·0237

Spiers and his colleagues have also applied methods of thermoluminesce dosimetry to animal bone containing $^{90}Sr+^{90}Y$ (Zanelli 1968; Zanelli, Darley, and Goldman 1971). In Table 8.13 are shown dose rates measured by the thermoluminescence technique using CaF_2:Mn as the dosemeter in different parts of the tibia of a beagle dog fed 4 μCi ^{90}Sr throughout the growth period.

Spiers and his colleagues consider that probably thermoluminesce methods afford the best means of studying dosimetry in animal bone, while in man bone scanning and appropriate calculations offer the best method of studying dosimetry from internal emitters. If it is possible to

TABLE 8.13

Dose rates to CaF_2: Mn in tibia of beagle D40M09
(Spiers et al. 1972, by courtesy of authors and publishers)

Diet: ^{90}Sr at 4 μCi day^{-1}, throughout growth period.

Specimen no.	Description		Average ^{90}Sr content* (μCi/g bone)	Dose rates	
	Trabeculation	Cortex		(rad day^{-1})	(g rad μCi^{-1} day^{-1})
1	Fine	Thin	0·022	0·95	43·3
2	Medium	Medium	0·021	0·72	34·1
3	Coarse	Medium	0·037	1·07	28·9
4	Little present	Thick	0·033	1·12	33·9
5	Medium	Medium	0·022	0·94	42·8
6	Fine	Medium	0·0165	0·86	51·9
Mean values:			0·0253		39·2 + 3·3

* averages obtained from measurements of both trabecular and cortical samples.

TABLE 8.14

Figures for dose response curve calculated by Mays and Lloyd (1972) for osteosarcoma incidence following single injections of ^{90}Sr (from the data of Moskalev et al. 1969, by courtesy of authors and publishers)

Injection (μCi kg^{-1})	Dead rats	Sarcoma rats	Incidence (sarcoma/dead)	Bone sarcoma rats	
				Average days injection to death	Rad 170 days before death
500	43	22	51·2%	223	5760
250	78	40	51·3%	356	7710
75–100	158	6	3·8%	435	3530
50	379	0	0	—	2480*
25	374	0	0	—	1240*
10	300	1	0·3%	500	480
5	300	1	0·3%	407	190
2·5	382	0	0	—	124*
0·5	383	0	0	—	25*
0·25	300	0	0	—	12*
0·005	300	0	0	—	0·2
0	722	0	0	—	0

* Dose for non-sarcoma levels calculated at control 520 − 170 = 350 days.

prove the equivalence of the theoretical method and thermoluminescence dosimetry in beagle or pig bone this should prove an important step in predicting radiological effects in human bone from animal experiment.

1.3.5. *Dose–response relationships following* ^{90}Sr *administration.* Mays and Lloyd (1972) have attempted to determine dose-response relationships for ^{90}Sr in experiments in mice, rats, dogs, and pigs using as their endpoint osteosarcoma induction. To facilitate interspecies comparisons and to correct for the escape of beta particles from the skeleton Mays chose *average skeletal dose* as a common dose parameter and computed it to the estimated start of tumour growth rather than to the time of death as he has done for all dosimetric discussions on the Utah dogs (see Chapter 6.4). The figures calculated for rats given a single injection are shown in Table 8.14 and the dose-response curves for mice and rats in Figs 8.21 and 8.22. The dose–response curve for dogs given a single injection is shown in Fig. 8.23 and that for pigs fed ^{90}Sr continuously in Fig. 8.24. The figures on which Fig. 8.23 is based are shown in Table 8.15. The data used for the mouse calculations comes from Finkel's CF 1/Anl mice, which as already discussed are not an ideal experimental animal in view of their high sarcoma incidence, but the same non-linear response was found by Mays for Nilsson's CBA mice, a strain with a low natural sarcoma incidence (Nilsson 1970). Mays (Mays and Lloyd 1972) found the incidence of bone sarcoma at low doses was consistently less than predicted from linear dose–response relationships which started at natural incidence. He states '"Linearity" is thus not supported but the low-dose data are fitted well either by practical threshold

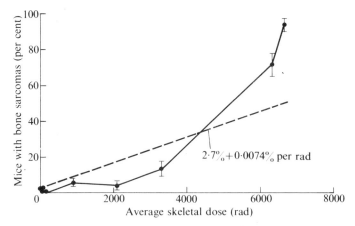

FIG. 8.21. ^{90}Sr dose–response in CF 1 female mice injected at 70 days of age. The data demonstrates a non-linear dose–response relationship. The indicated linear equation predicts 20 radiation induced plus 14 naturally occurring sarcoma cases (34 total predicted cases) in the 540 lower-dose mice injected with 1·3–88 μCi kg^{-1} (26–2090 rad), whereas they had 14 cases of bone sarcoma. The probability is only 10^{-4} for observing 14 or fewer cases if 34 are predicted. (Mays and Lloyd 1972, by courtesy of authors and publishers.)

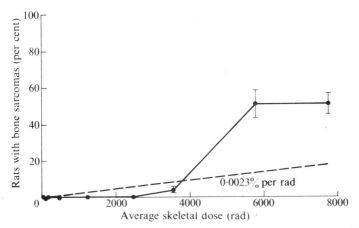

FIG. 8.22. ⁹⁰Sr dose–response in Moskalev's Wistar-derived rats injected at 3 months of age. The indicated linear equation predicts 38 sarcoma cases in the 2718 lower-dose rats injected with ⁹⁰Sr at 0·005–50 μCi kg⁻¹ (0·2–2480 rad), whereas they had only 2 bone sarcomas. The probability is only 10⁻¹³ for observing 2 or fewer cases if 38 are predicted. Although no bone sarcomas were observed among 722 control rats, it is possible that one or both of the 2 bone sarcomas observed in the 2718 lower-dose rats were spontaneous, rather than radiation-induced. (Mays and Lloyd 1972, by courtesy of author and publishers.)

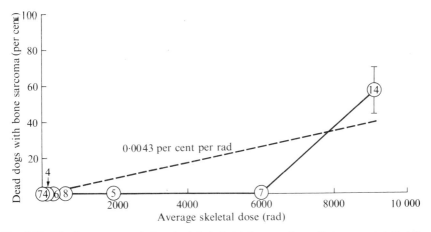

FIG. 8.23. ⁹⁰Sr dose–response in beagles injected at 1·4 years of age. Data are *not* plotted for the dogs injected with 63·6 μCi kg⁻¹ in 1966, since insufficient time has passed to establish their incidence point with reliability. Omitting this data point, the indicated linear equation (0·0043% per rad) predicts 2·5 sarcoma cases among the 30 deaths in the lower-dose dogs, injected with 0·57–32·7 μCi kg⁻¹ (105 rad, 316 rad, 635 rad, 1980 rad, and 6000 rad), whereas they had no bone tumours. The probability is 0·07 for observing 0 cases if 2·5 are predicted. A similar linear equation (0.0051% per rad) based on all of the dead dogs, including those injected with 63·6 μCi kg⁻¹, predicts 3·0 sarcoma cases for the non-tumour levels. The probability is 0·04 for observing 0 cases if 3 are predicted. A total of 74 life-span control beagles have died without bone tumours at the University of Utah and the natural incidence is 1–100 osteosarcomas per 10⁵ beagle deaths. (Mays and Lloyd 1972, by courtesy of authors and publishers.)

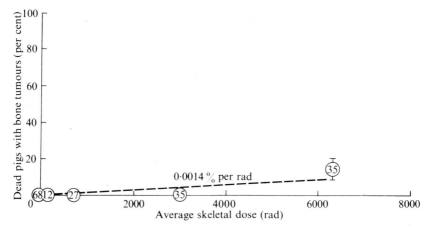

Fig. 8.24. ^{90}Sr dose–response in Pitman Moore miniature pigs continually exposed to dietary ^{90}Sr from conception to death. (Not plotted are the smaller number of parent pigs whose exposure started at 9 months of age.) The indicated linear equation predicts 1·8 cases of bone tumours among the 74 deaths in the lower-dose offspring pigs fed 1–25 μCi day^{-1} (130–3000 rad), whereas they had no bone tumours. The probability is 0·16 for observing 0 cases if 1·8 are predicted. No bone tumours were observed among the 68 control pigs which have died, and the spontaneous incidence of malignant bone tumours in miniature pigs is believed to be very low. (Mays and Lloyd 1972.)

TABLE 8.15

Figures for dose-response curve calculated by Mays and Lloyd (1971) for osteosarcoma incidence following a single injection of ^{90}Sr in Beagle dogs (Mays and Lloyd 1972, by courtesy of author and publishers)

Injection level	Injection (μCi kg^{-1})	Injected dogs	Dead dogs	Sarcoma dogs	Incidence sarcoma/dead =coma (%)	Bone sarcoma dogs	
						Years from Injection to death	Rad 1 year before death
5	97·9	14	14	8	57	4·02	9100
4·5*	63·6	12	2	2	100	2·77	3920
4	32·7	12	7	0	0	—	6000†
3	10·8	12	5	0	0	—	1980†
2	3·46	12	8	0	0	—	635†
1·7	1·72	13	6	0	0	—	316†
1	0·57	12	4	0	0	—	105†
0	0	13	6	0	0	—	0†

* The 4·5-level injected 16 March 1966: Older levels 1955–60.
† For levels without bone sarcoma, rads at 10 years.

TABLE 8.16

Lowest ^{90}Sr doses proven to induce bone sarcomas
(from Mays and Lloyd 1972, by courtesy of authors and publishers)

Animals	Administration	Days to estimated tumour start	Rads
*Mice	200 μCi kg^{-1} i.v. at 70 days of age	370	3300
Rats	88 μCi kg^{-1} i.p. at 3 months	265	3500
Rabbits	200 μCi kg^{-1} i.v. at 485 days	900	5700
Monkeys	170 μCi kg^{-1} orally at 3–7 years	870	2500
Beagles	64 μCi kg^{-1} i.v. at 504 days	655	3900
Pigs	125 μCi day^{-1} *lifetime* diet	570	6300

* It must be noted that this Table gives 200 μCi kg^{-1} as the lowest dose-level proven to produce osteosarcoma in mice. Finkel and Biskis (1968) state they obtained many tumours with 44 μCi kg^{-1} but Mays considers this apparent increase could be either real or random (Mays 1971, personal communication).

TABLE 8.17

Observed and predicted cases of osteosarcoma at low doses of ^{90}Sr
(from Mays and Lloyd 1972, by courtesy of authors and publishers)

Experiment	Observed cases	Threshold prediction	Linear prediction	P
^{90}Sr in 540 mice	14	14	34	10^{-4}
^{45}Ca in 300 mice	7	10	30	10^{-8}
^{90}Sr in 2718 rats	2	0	38	10^{-13}
^{90}Sr in 30 beagles	0	0	2·5	0·07
^{90}Sr in 74 pigs	0	0	1·8	0·16

relationships as Evans (Evans, Keane, and Shanahan 1969) has suggested for the human radium cases or "a sigmoid" relationship'. This suggests to Mays that beta irradiation, because of its low LET and therefore the possibility of cell recovery, is not highly effective in the induction of bone sarcomas provided that the dose rate is sufficiently low.

In Tables 8.16 and 8.17 Mays summarizes the results of his calculations. He gives in Table 8.16 the lowest ^{90}Sr dose level proven to induce bone sarcomas and in Table 8.17 the observed and predicted cases in the lower dose ranges. The observed cases with bone sarcomas at low doses are compared with the cases predicted by (a) the threshold model (natural incidence only) and (b) the linear model in which a radiation induced incidence (assumed proportional to dose) was added to the natural incidence. The P values shown in the last column are the probability that if the linear model was correct the observed cases could have occurred by chance. The number of all the animals listed is extremely small on which to base any expectation of tumour development at low doses. Very large populations must be examined to detect low dose-hazards.

Mays and Lloyd (1972) conclude that their present best estimates of the actual 50-year risk for man below 1000 rad from ^{90}Sr are 1 ± 1 sarcomas/10^6 person rad for a 'low dose' linear model, and 4 ± 4 sarcomas/10^{10} person rad^2 for a dose-squared model. These estimates are based on the observations of the radium-dial painters, where Rowland, Failla, Keane, and Slehney (1969–70, 1970–71) have recently found a dose-squared model best fits their results (Chapter 3.6.1), and the toxicity of ^{90}Sr relative to ^{226}Ra in mice and dogs.

Evidence from humans, mice, dogs, and pigs indicates in the opinion of Mays and Lloyd (1972) that children may be $\frac{1}{2}$ to 4 times as sensitive as adults to the induction of bone sarcoma, and suggests that the risk of leukaemia for children might well exceed the risk of bone sarcoma (Mays and Lloyd 1972). However in making this statement it is not clear whether they are separating the effects of a single injection from those of continuous ingestion. It would clearly appear from the evidence already discussed that the risk with continuous ingestion in all age groups of dogs and pigs was a haemoproliferative lesion rather than one affecting osteogenic tissue. It must not be forgotten that the swine dams on continuous feeding as well as their progeny developed haemoproliferative lesions.

It is important in discussion of dose-response relationships to differentiate between single and continuous exposure. A single injection of ^{90}Sr is unlikely to occur in real life. It has already been pointed out that continuous ingestion from contaminated food and water is the most likely ^{90}Sr hazard in man: therefore to calculate risks to man on the basis of single injection in dogs seems questionable, more especially when the hazard appears to be different depending on the exposure pattern. Osteosarcoma results from a single injection but a myeloproliferative disorder may follow from continuous ingestion.

A further difficulty in accepting Mays estimations of risks to man based on the Utah experiment on dogs is the fact that Spiers and his colleagues have shown that differences in bone trabeculation in man and dog lead to very different radiation doses to sensitive tissues (Section 1.3.4).

1.3.6. *Histopathology of skeletal tumours arising from ^{90}Sr irradiation*

1.3.6.1. *Osteogenic tissue.* Reference to Tables 8.2 and 8.3 indicates that following the single administration of ^{90}Sr to a variety of animals a high proportion of osteosarcomas result. Chondrosarcomas are noted by Skoryna and his colleagues in rats and one is noted in a monkey (Skoryna, Kahn, and Webster 1958; Skoryna and Kahn 1959; Casarett, Tuttle, and Baxter 1962). Following continuous ingestion or repeated injections osteosarcomas are also relatively common, except in swine when they were only found at post mortem in animals which had died of a haemoproliferative dyscrasia.

1.3.6.2. *Marrow tissue other than haemopoietic tissue.* The striking feature of the majority of experiments in which ^{90}Sr has been given to all species both by single and by continuous administration is the number of miscellaneous tumours that arise from mesenchyme elements in the marrow other than the haemopoietic. In many cases these tumours are well described and illustrated in photomicrographs. The nomenclature used by different workers is variable but they are clearly not describing the classical osteosarcoma where osteoid formation dominates the histological picture. Admittedly the diagnosis of bone tumours is controversial and there may often be a divergence of opinion among pathologists. However it would appear that the frequent mention of tumours other than straightforward osteosarcoma is remarkable.

TABLE 8.18

Skeletal tumours, Utah dogs, March 1970
(from Research in Radiobiology COO 119, 242, 1970)

	Myeloid leukaemia	Blood dyscrasia	Osteosarcoma	Angiosarcoma	Fibrosarcoma gingivae	Carcinoma sinuses skull
^{226}Ra			38			
^{90}Sr		3	8	2	1	1
^{228}Ra			36			
^{228}Th		2	36	1		
^{239}Pu	2		54			
Totals	2	5	172	3	1	2

In Table 8.18 are listed the tumours found in the Utah beagles given ^{228}Th, ^{228}Ra, ^{226}Ra, ^{239}Pu, and ^{90}Sr. There were in all 172 osteosarcomas, no fibrosarcomas, and only 3 haemangiosarcomas of bone. Two of the latter occurred in the ^{90}Sr dogs and 1 in a ^{228}Th dog (i.e. a dog receiving a radiation dose from a long-range alpha emitter that might be expected to irradiate marrow as well as osteogenic tissue). The beagles that inhaled ^{90}SrCl$_2$ had 7 angiosarcomas, 5 osteosarcomas, 3 fibrosarcomas, and 2 osteochondromas (McClellan, Boecker, Jones, Barnes, Chiffelle, Hobbs, and Redman 1972). These figures are significant when it is remembered that the most common bone tumour occurring anyhow in man, if not in beagles, is an osteosarcoma.

Kuzma and Zander (1957*a*) give no quantitative data about the types of tumours they induced with ^{89}Sr, which, like ^{90}Sr + ^{90}Y, emits beta particles of high energy, however they give a detailed description which confirms the other observations reported, namely that ^{89}Sr while inducing the classical bone-forming osteosarcoma will also give rise to (i) a highly cellular reticulum-cell sarcoma, (ii) a bizarre anaplastic type of sarcoma, and (iii) a highly and delicately vascularized fibroblastic type which closely mimics an angiosarcoma. These latter tumours were not seen in the rats given ^{45}Ca, which has a short-range low-energy beta radiation.

Barnes, Carr, Evans, and Loutit (1970) noted 5 of 11 bone tumours in CBA mice as haemangioendotheliomas, which they suggest arose from the reticuloendothelium of bone marrow. Nilsson (1970a) has made an elaborate histological study of the relationship of tumour type to the amount of ^{90}Sr given as a single injection and expressed per g of body weight. The result is shown in Table 8.19. It should be noted that he also has an appreciable

TABLE 8.19

Histological classification of bone tumours in mice following administration of ^{90}Sr
(*Nilsson 1970a, by courtesy of author and publisher*)

Type of tumours	Percentage distribution by dose expressed in μCi g^{-1}			
	1·6 n:219	0·8 n:292	0·4 n:90	0·2 n:8
Osteoblastic osteosarcomas	87·2	68·8	24·4	(50·0)
Fibroblastic osteosarcomas	11·0	29·8	71·1	(37·5)
Chondroblastic osteosarcomas	0·9	—	—	
Angiosarcomas	0·9	2·1	4·4	(12·5)

incidence of angiosarcomas and that he considers that the fibrosarcomas on histological evidence also arise from marrow reticulum rather than from osteogenic tissue on bone surfaces. (Nilsson 1962, 1970a; Sundelin and Nilsson 1968). Finkel and her colleagues (Finkel, Biskis, and Scribner 1959) using CF 1 mice, a strain known to have a natural incidence of osteosarcoma of at least 2 per cent, note a high incidence of haemangioendothelioma of bone marrow as well as of osteosarcoma following injection of ^{90}Sr.

No such incidence of tumours arising from marrow elements rather than from osteogenic tissue (Chapter 4.3.2) is reported for any radionuclide other than ^{90}Sr. As already explained their occurrence is probably accounted for by the long range of the ^{90}Sr + ^{90}Y beta particle.

1.3.6.3. *Haemopoietic tissue.* The precise nomenclature and definition of proliferative lesions and dyscrasias affecting haemopoietic tissue is a fruitful source of controversy (Dameshek 1951, 1956, 1970; Vaughan 1970b). Such a definition today should certainly include proliferation and dyscrasia of both myeloid and lymphoid elements, since stem cells of all circulating cell types may be found in the marrow (Loutit 1967). What consitutes leukaemia is also a matter hotly debated. The classical leukaemia is usually associated with an anaemia, which is leucoerythroblastic in character but not necessarily a leucocytosis. There is proliferation of young and often abnormal cells of one or other cell line in the marrow and infiltration by similar cells in organs elsewhere in the body, characteristically liver, spleen, and kidneys. This may lead to gross enlargement particularly of liver and spleen.

The pathological picture presented may vary greatly in the degree of severity. Reference to Table 8.3 indicates that in animals fed ^{90}Sr continuously the outstanding lesion is one affecting the haemopoietic marrow. McClellan records one instance of myeloid leukaemia in a dog that had a single inhalation of ^{90}SrCl$_2$. Moskalev and his colleagues (Moskalev, Streltsova, and Buldakov 1969) record it in rats and Nilsson in mice (Nilsson 1969).

All forms of haemoproliferation have proved a fatal condition in swine fed continuously (Clarke, Busch, Hackett, Howard, Frazier, McClanahan, Ragan, and Vogt 1972); myeloid leukaemia and myeloid hyperplasia are as common as osteosarcoma in dogs fed continuously. Clarke, Busch, Hackett, Howard, Frazier, McClanahan, Ragan, and Vogt (1972) say in their most recent report 'It is readily apparent that the primary response of miniature swine to the chronic ingestion of ^{90}Sr is manifested as leukaemogenic effects on bone marrow and the lymphoreticular systems and not as tumourigenic effects on bone.'

1.3.7. *Dysplasia.* Like all forms of radiation, that from strontium will

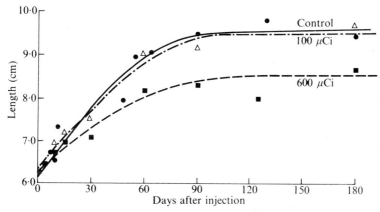

FIG. 8.25. Graph showing the increase in length of the tibia, plotted against time, for control rabbits and those which received 100 μCi kg^{-1} or 600 μCi kg^{-1} of ^{90}Sr. (Each point represents the mean of all the animals at that time-interval, generally about four.)

cause severe bone dysplasia. This is particularly apparent in young animals. ^{90}Sr, by whatever route it reaches the blood stream, is taken up particularly in sites of mineral accretion. In the young this means heavy concentration, as already shown in Fig. 8.2, beneath the epiphyseal plate and on the endosteal surface of the metaphysis and the periosteal surface of the diaphysis.

Figure 8.25 shows the effect on growth in length of the tibia of 600 and 100 μCi kg^{-1} ^{90}Sr injected into weanling rabbits. The cellular lesions underlying this inhibition of growth were analyzed in great detail by Macpherson and her colleagues (Macpherson, Owen, and Vaughan 1962). Figure 8.26 shows the number of nuclei undergoing disintegration in an area of the metaphysis in these rabbits adjacent to the area receiving the highest measured radiation dose; Fig. 8.27 shows the number of cells in mitosis in the same area. All cell types were affected. This cellular damage results in great thickening of the cartilage plate with failure of resorption of cartilage as shown in Fig. 8.28 and ultimate shortening of the limb. The damage to the osteogenic cells is associated with some damage to blood vessels as evidenced by leaking of red cells into the tissues. In the study of the rabbit's metaphyseal region this vascular damage was noted about 3 days after injection, and in the animals receiving the higher dose the whole blood supply of the metaphysis was interfered with so that ultimately the thickened plate became separated as a bar of dead bone. Rowland has described

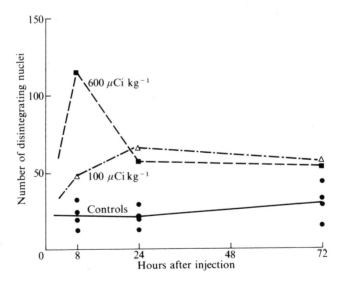

FIG. 8.26. Graph to show the number of nuclei undergoing disintegration in an area of the metaphysis 600 μ square in weanling rabbits injected with Sr^{90} (100 or 600 μCi kg^{-1}) and killed up to 3 days later. (Macpherson *et al.* 1962, by courtesy of the authors and editors.)

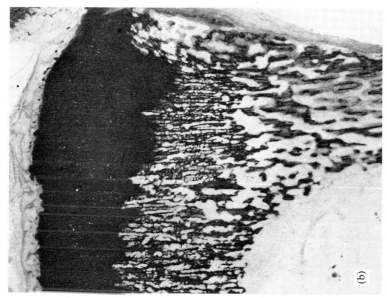

FIG. 8.28. The posterior side of the metaphysis in (a) a control rabbit (wearling + 30 days) and (b) a weanling killed 30 days after an injection of ^{90}Sr 600 µCi kg^{-1}, to show the thickening of the cartilage plate and the excessive number of cartilage remnants in the spongiosa. Azure II, × 175. (from Macpherson et al. 1962, by courtesy of authors and publishers.)

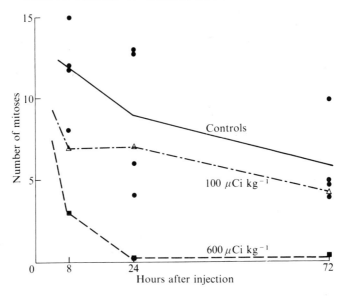

FIG. 8.27. Graph to show the number of cells in mitosis in the same area as in Fig. 8.25. (Macpherson *et al.* 1962, by courtesy of the authors and editors.)

vascular damage in adult dogs which have carried ^{90}Sr for a number of years as evidenced by plugged canals in cortical bone like those seen in long-standing radium poisoning (Rowland 1960*b*). These vascular changes in adult bone have been studied particularly by Jee and his colleagues (Jee, Bartley, Dockum, Yee, and Kenner 1969) who emphasize their importance in causing bone dysplasia which will frequently result in fracture. This vascular damage is associated with death of osteocytes and the presence of empty or plugged lacunae. However, the injury produced by ^{90}Sr is not as great as that produced by the alpha emitters. The observations were made on the bones of dogs in the Utah experiment. The amount of each radionuclide injected at each dose level was so adjusted that the desired retained μCi kg^{-1} was the same for all the radionuclides except ^{90}Sr in which case they were greater by a factor of 10 (Chapter 6.4).

9. Plutonium Isotopes

EARLY experimental work indicated that plutonium is one of the most effective skeletal carcinogens. The extreme toxicity of this radionuclide in minute quantities has meant that experimental work involving the use of plutonium isotopes has been limited to a few laboratories where adequate facilities are available. The most recent metabolic studies suggest that when the body burden is low the liver may be at greater risk than the skeleton through the effect of retained plutonium on the ageing mechanism, though bile duct tumours are also reported (Stover, Atherton, and Buster 1971). Further liver retention is proving rather greater than was originally expected.

There are 16 known isotopes of plutonium (Lederer, Hollander, and Perlman 1967). The majority decay by alpha particle emission; they have half-lives ranging from 0·18 seconds to 76 million years. Only two of these isotopes are at present recognized as of biological importance, ^{239}Pu and ^{238}Pu. It has been suggested that ^{237}Pu might have value in metabolic studies since it has a short half-life, but it differs from ^{239}Pu in other physical characteristics which mean that its biological characteristics are not comparable. There are also differences in the characteristic behaviour of ^{238}Pu and ^{239}Pu which, though less striking, make it unfortunate that earlier workers failed to discriminate between the two isotopes. ^{238}Pu is more soluble in tissue fluids than ^{239}Pu and therefore more readily translocated (Stuart 1970). The physical characteristics of the three isotopes are shown in Table 9.1.

TABLE 9.1

Physical characteristics of ^{239}Pu, ^{238}Pu, and ^{237}Pu

Pu isotope	Half-life	Mass for 1 μCi of α activity
239	24 360 years	16 μg
238	86·4 years	58 ng
237	45·6 days	25 ng

Plutonium 239, the most important of the plutonium isotopes, has a high cross-section for nuclear fission by thermal neutrons, the property which makes it important both as a nuclear fuel and a nuclear explosive. It is now produced in tonnage quantities from uranium 238. This will be increased as fast breeder reactors are brought into use. Since it is one of the most effective carcinogens known to man it probably presents the main hazard facing the modern world (Wick 1967, Cleveland 1970). The present maximum permissible body-burden of ^{239}Pu is 0·04 μCi. This figure may well prove too high (Mays, Taylor, Jee, and Dougherty 1970). Plutonium 238 is used to power both satellites and cardiac pacemakers, and both are possible sources of contamination. At least one such satellite is known to have burnt up and released measurable quantities of fall-out.

The chemistry of plutonium is complex. It may exist in several valence states though it is believed that once it reaches the blood stream it becomes tetravalent (Stover, Atherton, Keller, and Buster 1960; Taylor 1973).

The plutonium ion is readily hydrated giving rise to large polymeric particles. Ions, which undergo such extensive hydrolysis, will also form complexes readily. Plutonium cannot exist as a simple ion at a physiological pH and a colloidal or complexed form will inevitably be presented to the tissues.

Plutonium, reaching the blood stream, is largely retained in the liver or the skeleton. It is excreted extremely slowly. When it is present in monomeric form it is largely retained in the skeleton; when in polymeric form it is retained in the liver though it may then subsequently translocate from liver to skeleton. If present in a highly polymeric form of a narrow range of particle size only 3 per cent is found in the skeleton of mice and 84 per cent in the liver 6 days after intravenous injection but if given in a monomeric form, again of narrow particle size, 84 per cent is immediately found in the skeleton (Lindenbaum, Lund, Smoler, and Rosenthal 1968; Lindenbaum, Rosenthal, Russell, Moretti, and Smyth 1969). Only too little is known about the form in which ^{239}Pu translocates from the tissues to the blood stream. It gains access to the body of man most commonly from wounds or inhalation but may be absorbed from the gastrointestinal tract particularly in very young animals (Vaughan, to be published). Whatever the route of entry appreciable quantities may translocate to the skeleton (Buldakov, Kyubchanski, Moskalev, and Nifatov 1970; Vaughan, to be published).

1. Metabolism

In looking at experimental results in animals it must be remembered that different species handle plutonium differently. For instance in the dog and man what little is excreted is generally lost in the urine while in the rat it is lost in the faeces (Vaughan, to be published). On the other hand the relative loss in urine and faeces even in man is extremely variable both from one individual to another and in the same individual from day to day (Lister,

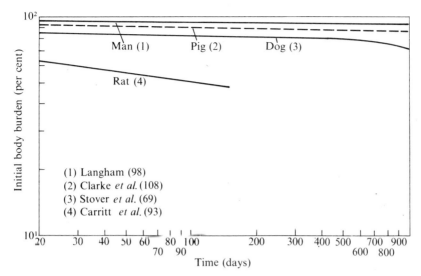

FIG. 9.1. ^{239}Pu retention in four species after i.v. administration. (Stara *et al.* 1971, by courtesy of authors and publishers.)

Morgan, and Sherwood 1963). ^{239}Pu retention in four species after intravenous injection is shown in Fig. 9.1 (Stara, Nelson, Della Rosa, and Bustad 1971).

1.1. *Man*

For obvious reasons the metabolism of plutonium in man is poorly documented. The accepted data on the metabolism of ^{239}Pu in man was, until very recently, that given by Langham and his colleagues (Langham, Bassett, Harris, and Carter 1950; Langham 1959; Langham, Lawrence, McClelland, and Hempelmann 1962). Sixteen patients, the majority of whom were extremely ill, were given an intravenous injection of ^{239}Pu or ^{238}Pu. Three patients received hexavalent ^{239}Pu, one received ^{238}Pu nitrate, and 12 ^{239}Pu citrate. The mean figures for these 16 patients gave the highest retention in the skeleton, 66 per cent, while the liver retained 23 per cent 150 days later. It is open to question how far such average figures are valid when the plutonium was given in such different chemical forms, to extremely ill patients. They have recently been reviewed in great detail by Durbin (1972). Her analysis makes little difference to the figures for total retention but results in a significant difference in organ retention. She concludes that in healthy humans, bone and liver will retain equal amounts of plutonium 15 years after exposure, and that at later intervals the liver may gain further at the expense of the skeleton. This fresh assessment of the old data is dependent on a variety of factors, the most important of

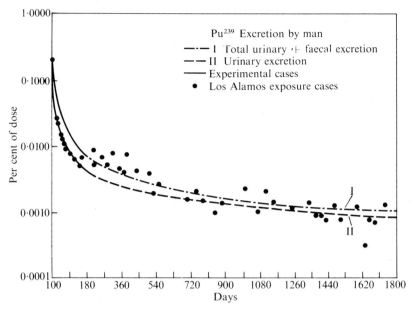

Fig. 9.2. Urinary and urinary and faecal excretion of plutonium by man over a five-year period following intravenous injection of plutonium (Langham 1959, by courtesy of author and publisher.)

which is that allowances were made for the fact that the patients were extremely ill. Abnormalities in the capacity of their plasma proteins to bind plutonium in their circulatory system, in their liver function, and in their dietary intake have been taken into account in making the new calculations. The classical curves for urinary and urinary and faecal excretion of plutonium in man from the Langham data are shown in Fig. 9.2 (Langham 1959). They demonstrate the high level of retention of this radionuclide once it has reached the blood stream.

Mays and his colleagues (Mays, Taylor, Jee, and Dougherty 1970) have also recently stated that limited data suggest to them that roughly half the ^{239}Pu reaching the human circulation from the lung or a puncture wound deposits in bone and the other half in the liver. In two plutonium workers detailed tissue analyses were made at autopsy (Foreman, Moss, and Langham 1960; Lagerquist, Borowski, Hammond, and Hylton 1968, 1969) and are shown in Tables 9.2 and 9.3.

In two additional autopsies of plutonium workers the vertebral concentrations were 5 and 19 per cent respectively of that in the liver. In two normal persons, whose ^{239}Pu was presumably all from fall-out, the vertebral concentration of plutonium averaged about 10 per cent of that in the liver (Magno, Kauffman, and Shleien 1967; Magno, Kauffman, and Grouly 1969).

TABLE 9.2

Case of Lagerquist et al. (1969) tissue concentration and extrapolated burden.
(Two contaminated puncture wounds and several inhalation exposures
during the 9-year prior to death)
(taken from Mays et al. 1970, by courtesy of authors and publishers)

Tissue	Organ mass (g)	Average concentration (dis/min/g)	Total content (dis/min)
Skeleton	10 000	0·13*	1300 (58%)*
Liver	1649	0·32	526 (23%)
Lung	1015	0·21	214 (10%)
Bronchial lymph nodes	3†	1·41	4 (0·2%)†
Kidney	289	0·002	0·6 (0·03%)
Spleen	120	0·004	0·5 (0·02%)
Remaining tissue	69 500	0·003‡	208 (9%)‡
Total	82 576		2253 (100%) (0·001 µCi)

* The skeleton was assumed to contain 7000 g of bone plus 3000 g of marrow, as in standard man. The skeletal samples consisted of pieces of sternum, rib, and vertebrae which probably contained higher concentrations of Pu than the skeleton as a whole. Thus, the skeletal content in this man may have been lower than 58%.
† The mass of the bronchial lymph nodes was visually estimated at 3 g. The 4 dis/min in bronchial lymph nodes averaged over the total 700 g of lymphoid tissue in standard man indicates an average lymphoid concentration of 0·006 dis/min/g, assuming negligible Pu in non-bronchial lymphoid tissue.
‡ The Pu concentration in the remaining tissue was assumed equal to the average of kidney, spleen, and a 32 g soft-tissue sample from the rib area.

The figures in Tables 9.2 and 9.3 were calculated from the original data by Mays and his colleagues (Mays, Taylor, Jee, and Dougherty 1970).

1.2. *Animals*

Much of the available knowledge of the biological behaviour of plutonium comes from experiments on animals when plutonium in various different forms has been injected into the blood stream (Buldakov, Lyubchanskii, Moskalev, and Nifatov 1970; Vaughan, to be published). However plutonium is likely only to reach the blood stream in man by translocation following other routes of entry and only too little is known about rates and methods of translocation.

The only experimental evidence comes from a comparison of the effects of intravenous and intramuscular injection of ^{239}Pu(NO$_3$)$_4$ into rabbits (Taylor 1969, Bleaney and Vaughan 1971). These experiments suggest that initially the plutonium is translocated in monomeric form in the blood

TABLE 9.3

Case of Foreman et al. (1960) tissue concentration and extrapolated burden.
(Intermittent exposure to Pu, primarily by inhalation, during the 12 year prior to death)
(from Mays et al. 1970, by courtesy of authors and publishers)

Tissue	Organ mass (g)	Average concentration (dis/min/g)	Total content (dis/min)
Skeleton	10 000*	1·4*	14 000 (36%)*
Liver	1950	9·9	19 300 (49%)
Lungs	850	4·8	4080 (10%)
Bronchial lymph nodes	10	125†	1250 (3%)‡
Muscle	30 000	0·01	300 (0·8%)
Heart	400	0·06	24 (0·06%)
Spleen	116	0·18	21 (0·05%)
Kidney	270	0·05	14 (0·04%)
Balance‡	26 400	0·01‡	264 (0·7%)‡
Total	70 000		39 253 (100%) (0·018 µCi)

* The skeleton was assumed to contain 7000 g of bone plus 3000 g of marrow, as in standard man. The skeletal samples consisted of pieces of rib, sternum, and vertebrae which probably contained higher concentrations of Pu than the skeleton as a whole. Thus, this man's skeletal content may have been lower than 36%.
† The 1250 dis/min in bronchial lymph nodes averaged over the total 700 g of lymphoid tissue in standard man indicates an average lympoid concentration of 1·8 dis/min/g, assuming negligible Pu in non-bronchial lymphoid tissue.
‡ The Pu concentration in the balance was assumed equal to that in the muscle.

stream but that later a more colloidal form, possibly transported by macrophages or lymphatics, reaches the marrow. All the experimental evidence suggests that plutonium is much more readily absorbed from the gut in young than in old animals of several different species (Vaughan, to be published).

Moskalev has reported death from osteosarcomas in rats following oral ingestion of plutonium citrate (Moskalev, Streltsova, and Buldakov 1969). The amount of plutonium found in the skeleton at autopsy was 2–3 times higher that that in the liver. In the case of ^{239}Pu nitrate at pH 3 applied to the skin Khodyreva noted 77·5 per cent of the absorbed plutonium was in the skeleton and only 7·5 per cent in the liver 14 days after application. Khodyreva suggests that the alpha particles have damaged the skin so allowing the plutonium to reach the tissue fluids (Khodyreva 1965).

Inhalation studies in dogs that survived at least 4·5 years after exposure have indicated that 10–15 per cent of the inhaled ^{239}PuO$_2$ was translocated in some form to the liver and 5 per cent to the skeleton. Ten to eleven

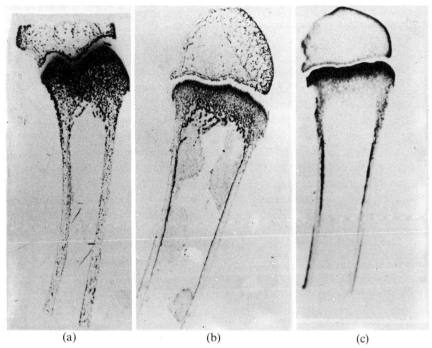

(a) (b) (c)

FIG. 9.3. Comparison of gross autoradiographs through femora of weanling rabbits injected with (a) ^{241}Am, (b) ^{239}Pu, and (c) ^{45}Ca and killed within 24 hours, exposed on Ilford ordinary plates. Note heavy uptake beneath the epiphyseal plate in all. Heavy endosteal uptake throughout the diaphysis in ^{239}Pu only and outlining of all trabeculae with both ^{241}Am and ^{239}Pu. Heavy uptake in marrow in ^{239}Pu only. Uptake on metaphyseal endosteal surface and diaphyseal periosteal surface with ^{45}Ca. Speckled uptake throughout the bone with ^{241}Am. (Williamson 1963, by courtesy of author.)

years post exposure the skeletal content was of the same order (Park, Bair, and Busch 1972). The lower skeletal retention previously described by Morrow and his colleagues (Morrow, Gibb, Davies, Mitola, Wood, Wraight, and Campbell 1967) may be due to the fact that the survival time of their dogs was only about eighteen months. Experimental studies on retention following intramuscular injection of ^{239}Pu(NO$_3$)$_4$ to rabbits, so simulating a wound, indicate that appreciable amounts of ^{239}Pu translocate to the skeleton over a period of time (Taylor 1969, Bleaney and Vaughan 1971).

2. Skeletal distribution

Plutonium is characteristically concentrated on endosteal 'bone surfaces'. Its distribution can be distinguished in an autoradiograph from that of either a volume-seeker like calcium or of another surface-seeker like americium. In Fig. 9.3 are seen autoradiographs of longitudinal sections of the femora of weanling rabbits injected with (a) ^{45}Ca, (b) ^{239}Pu, and (c) ^{241}Am. In the case of ^{241}Am there is concentrated deposition beneath the epiphyseal plate and on all bone surfaces. ^{45}Ca is concentrated also beneath the epiphyseal plate but only on the endosteal surface of the metaphysis and the periosteal surface of the diaphysis. There is also a diffuse deposition throughout mineralized bone, but in the case of ^{239}Pu there is no such diffuse distribution. ^{239}Pu is concentrated again beneath the epiphyseal plate but particularly on endosteal surfaces and in the marrow. This characteristic endosteal distribution is better seen in a cross-section of the femur (Fig. 9.4). In a young growing animal some of the heavy endosteal concentration becomes buried by new bone laid down subsequent to injection. The new bone contains very little plutonium as seen in Fig. 9.5. Little if any plutonium is found on the surfaces of Haversian systems or resorption cavities as seen in Fig. 9.6. and Fig. 9.7. Again

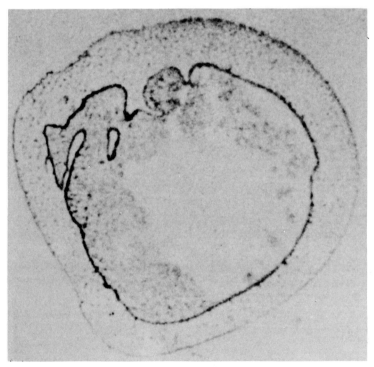

FIG. 9.4. Autoradiograph of cross-section of femur of young rabbit killed 24 hours after an injection of ^{239}Pu citrate (5 μCi kg^{-1}) exposed on Ilford 'Ilfex' for 5 weeks. Note heavy endosteal concentration and uptake in the marrow. (Williamson 1963, by courtesy of author and publisher.)

FIG. 9.5. Detailed autoradiograph from the vertebra of a 0.1 μCi kg^{-1} animal sacrificed 1200 days after ^{239}Pu injection. The dense wavey line represents the deposition at the original endosteal border; the bone below formed after injection contains a diffuse distribution of alpha activity, while the bone deposited before injection is free of activity. (Arnold and Jee 1962, by courtesy of authors and publishers.)

plutonium behaves differently from americium where deposition is particularly heavy on resorbing surfaces (see Fig. 9.8). The characteristic endosteal and marrow distribution has been analysed further by Vaughan and her colleagues (Vaughan, Bleaney, and Williamson 1967; Bleaney 1967, 1969a,b; Bleaney and Vaughan 1971).

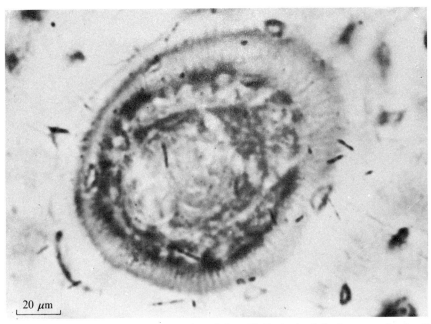

FIG. 9.6. Autoradiograph of a cross-section of the mid-diaphysis of the humerus of a dog 7 months old injected with ^{239}PuCl$_4$ (5 μCi kg^{-1}) and killed 4 hours later (Kodak AR 10, exposure 1 month) showing a few tracks behind the osteoid border and over the contents of an active Haversian system ($\times$ 780). (Herring *et al.* 1962, by courtesy of authors and publishers.)

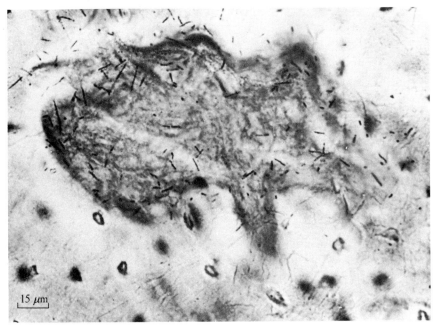

FIG. 9.7. Autoradiograph of a cross-section of the mid-diaphysis of the humerus of a dog 7 months old injected with ^{239}PuCl$_4$ (5 μCi kg^{-1}) and killed 4 hours later (Kodak Ar 10, exposure one month), showing scanty concentration of alpha tracks on irregular resorbing surface of an Haversian system ($\times$ 685). (Herring *et al.* 1962, by courtesy of authors and publishers.)

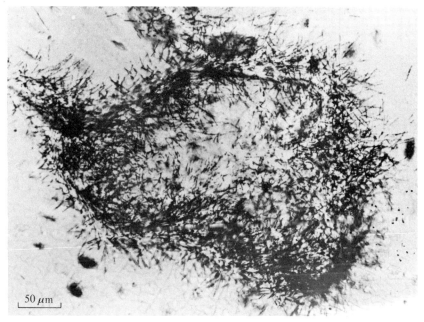

FIG. 9.8. Autoradiograph of a cross-section of the mid-diaphysis of the humerus of a dog 7 months old injected with ^{241}AmCl$_3$ (5 μCi kg^{-1}) and killed 4 hours later (Kodak AR 10 exposure, 5 weeks). Note heavy concentration of alpha tracks on irregular resorbing surface of an Haversian system (× 730). (Herring et al. 1962, by courtesy of authors and publishers.)

2.1. Bone surfaces

The term 'bone surface' is a geographical rather than a biological term. As already discussed (Chapter 2.2.3) a bone surface is made up of the osteogenic cells and the mineral/matrix of the hard bone. If ^{239}Pu(NO$_3$)$_4$ is given to rabbits by intravenous or intramuscular injection autoradiographic studies show that approximately 30–50 per cent of the endosteal plutonium is retained in or on the osteogenic cells. This is illustrated in Table 9.4 where the dose rates at different times have been measured following both intravenous and intramuscular injection. In Fig. 9.9 is seen a longitudinal section of marrow extruded from the mid-diaphysis of the femur with the osteogenic tissue on the left-hand surface. In Fig. 9.10 is an autoradiograph of the same section showing the heavy band of alpha tracks in the osteogenic cells. In Fig. 9.11. is seen the bone mineral/matrix surface devoid of cells. This relative distribution between cells and mineral/matrix interface is known to be maintained for at least 8 days. There is indirect evidence that it is maintained for at least 1 year in adult rabbits (Bleaney 1969a,b; Taylor 1969). The same pattern of distribution is found following both intravenous and intramuscular injection though the total amounts of

TABLE 9.4

The distribution of ^{239}Pu on the endosteal surface of bone of adult rabbits. Pu given as $^{239}Pu(NO_3)_4$ expressed in terms of dose rate $rad^{-1}\ day^{-1}\ \mu Ci^{-1}\ kg^{-1}$ of injected dose (adapted from Bleaney and Vaughan 1971)

Rabbit	Dose ($\mu Ci\ kg^{-1}$)	Dose rate. Rad day^{-1} $\mu Ci^{-1}\ kg^{-1}$ of injected dose at 5 μm from surface			Location of plutonium	P*
		Min.	Max.	Average		
Intravenous injection	1·25	7	32	20	Bone + endosteal cells	
		6	16	11	Bone only	44
		1	31	10	Endosteal cells	
Intravenous injection	1·25	12	42	21	Bone + endosteal cells	
		6	25	15	Bone only	30
					Endosteal cells	
Intramuscular injection	2·5	2	15	7	Bone + endosteal cells	
		1	7	4	Bone only	50
		0	1	1	Endosteal cells	
Intramuscular injection	2·5	3	14	7	Bone + endosteal cells	
		1	10	4	Bone only	42
		1	14	5	Endosteal cells	

* P is the percentage of the total plutonium on 'the endosteal surface of bone' which is in endosteal cells. This is calculated as $100 \times (1 - D_a/D_b)$ where D_a is the average dose rate from Bone only, and D_b is the average dose rate from (Bone + endosteal cells) given in the previous column.

plutonium, as might be expected, are different and vary differently with time. In these experiments the initial dose of ^{239}Pu given was 1·25 $\mu Ci\ kg^{-1}$ intravenously and 2·5 $\mu Ci\ kg^{-1}$ intramuscularly. In the case of the intramuscular injection the build up of ^{239}Pu in the skeleton was slow but at a maximum was as high as 26 per cent of the injected dose. This is perhaps important as slow translocation from lung or wound is the most likely way that ^{239}Pu will reach the human skeleton. A direct intravenous injection is a most unlikely event. Jee and his colleagues (Jee, Park, and Burggraaf 1969) have studied the 'residence time' of ^{239}Pu in the trabecular bones of beagles 1·5 months old* given single intravenous injections of ^{239}Pu citrate (0·3 and 0·015 $\mu Ci\ kg^{-1}$). They found that after 3 months most of

---

* The rib epiphysis in these dogs had not closed, so it is likely that the skeleton was less mature than in the 7-month old rabbits used by Bleaney, where all epiphyses examined are closed.

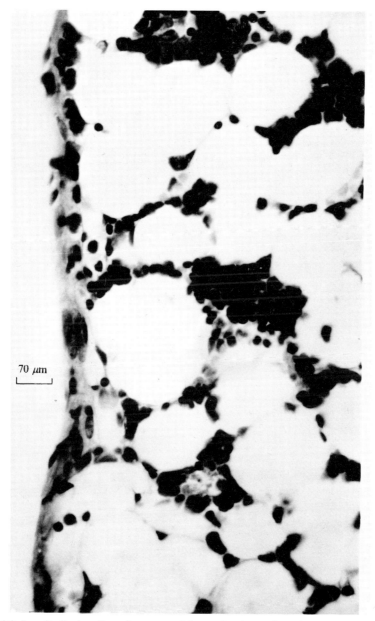

FIG. 9.9. Longitudinal section of marrow of femur showing endosteal mesenchyme cells on left side (× 128). (Bleaney and Vaughan 1971, by courtesy of authors and publishers.)

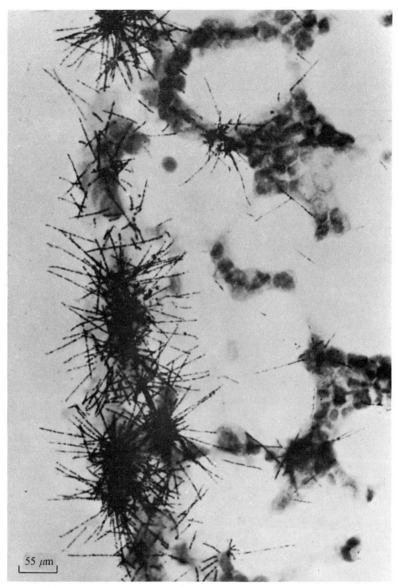

Fig. 9.10. Autoradiograph left on section of marrow of femur showing endosteal mesenchyme cells on left side containing ^{239}Pu. Rabbit injected (intramuscularly) with 2·62 µCi kg^{-1} ^{239}Pu(NO$_3$)$_4$, killed 4 weeks later. Long exposure, × 160. (Bleaney and Vaughan 1971, by courtesy of authors and publishers.)

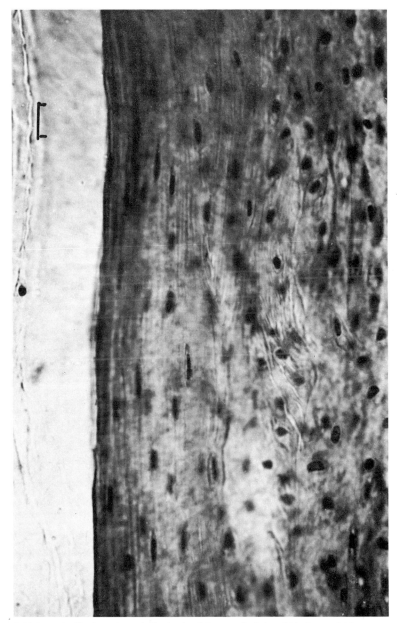

FIG. 9.11. Longitudinal section of diaphysis of femur of adult rabbit from which marrow has been extruded. The surface is devoid of endosteal cells. The marker line is 20 μ. (Bleaney and Vaughan 1971, by courtesy of authors and publishers.)

the plutonium originally located upon trabecular bone surfaces had relocated, much of it being buried within the bone as shown in Fig. 9.5. They associate this with the known rate of remodelling of trabecular bone in dogs of approximately the age group studied (Lee, Marshall, and Sissons 1965). In older dogs the rate is much slower (Amprino and Marotti 1964). Jee suggests that the long surface retention noted by Bleaney is possibly due to damage to the osteogenic cells by the high dose given. This may be true but long-term surface retention was still found following what amounts to a lower dose, reaching the blood stream from an intramuscular injection (Table 9.4). Unfortunately the rate of remodelling of adult rabbit trabecular bone is not known so any further comparison between the results of Bleaney and her colleagues and those of Jee is not at present possible. It can only be noted that even with the short surface residence-time described by Jee, doses of 0.015 μCi kg^{-1} were productive of osteogenic sarcoma in dogs. Further experiments are urgently needed to determine whether this relative distribution between cells and mineral/matrix occurs with other forms of plutonium reaching the skeleton either by inhalation or wounding and in other species and also at lower dose-levels. As far as possible fully mature animals should be used so that their trabecular bone remodelling time might be expected to approach that of adult man. In this connection it should be noted again that the measurements both of Sissons, Holley, and Heighway (1967) and Jowsey, Kelly, Riggs, Bianco, Scholz, and Gershon-Cohen (1965) indicate that in the trabecular bone of the ilium in man only 6–13 per cent of the bone surface is resorbing at any point in time. This would suggest that the residence time of plutonium on adult human trabecular bone surfaces might be considerably longer than in young dogs.* In very young animals plutonium may characteristically also be found in osteoclasts remodelling bone (Arnold and Jee 1962, Vaughan 1970a) as shown in Fig. 9.12. This active resorption is less easily demonstrated in adult bone.

2.2. *Marrow*

Plutonium is present in the marrow in a diffuse pattern, demonstrable as single scattered alpha tracks and also in aggregates. The relative proportion of diffuse and aggregated plutonium varies both with time after both intra-

---

* The points raised in this section are of considerable importance in assessing hazards from plutonium deposition. It must be recognized in considering the scanty facts available that it is difficult for technical reasons to make a really meaningful comparison of the data of Jee and Bleaney. The thickness of the sections and the injection doses were different, and Bleaney also used solid-state detection techniques which were equivalent to an exposure time for an autoradiograph of about six times the exposure time used by Jee. Regions of low plutonium deposition that show no tracks in Jee's autoradiographs might show tracks with longer exposure.

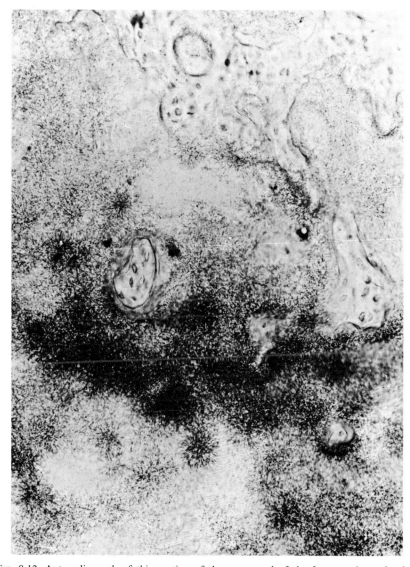

FIG. 9.12. Autoradiograph of thin section of the upper end of the femur at lower level of metaphyseal trabeculae of a weanling rabbit killed 8 days after an intravenous injection of 5 μCi kg^{-1} of Pu(NO$_3$)$_4$. AR 10, exposed 6 weeks. Note heavy concentration of Pu in the marrow at the end of the resorbing trabeculae, some of it probably in osteoclasts. (Vaughan 1970a.)

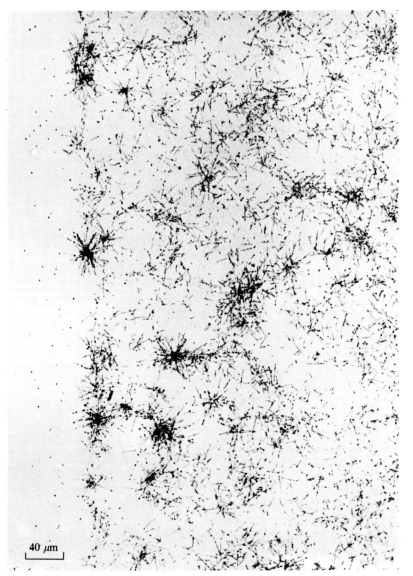

Fig. 9.13. Autoradiograph of marrow of rabbit 6–8 weeks old, killed 24 hours after injection of 1·25 μCi kg^{-1} ^{239}Pu. (AR 10, exposed 20 weeks, ×250.)

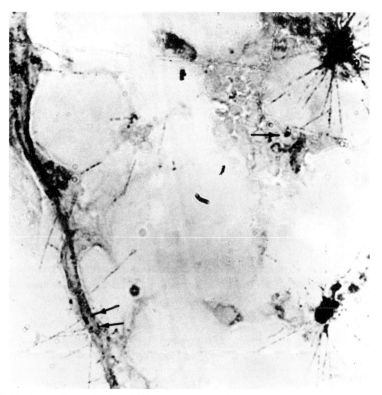

FIG. 9.14. Autoradiograph of rabbit bone marrow stained with Perl's reagent for iron prior to stripping film AR 10 application. No counterstain. The rabbit was injected intravenously with 1·15 μCi of ^{239}Pu(NO$_3$)$_4$ pH 4 and killed 16 weeks after injection. Note haemosiderin containing cells with ^{239}Pu aggregates. (Beno 1968, by courtesy of author and publisher.)

venous and intramuscular injection and also with the character of the solution injected (Vaughan, Bleaney, and Williamson 1967; Rosenthal, Marshall, and Lindenbaum 1968; Rosenthal and Lindenbaum 1969; Bleaney and Vaughan 1971). A typical distribution shortly after the intravenous injection of ^{239}Pu(NO$_3$)$_4$ is shown in Fig. 9.13. The aggregates are found in macrophages associated with haemosiderin (Arnold 1954, Beno 1968). This is illustrated in Fig. 9.14.

The relative amounts retained on the 'bone surface' and in the marrow are dependent on the physicochemical character of the plutonium in the blood stream. Rosenthal and her colleagues have shown that in mice monomeric plutonium is concentrated largely on the 'bone surface' while polymeric plutonium is largely in the marrow (Lindenbaum, Lund, Smoler, and Rosenthal 1968; Rosenthal, Marshall, and Lindenbaum 1968; Rosenthal and Lindenbaum 1969). What the distribution between bone and marrow will be when plutonium reaches the skeleton after inhalation or wounding is still obscure. The only available data comes from some experimental observations made by Taylor (1969) and Bleaney and Vaughan (1971), when ^{239}Pu(NO$_3$)$_4$ was injected intramuscularly into rabbits (Taylor 1969, Bleaney and Vaughan 1971). In these experiments the pattern of distribution changed with time in such a way that it suggested that initially plutonium was translocated largely in monomeric form but that later a more colloidal form was reaching bone marrow since the number of aggregates increased with time. The marrow dose rate at different times after injection is shown in Table 9.5 for two different experiments.

2.3. Binding of plutonium

2.3.1. Bone.
In discussing the constituents of bone matrix (Chapter 2. 2.3.2) the point was made that plutonium is bound strongly by certain glycoproteins that have been isolated from ox cortical bone. It is tempting to suggest that such binding accounts for the concentration of the radionuclide on the mineral/matrix of endosteal surfaces (Chipperfield and Taylor 1970, Bleaney and Vaughan 1971). It has not been possible to demonstrate the

TABLE 9.5

Diffuse dose rate in rad day^{-1} to marrow following intramuscular injection of ^{239}Pu into 7-month old rabbits. Dose rate is per µCi kg^{-1} of injected dose (from Bleaney and Vaughan 1971, by courtesy of authors and publishers)

Rabbit	Time after injection in days	First series		Second series	
		% of dose at injection site	Dose rate rad day^{-1} for 1 µCi kg^{-1}	% of dose at injection site	Dose rate rad day^{-1} for 1 µCi kg^{-1}
H 332 A	1			98	0·14–0·06
H 332 B	8			76	0·40–0·25
1688	8	103·7	0·08–0·04		
1673	28	105·2	0·12–0·03		
H 332 C	35			75	0·32–0·18
1669	49	59·4	0·12–0·03		
H 332 D	56			52·4	0·48–0·14
1671*	63	—	0·51–0·23		
H 332 E	112			47	0·70–0·33
1670	112	61·9	1·00–0·53		
1672	280	33·3	1·04–0·68		
H 332 F	288			29·3	0·20–0·10
H 332 G	365			34·8	0·12–0·10
1674	365	67	0·46–0·12		
H 332 H	561			50·9	0·12–0·05

* Killed with lung infection.
The % of the injected dose at the injection site at death is from Taylor (1969).

presence of glycoproteins on the endosteal mineral-matrix interface though it is known that there is a sialoprotein present in hypertrophic cartilage cells at the epiphyseal plate where plutonium is also concentrated (Williamson and Vaughan 1967). Mesenchymal cells are also known to be rich in glycoproteins though they have not been characterized in detail (Cooke 1968). These may account for the presence of plutonium in the osteogenic cells. Binding to collagen as Table 2.3. shows is not significant. There is probably some plutonium associated with the crystals of calcium hydroxyapatite but the importance of this is still not clear. Chipperfield and Taylor have shown thst if plutonium is mixed with bone mineral and bone sialoprotein, or chondroitin sulphate protein complexes, deposition of plutonium into bone mineral is inhibited (Chipperfield and Taylor, personal communication, 1971).

2.3.2. *Marrow.* There is no explanation for the diffuse distribution of plutonium in the marrow. The aggregates are always found in macrophages in association with haemosiderin (Vaughan, Bleaney, and Williamson 1967; Beno 1968). Whether the plutonium is taken up by the macrophages as part of their usual scavenging activities or whether it is bound to some part of the haemosiderin complex is not at present clear. The close association of plutonium with iron and iron-carrying complexes elsewhere in the body makes the latter a likely hypothesis (Boocock and Popplewell 1965; Stevens, Bruenger, and Stover 1968; Turner and Taylor 1968; Taylor 1969; Boocock, Dampure, Popplewell, and Taylor 1970; Stover, Bruenger, and Stevens 1970; Stover, Atherton, and Buster 1971).

3. Effects on man

Nothing is known of the clinical effects of plutonium in man. There is no evidence that it has yet produced cancer (Langham, Lawrence, McClelland, and Hempelmann 1962; Newton, Heid, Larson, and Nelson 1967; Langham 1970). Langham's most recent report (Langham 1970) contains individual data on each of 37 plutonium workers exposed during the original Manhattan project or in subsequent AEC operations. Their estimated systemic body-burdens range from 1 to 13·5 times and averaged 3 times the permissible 0·04 μCi level of ^{239}Pu. Periods after first exposure ranged from 4 to 24 years and averaged 14 years. No tumours attributable to ^{239}Pu have been reported in this group but the period is still relatively short for a tumour like osteosarcoma with a long latent period.

There is however extensive experimental data on mice, rats, and dogs. In view of its distribution pattern already described it is not surprising to find its skeletal toxicity is high since it is concentrated particularly in or adjacent to the sensitive osteogenic cells themselves.

4. Effects on animals

4.1. *Neoplasia*

Plutonium is known to induce tumours of osteogenic tissue, of haemopoietic marrow, and of epithelium closely applied to bone.

4.1.1. *Osteogenic tissue.* With the exception of ^{228}Th, another surface-seeker, ^{239}Pu is the most carcinogenic of the bone-seeking radionuclides, producing an extremely high incidence of osteogenic sarcomas at a relatively low body-burden (Brues 1949; Finkel 1953; 1956, 1959a,b; Finkel and Biskis 1962, 1968; Dougherty and Mays 1969; Moskalev, Streltsova, and Buldakov 1969; Mays, Taylor, Jee, and Dougherty 1970). This high carcinogenicity is presumably associated with the fact that plutonium is concentrated both in the sensitive endosteal cells themselves and on the mineral/matrix surface for at least one year after both intravenous and intramuscular injection (Bleaney and Vaughan 1971), as already described in Section 2.1.

4.1.1.1. *Site.* The majority of plutonium induced tumours arise in spongy bone particularly in the spine, which is not surprising considering the location of the radionuclide on endosteal surfaces of trabecular bone. The distribution of the Utah beagle tumours associated with a plutonium burden is shown in Fig. 9.15. (Jee, Stover, Taylor, and Christensen 1962).

4.1.1.2. *Histopathology.* The histopathology of the osteogenic sarcomas occurring in the Utah beagles has been described by Jee and his colleagues (Jee, Stover, Taylor, and Christensen 1962). The tumours varied over a wide range of sizes from 9×10 cm to microscopic lesions of around 400 μ. The character of the tumours was also extremely variable both within each tumour itself and from lesion to lesion. Some were largely osteoblastic with much bone formation and others were more fibroblastic. A few showed cartilage formation. Only one typical fibroblastic sarcoma, arising in a vertebra, was recorded. Jee emphasized that all the tumours seem to arise from osteogenic tissue on bone surfaces and confirms the observation of other workers that gross tissue damage is not a necessary prelude to bone tumour production.

Moskalev and his colleagues (Moskalev, Streltsova, and Buldakov 1969) consider that, on the whole, ^{239}Pu causes more osteolytic than osteoblastic tumours in rats, though they state that other Russian workers find the reverse (Buchtoyarova and Lemberg 1969). They suggest that the difference may be due to the fact that the latter used ^{239}Pu(NO$_3$)$_4$ while they themselves had used ^{239}Pu citrate. This description of plutonium-induced bone tumours does not differ from that for other radionuclides causing bone tumours except that tumours arising from marrow elements, such as angiosarcoma and reticuloendothelial sarcoma, are not mentioned.

4.1.2. *Haemopoietic marrow.* In view of the deposition of ^{239}Pu in marrow,

PLUTONIUM ISOTOPES

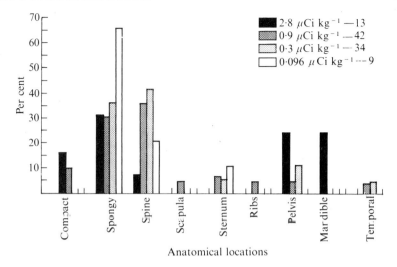

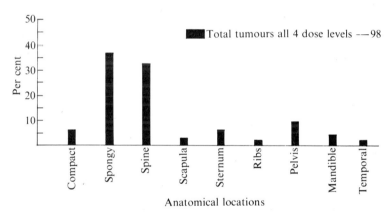

FIG. 9.15. Anatomic distribution of osteogenic sarcoma following intravenous injection of ^{239}Pu. (Jee *et al.* 1962, by courtesy of authors and publishers.)

malignant blood dyscrasia might be anticipated. Only 2 cases have occurred in the Utah beagles, against 52 dogs with osteosarcomas. Bensted and his colleagues (Bensted, Taylor, and Sowby 1965), in a group of 26 rats which survived 11 or more months following an intravenous injection of ^{239}Pu(NO$_3$)$_4$ given in divided doses, recorded 3 cases of myeloid leukaemia occurring 46–54 weeks after the last injection. The strain of rat had previously been free of leukaemia so it was attributed to the radionuclide. Moskalev also notes a relatively high incidence of tumours of the blood-forming organs in rats compared with osteosarcoma (Moskalev, Streltsova, and Buldakov 1969) following injections of ^{239}Pu citrate. They are described as 'leukoses of the reticulosis type; haemocytoblastoses with extensive leukaemia infiltrates in the bone marrow, liver, spleen, adrenals, kidneys, lymphatic nodes, and lungs whereas pulmonary lymphosarcomas predominated in the controls'. He emphasizes that these neoplasms of the haemopoietic tissues are of great importance from the point of view of radiation hazards.

Finkel and Biskis (1962) tabulate a large number of reticular endothelial tumours in mice following ^{239}Pu citrate injection but an almost equally large number were found in their controls and the CF 1/Anl mouse used in these experiments is known to develop lymphoma (Lisco, Rosenthal, and Vaughan 1971).

4.1.3. *Epithelium closely applied to bone.* In mice Finkel and Biskis (1962) record one nasal epidermoid carcinoma. In the Utah beagles one sinus carcinoma is noted. This occurred 2093 days after injection of 0·0996 μCi kg^{-1}. The skeletal dose was calculated to be 210 rad. A photomicrograph of the wall of the frontal sinus of an adult beagle is shown in Fig. 9.16. It is clear that the basal layer of the mucosa applied to bone is well within the range of a typical alpha particle (Taylor, Dougherty, Shabestari, and Dougherty 1969). Experience in the case of both ^{226}Ra and ^{90}Sr has

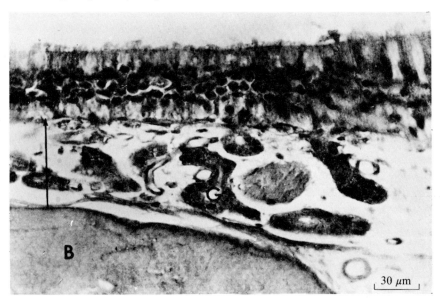

FIG. 9.16. Photomicrograph of the wall of the frontal sinus of an adult beagle showing the proximity of the epithelial components of the mucosa to the underlying bone (B) and tubular glands (G)—a typical alpha range. H and E, × 474. (Taylor *et al.* 1969, by courtesy of authors and publishers.)

shown that the latent period for the development of sinus and mastoid carcinomas is long and that on the whole they tend to develop in animals with a relatively low body-burden so it is possible that more such tumours may develop in the lower-dose beagle dogs.

4.2. *Dysplasia*

Bone. Apart from inducing osteogenic sarcomas, plutonium will cause great damage to bone as was first described by Heller (1948). The effects are listed as (i) generalized osteoporosis, (ii) increased sclerosis along epiphyseal growth lines, (iii) bone necrosis, and (iv) marked thinning of the cortex of the long bones (Langham and Carter 1951). Multiple fractures have been a marked finding in the Utah beagles. The incidence was highest in the ribs and the processes of the vertebrae. They caused little pain and healed readily. A radiograph of a rib cage showing many healed fractures is shown in Fig. 9.17. Between 1 and 3 μCi kg^{-1} of ^{239}Pu citrate gave the highest incidence of fractures; none were seen with less than 0·3 μCi kg^{-1} (Jee, Stover, Taylor, and Christensen 1962). In pigs, the younger the animal the more severe the dysplasia (Bustad, Clarke, George, Horstman,

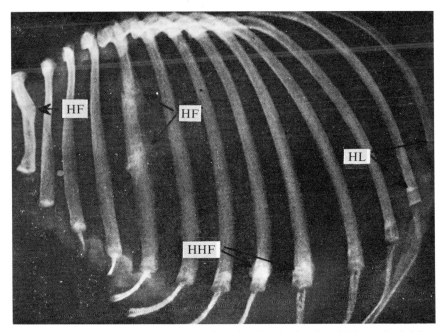

FIG. 9.17. Radiograph taken 1192 days post-injection showing hot lines (HL), two healed 'hot line' fractures (HHF), and three other healed fractures (HF). The dog received 2·69 μCi kg^{-1} ^{239}Pu by a single intravenous injection. (Taylor *et al.* 1962, by courtesy of authors and publishers.)

McClellan, Persing, Seigneur, and Terry 1962). The effect on growth has been carefully analysed by Fabrikant and Smith (1964). They compared the effect on growth in rats of ^{239}Pu, ^{241}Am, and ^{32}P and found ^{239}Pu effective in smaller doses than the other two radionuclides. Growth was considerably reduced both in rate and in absolute amount. Clarke, who has studied the histopathology, attributes the more deleterious effects of plutonium to the extreme degree of fibrosis of the medullary blood vessels which results in bone necrosis. He found the damage done by plutonium greater than that caused by radium or strontium (Clarke 1962).

Marrow. High doses of plutonium will cause severe aplasia of marrow which in view of the marrow deposition of the radionuclide is not surprising (Langham and Carter 1951). Dougherty has also shown, in the Utah beagles, that ^{239}Pu is approximately 4–8 times as effective as ^{226}Ra in depressing the total white-cell count, particularly that of granulocytes in the peripheral blood (Dougherty and Rosenblatt 1969). Clarke found the same relative effects of ^{226}Ra and ^{239}Pu in miniature swine (Clarke 1962).

5. Dosimetric considerations

The result of the Utah experiment on dogs (already discussed in Chapter 6.4, Table 6.4) indicates the extreme toxicity of ^{239}Pu. Osteosarcomas occurred in the dogs at an injection dose as low as 0·0157 μCi kg^{-1} and an average skeletal dose of 75 rad. It is of some interest that Moskalev and his colleagues report an average radiation dose of 57 rad in rats fed ^{239}Pu citrate that died with osteosarcomas (Moskalev, Streltsova, and Buldakov 1969). The maximum permissible skeletal burden for man is at present as low as 0·04 μCi. Mays and his colleagues (1970) have recently suggested that it may prove necessary to reduce this figure even more, basing this proposition largely on their results with the Utah beagles and a few figures for retention in human subjects which suggest approximately equal retention in liver and skeleton.

Liver tumours occurred in the Utah beagles only at low dose-levels indicating that the dogs at the higher levels were 'protected' against liver tumours by premature death from radiation-induced bone cancer. It is therefore feared that human individuals with low burdens who escape osteosarcomas may later die with bile duct carcinomas (Mays, Taylor, Jee, and Dougherty 1970; International Commission on Radiological Protection 1972).

Mays and his colleagues (Mays, Taylor, Jee, and Dougherty 1970) have recently produced some speculative figures for the risk to bone and liver from ^{239}Pu on the basis that long-term retention is roughly the same in bone and liver. For a constant 0·02 μCi ^{239}Pu in bone and 0·02 μCi ^{239}Pu in liver the 50-year doses to man are 14 rad and 57 rad averaged over these respective organs. Probabilities of radiation induced tumours from these doses they calculate, whatever the model used, will be higher in the liver than in the skeleton.

The only measurements of actual radiation doses received by sensitive osteogenic cells and marrow cells are those of Bleaney (1967, 1969a,b) and Bleaney and Vaughan (1971). These were obtained on rabbits given either intravenous or intramuscular injections of plutonium nitrate. Some representative figures for the dose to cells on the surface have already been given in Table 9.4, but Table 9.5 shows how the marrow dose may vary. In the latter case two separate experiments are shown which indicate that results appear to be variable. The rabbits used were of the same stock and the solutions used were comparable, but the amount of plutonium translocating to bone from the site of the intramuscular injection differed in the two groups. There are also of course great approximations used in expressing a radiation dose from alpha particles in rad. The results in both tables can only be taken as an indication of the relative dosage pattern to bone and marrow under the particular experimental conditions in the rabbit.

Apart from measurements and calculations of radiation dose, based on animal experiments, it must be emphasized that all the data available on trabecular bone turnover in adult man suggests that it is much slower than it is in young adult dogs. Any plutonium, therefore, reaching the bone surfaces and the marrow by translocation from sites elsewhere in the body of man is likely to remain in contact with cells capable of malignant transformation. Calculations based on animal data are likely to minimize the hazard. It can only be concluded that osteosarcoma induction in man presents a serious hazard at extremely low body-burdens. This is not the place to discuss the question of removal of plutonium. It has been reviewed elsewhere (Vaughan, to be published). It must however be said that there is probably at present nothing that will remove plutonium once it is fixed in the skeleton. It can however be chelated in the blood stream. It is therefore essential to administer chelating agents, DDTA or TTPA, immediately contamination is known to have occurred. Further it is clearly of extreme importance to obtain as much data as possible from persons exposed to plutonium during their lifetime and from detailed autopsy studies at death. The Hanford Plutonium registry in the future should give us as much important scientific information as the follow-up of the radium cases. Any patient in whom osteosarcoma is diagnosed should be questioned about possible exposure in the past to radionuclides.

10. Americium

AMERICIUM 241, one of the actinide elements, is used in industry and therefore presents an industrial hazard. Its metabolic behaviour as a bone-seeker has been studied particularly in relation to that of plutonium, its immediate chemical neighbour in the periodic table. All the evidence suggests that there are considerable differences between tetravalent plutonium (the valency in which plutonium is usually present in the blood stream) and the trivalent actinides, in their reactions with biological systems.

1. Metabolism in relation to the skeleton

The plasma clearance of americium is not very different from the plasma

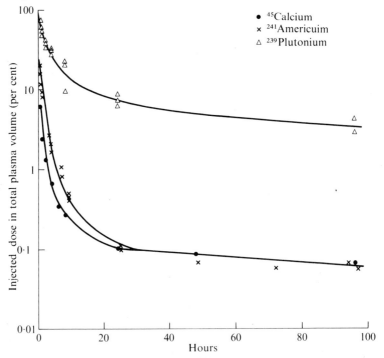

FIG. 10.1. Concentration of ^{45}Ca, ^{239}Pu, and ^{241}Am in the plasma of the rat during the first 4 days after intravenous injection of the radionuclide in 0.02 HNO$_3$ (Taylor 1962, by courtesy of author and publisher.)

clearance of calcium and is therefore much more rapid than that of plutonium. It is shown in the rat in Fig. 10.1. It is not greatly influenced by the chemical form of the injected material (Turner and Taylor 1968) which is extremely important in the case of plutonium. A comparison of the uptake of ^{241}Am and ^{239}Pu in the skeleton of rats indicates that the initial uptake of ^{239}Pu in the skeleton was about 1·4 times greater than that of ^{241}Am while the rate of loss from the skeleton was almost the same for the two

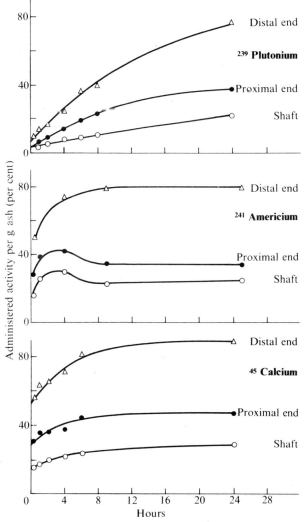

FIG. 10.2. Concentration of ^{45}Ca, ^{239}Pu, and ^{241}Am in the proximal, shaft, and distal regions of the rat femur during the first 24 hours after intravenous administration of the radionuclide in 0·02 M HNO$_3$ (Taylor 1962, by courtesy of author and publisher.)

nuclides (Taylor, Sowby, and Kember 1961). The pattern of uptake for ^{45}Ca, ^{239}Pu, and ^{241}Am is illustrated in Fig. 10.2. In the rat about 40 per cent is retained in the skeleton (Bensted, Taylor, and Sowby 1965). There is no data for the metabolic picture of ^{241}Am in man. In the dog initial retention in the skeleton following an injection of 0·002 to 0·3 μCi kg^{-1} ^{241}Am citrate is calculated to be 30 per cent of the injected dose while 50 per cent is in the liver. Four years later the figures are 23 per cent and 38 per cent respectively. Retention is to some extent dose dependent; after about 100 days higher doses are followed by a sharp decrease in liver americium and an increase in skeletal americium (Lloyd, Mays, Taylor, and Atherton 1970). The recognized high uptake in the thyroid gland in dogs is at present not understood. There is a preliminary report of short term metabolic experiments in the adult baboon. The animals were sacrificed one and three months post injection. At the end of 1 month the liver retained more radionuclide than other organs with decreasing concentrations in the order of bone, lung, aorta, and kidney. Among various

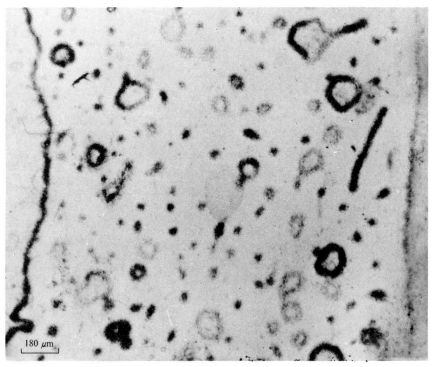

FIG. 10.3. Autoradiograph of a cross-section of cortical bone of a young dog killed 4 hours after injection of 5 μCi kg^{-1}AmCl$_3$. Note heavy periosteal uptake (on left) compared with endosteal uptake on right, also heavy irregular uptake round some Haversian canals and not others. ($\times$ 60, Ilford K, exposure 4 weeks.)

bones the concentration of ^{241}Am varied within a factor of 4, that in the vertebrae being the highest and that in the small long bones the lowest. The skull contained 19 per cent of the total activity and was 21 per cent of the skeletal wet weight (Rosen, Cohen, and Wrenn 1972).

2. Skeletal distribution

Americium is a surface seeker but its pattern of surface distribution differs from that of plutonium as shown in Fig. 9.2. There is as much americium, if not more, on periosteal surfaces as on endosteal surfaces (Durbin 1962). There is a heavy and early deposition in the young animal beneath the epiphyseal plate and at all ages there is heavy deposition on resorbing surfaces. In Fig. 10.3 is shown an autoradiograph of a cross section of dog cortical bone at a low manification following an injection of ^{241}Am. This demonstrates heavy but uneven concentration round some Haversian canals or resorption cavities. A high-power view of a resorption cavity shown in Fig. 9.7 shows how heavy this concentration is in the case of americium compared with that (illustrated in Fig. 9.6) of a resorption cavity in the cortical bone of a dog injected with plutonium.

Differences in the binding of plutonium and americium by the glycoproteins of bone (already discussed in Chapter 9) provide the probable explanation for these differences in distribution, though at present there is no information as to whether the different glycoproteins are found on bone surfaces. Americium is not found in the marrow. It has been reported in high concentration in the dental pulp of mice (Hammarstrom and Nilsson 1970a). It is of interest that this area stains strongly with periodic acid Schiff stain suggesting the presence of glycoproteins (Hammarstrom and Nilsson 1970b). Dysplasia of the jaw is often severe (Taylor and Bensted 1969).

3. Clinical effects

Comparative studies on the carcinogenic action of plutonium and americium have been made by Bensted and his colleagues (Bensted, Taylor, and Sowby 1965) using rats. About 80 per cent of the rats injected with ^{239}Pu which survived a year or more after injection developed osteogenic sarcomas whereas only 20 per cent of those given ^{241}Am developed bone tumours. Americium was given as a single injection of 2·5 μCi kg^{-1} and plutonium either as a single dose of 2·95 μCi kg^{-1} or 3 μCi kg^{-1} given in divided doses. Only 1 rat receiving ^{241}Am developed leukaemia while three ^{239}Pu rats died with a picture of myeloid leukaemia. An earlier report when much higher doses of both plutonium and americium were administered to rats gave a rather different result (Langham and Carter 1951). A much lower tumour incidence was found, presumably because the death of the relevant cells at risk from a high radiation dose reduced the potential tumour population. Langham and Carter (1951) used doses varying from 32–63 μCi kg^{-1} while Bensted, Taylor, and Sowby (1965) employed a dose of 2·5 µCi kg^{-1} ^{241}Am or 2·95 µCi kg^{-1} ^{239}Pu as a single injection or a total of 3 μCi kg^{-1} ^{239}Pu given in divided doses.

Since both ^{241}Am and ^{239}Pu are alpha emitters it appears probable that the difference in their carcinogenicity is explained by their different distribution on 'bone surfaces' and in the marrow.

^{239}Pu is largely concentrated on endosteal surfaces which have a much greater area than periosteal surfaces while ^{241}Am is more heavily concentrated on periosteal surfaces than ^{239}Pu and does not appear in marrow where plutonium is found characteristically both diffusely and in scattered macrophages.

Americium has recently been added to the radionuclides administered intravenously to beagles in the Utah experiments. No results on tumour incidence in bone are yet available.

11. Thorium Isotopes

The decay series of the radioisotopes in the thorium series has been shown in Table 6.2. The behaviour of one of the daughter products ^{224}Ra (or thorium X) has already been discussed (Chapter 7.4). Other isotopes in the series which have some interest as bone-seeking radionuclides are ^{228}Ra (or mesothorium) and ^{228}Th (or radiothorium).

1. ^{232}Thorium

^{232}Th has become of considerable biological interest and importance since colloidal thorium dioxide was introduced as a contrast medium in diagnostic radiology in 1928 and was subsequently used widely under the trade name of Thorotrast. Thorotrast is a colloidal thorium dioxide preparation containing about 0·20 to 0·25 g (or 22–28 nCi) of ^{232}Th per ml (Dudley 1967).

1.1. *Distribution*

Following injection, thorium dioxide itself may appear on bone surfaces (Jee, Dockum, Mical, Arnold, and Looney 1967). It also has important bone-seeking daughter products: ^{224}Ra, ^{228}Ra (mesothorium), and ^{228}Th. Characteristically the injected Thorotrast tends to become encapsulated in fibrous tissue, particularly in the liver. Some of the radioactive daughter products may however escape from such encapsulated lesions into the blood stream and so reach both bone and marrow (Marinelli and Lucas 1962, Marinelli 1964, Dudley 1967).

1.1.1. *Bone*. Jee and his colleagues describe Thorotrast as present in macrophages on bone surfaces, but the appearance of their autoradiographs and knowledge gained from a study of ^{239}Pu distribution already discussed (Chapter 9.2.1.9.2) suggests that, like plutonium, thorium may well be present in osteogenic cells as well as in macrophages and possibly also be associated with the glycoproteins of the bone matrix, since thorium was shown to be strongly bound by the carboxyl groups in bone sialoprotein by Peacocke and Williams (1966). Like plutonium, it may well prove to be bound even more strongly to other more recently isolated glycoproteins (Chipperfield and Taylor 1970). In Fig. 11.1 is seen an autoradio-

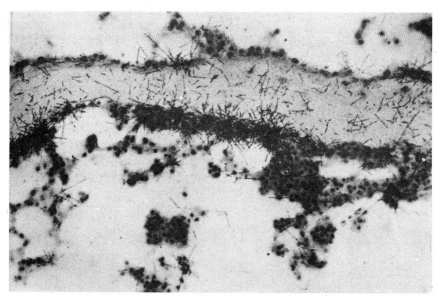

Fig. 11.1. A detailed autoradiograph of a portion of trabecular bone from patient AD with numerous macrophage thorium dioxide aggregates lining the ventral surface and the corresponding increase in alpha tracks in adjacent bone. Note also aggregates and alpha tracks in the marrow. (× 163) (Jee et al. 1967, by courtesy of authors and publishers.)

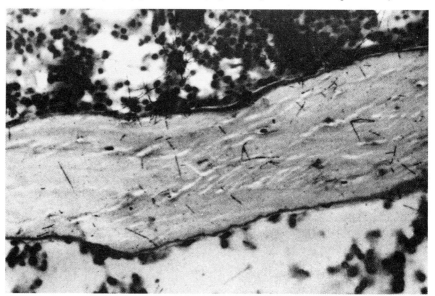

Fig. 11.2. A detailed autoradiograph of a trabecula from patient AD showing the absence of macrophages lining its bone surface and few alpha tracks on bone surface and within the bone. Note aggregates and alpha tracks in the marrow. (Jee et al. 1967, by courtesy of authors and publishers.)

graph from the bone of a woman who had received Thorotrast 19 years previously. It shows a trabecula with a heavy concentration of alpha tracks on the bone surface and diffuse alpha tracks throughout the bone. Jee and his colleagues (1967) considers the heavy concentration of alpha tracks is due to thorium dioxide aggregates and that the diffuse tracks within the bone are due to decay products that have translocated into the bone mineral. Trabeculae with a heavy surface distribution appear to have a much higher diffuse distribution of alpha tracks within the bone than trabeculae with a light surface distribution as shown in Fig. 11.2. The precise interpretation of such autoradiographs taken at one point in time would, however, appear to be difficult.

1.1.2. *Marrow.* Thorotrast has been found in marrow macrophages and surrounding marrow may also have a diffuse distribution of single alpha particles (Jee, Dockum, Mical, Arnold, and Looney 1967).

1.2. *Clinical effects*

The fact that Thorotrast may deposit both in marrow and on bone surfaces would lead to the expectation that severe marrow and bone injury might result. This has not proved to be the case. The characteristic lesions occur in the liver and spleen where the large deposits of Thorotrast in the reticulo-endothelial cells results in severe cirrhosis and the development particularly of malignant liver tumours (da Silva Horta 1967). Many large scale surveys of Thorotrast patients have been carried out; only the data relevant to the skeleton is reviewed here (Looney 1960; Faber 1968; Silva Horta, Abbatt, Motta, and Roriz 1965; Takahashi, Kitabatake, Yamagata, Miyakawa, Masuyama, Mori, Tanaka, Hibino, Miyakawa, Kaneda, Okajima, Komiyama, Koga, Adachi, Hashizume, and Hashimoto 1965; Abbatt 1967; Faber and Johansen 1967).

1.2.1. *Marrow.* At least 15 cases of leukaemia are reported following administration of Thorotrast (Faber 1968; Silva Horta, Motta, Abbatt, and Roriz 1965; Takahashi, Kitabatake, Yamagata, Miyakawa, Masuyama, Mori, Tanaka, Hibino, Miyakawa, Kaneda, Okajima, Komiyama, Koga, Adachi, Hashizume, and Hashimoto 1965; Faber and Johansen 1967). Their evaluation is difficult since many of them also received external irradiation. The Danish series includes 6 cases of acute leukaemia and one of myeloid leukaemia. Faber and Johansen also list 6 other blood dyscrasias, largely affecting the red cell elements with associated myelosclerosis. Ten non-leukaemic blood dyscrasias have recently been listed by Abbatt (1967). Some of these were probably of the De Guglielmo type, which many people would recognize today as allied to leukaemia, i.e. a myeloproliferative disorder (Vaughan 1970*b*). Those diagnosed on clinical findings, only, as aplastic anaemia, may well have been aleukaemic leukaemias. The latent period is often recorded as 15–17 years, which is interesting when compared with the relatively short period recognized as usual for the latent period for leukaemia due to external irradiation.

1.2.2. *Bone.* No instance of osteosarcoma in patients who have received Thorotrast has been reported nor any form of injury to bone.

1.3. *Animals*

Few experimental studies with Thorotrast using animals have been carried out. As a result of injecting some mice with Thorotrast, and others with a non-radioactive contrast medium, Bensted (1967) questioned whether all the tumours found in soft tissues were due to radiation alone or whether other factors might be involved. At high dose-levels of ionium-enriched Thorotrast there were many deaths from anaemia and six out of 10 mice given a lower dose died with myeloid leukaemia. Since all these mice came from the same cage Bensted himself suggests a possible viral origin for the leukaemia. On the other hand, no leukaemia occurred in 50 control mice. The experiments would appear to merit repetition.

Guimaraes and Lamerton (1956) injected 0·1 ml of Thorotrast into mice. They found some accumulation of Thorotrast, particularly in liver and spleen and some in the lung. A very small amount was present in the bone marrow but no lesions of bone or marrow were found. Jee and his colleagues (Jee, Dockum, Mical, Arnold, and Looney 1967) injected one dog weighing 30 pounds with 30 ml of thorium dioxide containing a ^{232}Th : ^{228}Ra : ^{224}Ra ratio of 1·0 : 0·1 : 0·4. The dog was sacrificed after 51 days. The total skeletal activity of ^{228}Th as a per cent of injected ^{232}Th was 2·9 per cent. A diffuse and hot spot distribution was found in both cortical and cancellous bone.

1.4. *Dosimetric considerations*

The dosimetry, i.e. the measurement of radiation dose per unit mass of tissue expressed in rad, that results from retention of Thorotrast is extremely complicated for a number of reasons listed by Dudley (1967). First, Thorotrast has been administered in many different ways, into the blood stream, into the kidney, into arteries, perivascularly, etc., and each route will give a different distribution. Secondly, the colloid is deposited non-uniformly giving a wide range of dose rates to different regions of the same tissue and this distribution will change with time. Thirdly, there may be considerable self-absorption of the alpha rays (range about 50 μm) in the colloid itself so that the tissue dose is somewhat ambiguously related to Thorotrast concentration. Fourthly, thorium is the first number of a chain of many radionuclides. Fifthly, the radionuclides escape in varying degree

THORIUM ISOTOPES 237

from the colloidal particles and translocate in the body according to their chemical and physical properties.

Further, Parr (1967) has pointed out that the idea of a typical Thorotrast patient is untenable. In a series of 15 patients Rundo (1957) observed that the estimated ^{232}Th content of the spleen varied between 8 and 75 per cent of that of the whole liver. Data from one individual cannot necessarily be used to estimate radiation dose in another. In Table 11.1 are shown the thorium series activity and dose rates in various organs including bone and marrow in an individual who was said to have been given 50 ml Thorotrast in the femoral artery at the age of 35 and who died about 20 years later (Dudley 1967). In looking at these figures the great variability of distribution on both bone surfaces and in marrow already described must be remembered. The radiation dose in certain sites will be much higher than in others. In the case of Thorotrast and its daughter products, ^{224}Ra and ^{228}Th particularly, this radiation dose will be received by the sensitive osteogenic cells themselves. No detailed study of the radiation dose received at the microscopic level has been made.

2. ^{228}Thorium (Radiothorium)

There is no human data on the behaviour of ^{228}Th except what can be derived from the study of Thorotrast patients already discussed. Theor-

TABLE 11.1

Thorium series activity and dose rate in various organs
(50 ml thortrast injected into blood)
(Dudley 1967, by courtesy of authors and publishers)

Organ	Wet weight (g)	Nanocuries in organ at equilibrium								Alpha dose-rate from all isotopes (rad year^{-1} averaged over organ)
		^{232}Th	^{228}Ra	^{228}Th	^{224}Ra	^{220}Em ^{216}Po	^{212}Pb	^{212}Bi	^{208}Tl ^{212}Po	
Whole body	70 000	1250	625	625	556	500	500	500	500	
Liver	1700	950	425	420	378	302	190	130	130	62
Spleen	150	120	60	58	52	42	26	18	18	81
Red marrow	1500	160	80	32	29	23	26	40	40	16
Lungs	1000	6	3	2	2	24	24	24	24	10
Blood	5400	0	4	4	11	44	72	150	150	6
Kidneys	300	1·5	0·3	0·6	2	1·6	2	3	3	4
Skeleton (marrow free)	7000	6·5	16·5	21·5	43	40	40	40	40	3

etically the luminous paint ingested by the dial painters might have been contaminated with ^{228}Th but, as already discussed, experiments by Maletskos and his colleagues (Maletskos, Keane, Telles, and Evans 1969) using ^{234}Th(SO$_4$)$_2$ suggest that thorium is very poorly absorbed from the gastrointestinal tract so that little if any ^{228}thorium will have reached the skeleton.

Experiments have however been carried out on beagle dogs at Utah given a single intravenous injection of ^{228}Th. Mention has already been made of these results (Chapter 6.4).

2.1. Animal experiments

2.1.1. *Metabolism.* The metabolism of ^{228}Th and its decay series in adult beagles is similar but not identical to that of ^{239}Pu (Stover, Atherton, Keller, and Buster 1960; Stover, Atherton, Buster, and Bruenger 1965; Stover, Atherton, Buster, and Keller 1965). The difference is dependent on two facts: (i) the Pu IV ion is smaller than Th IV ion and consequently forms complex ions to a greater extent than Th IV; (ii) because of the differences in half-periods, the number of ^{239}Pu atoms is about 10^4 that of ^{228}Th at each dose level. Like ^{239}Pu, ^{228}Th is concentrated in the skeleton and only very slowly excreted.

2.1.2. *Distribution.* Again like plutonium, ^{228}Th is concentrated on endosteal surfaces. This is thought to be dependent on its binding by bone glycoproteins as already discussed in relation to thorotrast. It is not yet known whether this surface concentration occurs partly in osteogenic cells and partly on the mineral/matrix surface. It is also found both in the diffuse form and in aggregates in the marrow.

2.1.3. *Effects*

2.1.3.1. *Neoplasia.* The results of the comparative experimental study on beagles at Utah show that ^{228}Th produces more osteosarcomas in a shorter time than ^{239}Pu, ^{228}Ra, ^{224}Ra, or ^{90}Sr, as shown in Fig. 6.10 (Dougherty and Mays 1969). In the latest report (COO 119, 242, 1970) there were 36 osteosarcomas, 1 haemangiosarcoma, 1 severe anaemia, and 1 pancytopaenia. This suggests that the osteogenic cells were most severely affected but that the marrow was also subject to severe radiation insult. Leukaemia might well be expected (Vaughan, Bleaney, and Williamson 1967).

2.1.3.2. *Dysplasia.* Interesting results on fracture incidence due to ^{228}Th have been reported in the Utah beagles. Significant numbers of pathological fractures were induced by a single intravenous injection of 0·3 μCi kg^{-1} or 0·9 μCi kg^{-1} ^{228}Th. Fractures were not observed at the next highest level, 2·7 μCi kg^{-1}, because of early death from radiation induced nephritis or blood dyscrasia or from both. Only 1 fracture occurred at 0·1 μCi kg^{-1} level in nine dogs and no fractures were observed below this dose. This lowest level at which a fracture occurred was approximately six times higher than the dose which has so far produced osteosarcoma. The total number of fractures per animal varied from 4 to 36 at the 0·9 μCi kg^{-1} dose and from 0 to 19 at the 0·3 μCi kg^{-1} dose. The high number of long-bone fractures was one of the characteristics of the 0·9 μCi kg^{-1} dose level. The tendency towards fracture repair was low and the amount of pain was slight. Though the dogs were thought to be fully grown at injection it subsequently was apparent that the rib epiphysis had not closed in all the animals. This resulted in interference with growth at the costalchondral junction resulting both in dead bone and abnormal bone formation and remodelling producing the so called 'hot line'. This zone was the most susceptible fracture site in the rib cage. In comparison to ^{226}Ra, ^{228}Ra, ^{239}Pu, and ^{90}Sr these ^{228}Th fractures most nearly resembled those of ^{228}Ra except for the unique 'hot line' fractures which were similar to those of ^{239}Pu (Taylor, Jee, Christensen, Rehfeld, and Nebeker 1965).

J. H. Dougherty (1962) in an early report on the Utah beagles reported that radiothorium produced the most sustained red-cell depression initially though there was subsequently some recovery. No account of marrow change is available except that in their account of the fracture incidence Taylor and his colleagues remark on marrow aplasia near the hot-line fractures (Taylor, Jee, Christensen, Rehfeld, and Nebeker 1965).

2.1.4. *Dosimetric considerations.* The extreme toxicity of ^{228}Th is due to the high energy of its alpha particle and to the fact that it is concentrated in or adjacent to the osteogenic cells at carcinogenic risk.

In 1964 Mays and Tueller compared the range of the alpha particles of ^{239}Pu and ^{228}Th and pointed out that the somewhat longer range of ^{228}Th might account for its greater toxicity. It might well account also for its bone marrow effects. A further enhancing factor will be derived from its daughter ^{224}Ra which is known to be a powerful inducer of osteosarcomas in its own right (Chapter 7.11). If Dougherty and Mays' calculations (1969) are accepted, the relative biological efficiency of ^{228}Th relative to ^{226}Ra is 8 (Chapter 6.4).

12. ^{32}P

The physical and biological characteristics that make ^{32}P a theoretical radiogenic risk to the skeleton have already been discussed in Chapter 6.5. There is clinical evidence in man and experimental evidence in animals that this risk may have practical importance.

1. Clinical experience with ^{32}P

^{32}P has been used extensively since 1939 for the treatment of patients with 'polycythaemia vera'. A survey of the results of treatment made in 1968 (Vaughan 1970b) certainly suggested that ^{32}P must be under suspicion as a leukaemogenic agent. Some of the reported cases at that time are shown in Tables 12.1 and 12.2. Such a tabulation is unsatisfactory since it inevitably simplifies what is an extremely complex situation, and furthermore it has been difficult to determine how far some of the recorded cases appear in more than one publication.

TABLE 12.1

Terminal leukeamia or myelosclerosis or fibrosis in polycythemia vera

Reference	Date of reference	Total cases	Dead	Leukaemia Acute	Leukaemia Chronic	Myelo-fibrosis	Irradiation
Videbaek	1950	125	76	1	1		No
Chievitz and Thiede	1962	111	50+?	0	0		No
Perkins et al.	1964	127	44	1			No
Wiseman et al.	1951	450	40	12			^{32}P
Stroebel	1954	241	39	8		9?	^{32}P
Wasserman	1954	270	64	7			^{32}P and X-ray
Lawrence	1955	?	73	16			^{32}P
Calabresi and Meyer	1959	100	28		6	2	^{32}P
Ledlie*	1960	117	21	5	1		^{32}P
Halnan and Russell	1965	107	43			4	^{32}P and X-ray
Szur and Lewis	1966	169	?	4		12	^{32}P
Watkins et al.	1967	81	46	2			^{32}P
Harman and Ledlie*	1967	132	65	10	Marrow failure	12	^{32}P

* These two references probably contain some of the same cases.

TABLE 12.2

Frequency of acute leukaemia by diagnostic group, ionizing radiation, and chemotherapy treatment*
(from Modan and Lilienfeld 1965, by courtesy of authors and publishers)

Method of radiation treatment for polycythaemia vera	Patients receiving no chemotherapy			Patients receiving chemotherapy		
	Total patients	Patients developing acute leukaemia		Total patients	Patients developing acute leukaemia	
		No.	%		No.	%
No radiation treatment	101			32	1	3·1
X-ray	52	4	7·7	27	3	11·1
^{32}P	126	13	10·3	102	12	11·8
X-ray plus ^{32}P	42	4	9·5	30	8	26·7
Total	321	21	6·5	191	24	12·6

It would appear from these tables however that acute myeloid leukaemia is probably more common in patients who receive radiation, i.e. ^{32}P or X-rays, than in those who receive either no therapy as in Chievitz and Thiede's series (1962) or only chemotherapy as in the patients treated by Perkins (Perkins, Israels, and Williamson 1964). The development of acute leukaemia however continues to be reported in patients who have received chemotherapy only (Logue, Gutterman, McGinn, Lazlo, and Rundles 1970).

It is still disputed therefore as to whether ^{32}P has prolonged the expectation of life of such patients and also increased the incidence of so called 'acute leukaemia' and myelosclerosis. It has probably done both but certainly many physicians regard it as the treatment of choice for several reasons including the comfort of the patient (Lawrence, Winchell, and Donald 1969; Millard, Kay, and Lawler 1969; Campbell, Emery, Godlee, and Prankerd 1970; Höfer and Kühböck 1970). It is also disputed as to whether primary polycythaemia is always due to abnormal activity of all marrow elements or whether initially at least only the red cell stem-cells are affected.

Lawrence (Lawrence, Winchell, and Donald 1969), having recently reviewed 181 cases treated with ^{32}P alone or ^{32}P combined with X-rays, is still of the opinion he put forward in 1955 that the incidence of acute leukaemia in irradiated patients may be primarily a result of prolonged survival rather than radiation dose, the development of significant splenic myeloid metaplasia and acute leukaemia-like states being part of the evolutionary history of the disease, with which Millard and her colleagues (Millard, Kay, and Lawler 1969) would agree. Modan and Modan (1970) however indicated, following an analysis of a retrospective follow up of a large group of patients, that primary polycythaemia should be separated into benign erythrocytosis when the red cell element alone was involved and the form in which all cell types were involved and there was associated generalized myeloid hyperplasia evidenced by splenic enlargement before treatment.

In Table 12.3 they have (1970) separated the cases of benign erythrocytosis from polycythaemia vera. In both groups there was a higher number

TABLE 12.3

Cause of death in patients with polycythaemia vera and with benign erythrocytosis (Modan and Modan 1968, by courtesy of author and publishers)

Cause of death	No radiation				Radiation			
	Polycythaemia vera		Benign erythrocytosis		Polycythaemia vera		Benign erythrocytosis	
	No.	%	No.	%	No.	%	No.	%
Acute leukaemia	1	0·8	—	0·0	44	11·6	3	7·9
Other myeloproliferative disorders	8	6·0	—	0·0	70	18·5	—	0·0
Thromboembolic and haemorrhagic events	45	33·8	30	33·0	60	15·8	16	42·2
Malignant tumours	8	6·0	10	11·0	31	8·2	6	15·8
Other causes	32	24·1	13	14·3	34	9·0	4	10·5
Total deaths	94	70·7	53	58·2	239	63·1	29	76·3
Total patients	133	100·0	91	100·0	379	100·0	38	100·0

of deaths from acute leukaemia in irradiated than in non-irradiated patients, no distinction being made as to the type of radiation, but the difference between the two groups was apparent when other myeloproliferative lesions, myeloid metaplasia, chronic leukaemia of all types, myelofibrosis, and aplastic anaemia were involved. No irradiated case of benign erythrocytosis died with such lesions while 18·5 per cent of polycythaemia cases did so. Other causes of death did not appear to differ in the two groups. Campbell and his colleagues (Campbell, Emery, Godlee, and Prankerd 1970) have also emphasized that patients who initially have a small spleen have a better prognosis when treated with ^{32}P than those with a large spleen. The survival time in the first group was 12·5 years, two dying with acute leukaemia and none with myelosclerosis. In the second group the survival time was 7·5 years, 6 patients dying from acute leukaemia, 5 from myeloid metaplasia and myelosclerosis. There was no relationship between the amount of ^{32}P given and the leukaemia incidence. Campbell and his colleagues suggest that the development of leukaemia in undifferentiated polycythaemia vera may be related more to the type of patient than to the amount of radiation.

Millard, who has made an intensive study of the chromosome changes in marrow cells during treatment with ^{32}P, concludes that progression into the leukaemic phase of the disease is associated, in some cases, with gross chromosomal abnormalities, such as a shift of the stem-line chromosome number and bizarre chromosome 'markers'. In other cases, some of whom have not been irradiated for several years, the chromosomal changes are less pronounced (Millard 1965; Millard, Kay, and Lawler 1969). Unfortunately no study is available for chromosome changes occurring in the marrow cells of polycythaemia patients treated with venisection or non-radiogenic drugs. She considers, however, that irradiation producing chromosomal damage is one component of leukaemogenesis but that there are other factors probably of equal importance. It would perhaps appear that at present, when the underlying pathology of polycythaemia vera itself is not clearly understood, it is difficult to interpret the effects of radiation with any precision. There is however no question that ^{32}P is a valuable therapeutic agent. Hitherto there have been no reports of osteo-sarcomas arising in patients treated with ^{32}P (Millard, Kay, and Lawler 1969) but survival time is perhaps not long enough. It must be noted however that myelosclerosis, which suggests involvement of osteogenic as well as marrow tissue, is frequently reported as a terminal event (Szur and Smith 1961; Szur and Lewis 1966; Campbell, Emery, Godlee, and Prankerd 1970). Clearly careful macroscopic and microscopic examination of all parts of the skeleton, in patients who have received large doses of ^{32}P, should be made.

2. Animal experiments

Extensive experiments on animals using ^{32}P have not been carried out.

3. Marrow

Rats, given a lethal dose 3600 μCi kg^{-1} of ^{32}P, showed a depression of all the cellular elements of the marrow (Kligerman 1950). Early experiments on monkeys showed that ^{32}P lowered the absolute numbers of red cells, lympohocytes, and granulocytes in the circulating blood. Doses of 0·76 to 0·71 mCi per pound of body weight were well tolerated but 1·04 mCi proved fatal (Scott and Lawrence 1941); the former dose is equivalent to about 114 mCi given in one dose, i.e. 10 times that usually given therapeutically.

4. Skeleton

Bone growth in rats is affected with doses as low as 132 μCi kg^{-1} but the effect is temporary (Kligerman 1950).

Osteosarcomas have been produced in rats: 1 μCi g^{-1} was a carcinogenic dose though many rats at this level died of haemopoietic failure within the first 10 weeks (Bensted, Blackett, and Lamerton 1961). Increasing the dose increased the number of tumours. If the total dose of ^{32}P was divided between 5 injections at 2 weekly intervals and injected into rapidly growing rats, there was a reduction in the maximum radiation-dose rate to bone compared with a single injection, but a larger volume of bone was irradiated since much of the ^{32}P deposited, at succeeding injections, in the region of new bone formed since the previous injection. Such fractionation of the administered ^{32}P increased the number of tumours produced and also increased the generalized tissue damage. A comparison of the number of tumours produced by single and by repeated injections is shown in Fig. 12.1.

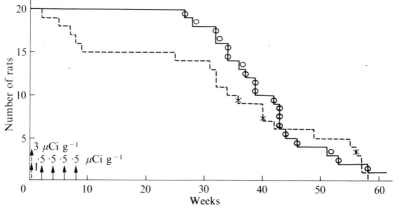

FIG. 12.1. Time of sacrifice or death of rats following 3·0 μCi kg^{-1} of ^{32}P as a single injection and 1·0 μCi g^{-1} + 4 × 0·5 μCi g^{-1} × and 0 indicates rats with bone tumours. (Bensted *et al.* 1961, by courtesy of authors and editors.)

Since ^{32}P clearly is able to induce osteosarcomas in rats it is extremely important to check the skeletons of patients treated with ^{32}P who come to autopsy, though the dose of ^{32}P given to the rats at 3·0 μCi g^{-2} was considerably larger than that given to patients per gram of body weight. The average dose of ^{32}P is 5 mCi though this of course may be repeated.

5. Dosimetric considerations

Millard and her colleagues have tried to estimate the bone-marrow dose from ^{32}P assuming that the greater part comes from bone. Following Spiers

(Spiers 1968a,b), the mean dose to the bone marrow from the trabecular bone has been calculated as 75 rad per mCi ^{32}P with an initial dose rate of about 0·8 rad per hour. Since there is some non-uniformity of distribution of dose to the bone marrow the dose may vary by a factor of perhaps two according to the distance of the marrow from the bone. Making other assumptions it is estimated that the additional radiation dose to the marrow is about 10 rad per mCi of ^{32}P. Therefore the mean dose to the bone marrow is about 85 rad per mCi of ^{32}P; since the average dose of ^{32}P is 5 mCi the total dose to the bone marrow per treatment is about 425 rad. This initial dose rate will of course decrease with time but it is considered by Millard to be adequate to account for the amount of chromosomal damage observed.

It should be noted that the response of different patients to the same administered dose of ^{32}P is extremely variable.

No calculations have been made of the dose received by osteogenic tissue.

13. Other Bone Seeking Radionuclides

THERE are many other bone seeking radionuclides about which less is known than those already discussed: among them cerium, curium, neptunium, einsteinium, and californium are likely to have practical importance in the future. This is particularly true of californium. It has recently been estimated that ^{252}Cf production requirements by 1980 may be of the order of several hundred grams per year (Seaborg 1968; Crandall 1968). This is largely because ^{252}Cf decays by spontaneous fission in about 3 per cent of its radioactive disintegrations and therefore, because of the accompanying emission of fission neutrons it may provide a portable neutron source of very small size with virtually unlimited applications. Each radioactive disintegration of ^{252}Cf by fission (3·1 per cent) releases 185 MeV and each disintegration by alpha emission (96·9 per cent) releases 6·11 MeV alpha plus 0·10 MeV recoil energy. Approximately half of the total alpha + fission dose rate is accounted for by each of these two modes of ^{252}Cf decay. ^{249}Cf decays by alpha emission only. It releases 5·83 MeV of alpha plus 0·10 MeV of recoil energy per disintegration (5·93 MeV). The dose rate therefore of ^{252}Cf is nearly twice that of ^{249}Cf. The present scanty evidence suggests that tissues at risk from californium are liver, skeleton, kidney and thyroid (Lloyd, Mays, Taylor, and Williams 1972). (Atherton and Lloyd 1972) Californium has recently been introduced into the Utah study on beagle dogs. The relative bone to bone distribution of ^{249}Cf in the skeleton of two dogs sacrificed at 7 and 21 days after injection was similar to that of injected ^{241}Am and ^{239}Pu. About one week following injection about 20 per cent of the injected californium was deposited in the liver and about 60 per cent remained in non liver tissue, mainly skeleton (Lloyd, Mays, Taylor, and Williams 1972). Concentration in the plasma decreased very rapidly during the first few hours after injection. At 24 hours less than 1 per cent of the injected dose was still circulating (Stevens and Bruenger 1972). The effects on the blood picture soon after injection are much more marked following ^{252}Cf than following ^{249}Cf due presumably to the added dose rate from fission fragments on bone surfaces from ^{252}Cf (Dougherty 1971). The available evidence at present suggests that the metabolic behaviour of californium resembles that of americium rather than that of plutonium. Little is known at present about its long term toxic effect but

OTHER BONE SEEKING RADIONUCLIDES

it certainly presents a skeletal hazard.

Levdik and his colleagues (Levdik, Lemberg, Buldakov, Lyubchanskii, and Persternikov 1972) have given ^{237}Np as both the oxide and the nitrate by intravenous injection to rats in doses ranging from 2·0 to 0·017 μCi kg^{-1} and induced osteosarcoma. If the ^{237}Np was given by intratracheal tube some osteosarcoma also occurred. They consider that this radionuclide is more carcinogenic than ^{239}Pu.

Though there may be differences in details of the metabolic behaviour of these radionuclides, evidence suggests that they are distributed in bone broadly like plutonium and americium, i.e. they are found on 'surfaces' rather than throughout the mineral (Durbin 1962; Menot, Masse, Morin, and Lafuma 1972). It is however of interest that recently Lafuma and his colleagues have found 'some migrating elements in deep bone, closely associated with the osteocytes and their canaliculi' (Menot, Masse, Morin, and Lafuma 1972). This migration may occur as the result of their binding to macromolecules in plasma (Owen and Triffitt 1971). There is some evidence from the work of Taylor and Chipperfield that certain of the glycoproteins of bone as well as plasma proteins are capable of binding americium, plutonium, and curium as discussed in Chapter 2.2.3.2. The binding for plutonium is stronger than that for americium while curium is again bound to different glycoproteins than either of the other two. No observations have yet been made as to the relationship of other radionuclides to bone matrix elements but it would appear possible that such binding may account for their presence on bone surfaces.

14. Conclusions

THIS review of the effects of irradiation on the skeleton emphasizes the wisdom of Evans' law: 'Everything is more complicated than most people think' (Evans, Keane, Kolenkow, Neal and Shanahan 1969). It is certainly rash, but it may be helpful, to see if any conclusions can be reached.

(1) Degenerative changes have not been seen when the radiation dose is less than the carcinogenic dose so that in the case of both external and internal radiation interest is concentrated on the carcinogenic effects of radiation.

(2) The cell at carcinogenic risk from radiation is the proliferating cell since it can carry forward injury from an ionizing event by later division. Such proliferating cells in the case of the skeleton are found in (i) osteogenic tissue on bone surfaces, (ii) the marrow, and (iii) epithelium closely applied to bone.

(3) Until the mechanisms involved in carcinogenesis from any cause are understood it is unlikely that the mechanisms of radiation carcinogenesis will be understood. It would appear however that radiation may prove a useful experimental tool in the study of such mechanisms. For instance the relationship of radiation and viruses in the aetiology of both leukaemia and osteosarcoma, at present under intensive study, has indicated the importance of the relationship between a cell and its environment.

(4) External irradiation of the whole skeleton or the greater part of the skeleton in the adult may result in malignant myeloproliferative disorders, usually described as 'leukaemia'. However, since the latent period of osteosarcoma is much longer than that for leukaemia, it may well be that longer study of irradiated populations may show that those who live long enough may develop osteosarcoma as well as other malignancies.

(5) The evidence from leukaemia studies at Hiroshima and Nagasaki suggests that the RBE of fast neutrons to gammas is about 5 with no indication of a threshold.

(6) External irradiation of part of the skeleton only will be likely to

induce a bone tumour in the adult. In the child, for reasons which are not at present clear, benign lesions such as exostoses are relatively common, associated with disturbances of bone growth, but bone malignancies are rare. A radiation dose above 2000 rad is at present thought to be carcinogenic if X-rays or γ-rays are involved. Fast neutrons are carcinogenic at a lower dose in experimental animals.

(7) Any analysis of the effects of internal irradiation from radionuclides deposited in the skeleton is complicated by a wide range of interacting factors. Experimental results have hitherto been largely concerned with the effects of a single intravenous administration, but in real life contamination is far more likely to arise from continuous ingestion, either by inhalation, ingestion, or wounding.

(8) In general terms it may be said that alpha-emitting radionuclides are more carcinogenic than beta emitters because of the higher LET of the radiation. Furthermore, radionuclides such as ^{239}Pu and ^{228}Th that deposit on bone surfaces adjacent to or within or on proliferating cells are more carcinogenic than radionuclides, such as ^{226}Ra, that are deposited throughout the bone mineral, even though all three are alpha emitters.

(9) The short range of the alpha-emitting radionuclides results in the induction of tumours of osteogenic tissue predominantly, while the long range of the beta radiation from ^{90}Sr+^{90}Y results in a wide variety of marrow tumours. If the ^{90}Sr+^{90}Y is continuously ingested the characteristic response is a myeloproliferative malignancy.

(10) Apart from an estimation of the dose received by external irradiation of the foetus *in utero* there is no satisfactory data on the response to low levels of radiation. It still remains to be determined, in the case of both external and internal radiation, whether there is a threshold, perhaps determined by life span, below which the probability of malignant transformation can be ignored.

(11) Now that techniques have been developed which enable an estimate to be obtained of the radiation dose delivered to sensitive tissues by different radionuclides in different species it will be important to determine such doses in different species and in man. Such determinations should enable experimental results which are obtained on animals to be extrapolated to man with more confidence than when average dose to the bone mass is used.

Bibliography

Bold numbers in brackets indicate page numbers on which the reference occurs.

Note References to papers in *Biomedical Implications of Radiostrontium Exposure* are based on papers circulated at the time of the meeting. The published volume may differ in detail. It is not yet available.

ABBATT, J. D. (1967). Leukaemia and other fatal blood dyscrasias in thorium dioxide patients. *Ann. N. Y. Acad. Sci.* **145**, 767–75. **[235]**

ADRIAN COMMITTEE (Committee on Radiological Hazards to Patients). *Interim report.* Her Majesty's Stationery Office, 1959. **[68]**

ADRIAN COMMITTEE (Committee on Radiological Hazards to Patients). *2nd report.* Her Majesty's Stationery Office, 1960. **[68]**

ALBEE, F. H. (1920). Studies in bone growth: an experimental attempt to produce pseudoarthrosis. *Am. J. med. Sci.* **159**, 40. **[138, 146]**

ALDERSON, M. R. and JACKSON, S. M. (1971). Long-term follow-up of patients with menorrhagia treated by irradiation. *Br. J. Radiol.* **44**, 295–98. **[78]**

AMPRINO, R. and MAROTTI, G. (1964). A topographic quantitative study of bone formation and reconstruction, in *Bone and tooth* (ed, H. J. J. Blackwood). Pergamon Press, Oxford. 21–33. **[21, 216]**

AMSTUTZ, H. C. (1969). Multiple osteogenic sarcomata. Metastatic or multicentric. *Cancer N. Y.* **24**, 923–31. **[80]**

ARDRAN, G. M. and KEMP, F. H. (1958). Radium poisoning. Two case reports. *Br. J. Radiol.* **31**, 605–610. **[120]**

ARKIN, A. M., PACK, G. T., RANSOHOFF, N. S., and SIMON, N. (1950). Radiation-induced scoliosis. A case report. *J. Bone Jt Surg.* (Am. ed.) **32**, 401–4. **[83]**

ARNOLD, J. S. (1954). *Second Annual Conference of Plutonium, Radium, and Mesothorium.* Salt Lake City, Utah. T.I.D. 7639. (ed. Stover, C. N.). pp. 112–29. **[219]**

—— and JEE, W. S. S. (1962). Pattern of long-term skeletal remodelling revealed by radioautographic distribution of plutonium 239 in dogs. *Hlth Phys.* **8**, 705–7. **[209, 216]**

—— STOVER, B. J., and VAN DILLA, M. A. (1955). Failure of ^{90}Y to escape from skeletally-fixed ^{90}Sr. *Proc. Soc. exp. Biol. Med.* **90**, 260–3. **[154]**

ATHERTON, D. R. and LLOYD, R. D. (1972). The distribution and retention of ^{249}Cf in beagle soft tissue. *Health Physics* **22**, 675–77. **[246]**

ATKINSON, P. J. (1967). Variation in trabecular structure of vertebrae with age. *Calcif. Tissue Res.* **1**, 24–32. **[16]**

AUB, J. C., EVANS, R. D., HEMPELMANN, L. H., and MARTLAND, H. S. (1952). The

BIBLIOGRAPHY

late effects of internally deposited radioactive materials in man. *Medicine, Baltimore* **31**, 221–329. [**111, 118, 122**]

AWA, A. A., HONDA, T., SOFUNI, T., NERIISHI, S., YOSHIDA, M. C., and TAKASHI, M. (1971). Chromosome-aberration frequency in cultured blood-cells in relation to radiation dose of A-bomb survivors. *Lancet* ii, 903–5. [**2**]

BACHRA, B. N. (1967). Some molecular aspects of tissue calcification. *Clin. Orthop.* **51**, 199–222. [**29**]

BACKWINKEL, K. D. and DIDDAMS, J. A. (1970). Hemangiopericytoma. Report of a case and comprehensive review of the literature. *Cancer, N.Y.* **25**, 896–901. [**28**]

BAENSCH, W. (1927). Knochenschädigung nach Rontgenbestrahlung. *Fortschr. Geb. RöntgStrahl.* **36**, 1245–7. [**88**]

BARDEN, S. P. (1943). Healing of radiation fractures of the necks of the femora, with a report of a case. *Radiology* **41**, 389–94. [**88**]

BARNES, D. W. H., CARR, T. E. F., EVANS, E. P., and LOUTIT, J. F. (1970). ^{90}Sr induced osteosarcomas in radiation chimaeras. *Int. J. Radiat. Biol.* **18**, 531–7. [**29, 44, 196**]

—— and KHRUSCHOV, N. G. (1968). Fibroblasts in sterile inflammation: study in mouse radiation chimaeras. *Nature, Lond.* **218**, 599–601. [**23, 46**]

—— and LOUTIT, J. F. (1967). Haemopoietic stem cells in the peripheral blood. *Lancet* **2**, 1138–41. [**46**]

—— EVANS, E. P. and LOUTIT, J. F. (1971). Local origin of fibro-blasts deduced from sarcomas induced in chimaeras by implants of pliable discs. *Nature, Lond.* **233**, 267–8. [**23, 46, 49**]

BASERGA, R., LISCO, H., and CATER, D. B. (1961). The delayed effects of external gamma irradiation on the bones of rats. *Am. J. Path.* **39**, 455–72. [**41, 79, 92**]

BATT, R. C. and HAMPTON, A. O. (1940). Spontaneous subcapital hip fractures occurring in tabes dorsalis. *J. Bone Jt Surg.* **22**, 137–46. [**88**]

BAUDISCH, E. (1960). Radiation damage to the ribs and clavicles in patients with cancer of breast. *Strahlentherapie* **113**, 312–18. [**88**]

BECK, A. (1922). Zur Frage des Rontgensarkoms zugleich ein Beitrag sur Pathogenese des Sarkoms. *Münch. med. Wschr.* **69**, 623–4. [**79**]

—— (1924). Zur Frage des Rontgensarkoms, zugleich ein Beitrag sur Pathogenese des Sarkoms, *Arch. klin. Chir.* **133**, 191–5. [**79**]

BENDER, M. A. and GOOCH, P. C. (1966). Somatic chromosome aberrations induced by human whole body irradiation: The 'Recuplix' Criticality Accident. *Radiat. Res.* **29**, 568–82. [**2**]

BENNETT, B. G. (1972). Global ^{90}Sr fall-out and its occurrence in diet and man. To be published in *Biomedical implications of radiostrontium exposure* (ed. Goldman, M. and Bustad, L. K.). Division of Technical Information, U.S. Atomic Energy Commission, Oak Ridge, Conference CONF. 710201. [**56, 57, 58**]

BENO, M. (1968). A study of 'haemosiderin' in the marrow of the femur of the normal young adult rabbit compared with that in rabbits four months after an intravenous injection of ^{239}Pu(NO$_3$)$_4$. *Br. J. Haemat.* **15**, 487–95. [**219, 221**]

BENSTED, J. P. M. (1967). Experimental studies in mice on the late effects of radioactive and non-radioactive contrast media. *Ann. N. Y. Acad. Sci.* **145**, 728–37. [**235**]

—— BLACKETT, N. M., and LAMERTON, L. F. (1961). Histological and dosimetric considerations of bone tumour production with radioactive phosphorus. *Br. J. Radiol.* **34**, 160–75. [**244**]

—— TAYLOR, D. M., and SOWBY, F. D. (1965). The carcinogenic effects of americium 241 and plutonium 239 in the rat. *Br. J. Radiol.* **38**, 920–5. [**223, 230, 231, 232**]

BERDON, W. E., BAKER, D. H., and BOYER, J. (1965). Unusual benign and malignant sequelae to childhood irradiation including unilateral hyperlucent lung. *Am. J. Roentg.* **93**, 545–56. [**84**]

BERENBLUM, I. and TRAININ, N. (1963). New evidence on the mechanism of radiation leukaemogenesis, pp. 41–56 in *Cellular basis and aetiology of late somatic effects of ionizing radiation* (ed. R. J. C. Harris). Academic Press, London. [**10**]

BERG, N. O., LANDBERG, T., and LINDGREN, M. (1966). Osteonecrosis and sarcoma following external irradiation of intracerebral tumours. *Acta radiol.* **4**, 417–36. [**81, 82**]

BICKEL, W. H. (1970). In discussion pp. 663–4. *J. Bone Jt Surg.* (Am. ed.) **52**a. [**82**]

BINGHAM, P. (1968). Studies of the cells of the osteogenic connective tissues using autoradiographic techniques. *D. Phil. Thesis, Oxford.* [**25**]

BINGHAM, P. J., BRAZELL, I. A., and OWEN, M. (1969). The effect of parathyroid extract on cellular activity and plasma calcium levels *in vivo. J. Endocr.* **45**, 387–401. [**24, 26**]

BIZZOZERO, O. J., JOHNSON, K. G., and CIOCCO, A. (1966). Radiation-related leukaemia in Hiroshima and Nagasaki 1946–64. *New Engl. J. Med.* **274**, 1095–101. [**62, 65, 67**]

BLACKBURN, J. and WELLS, A. B. (1963). Radiation damage to growing bone; the effect of X-ray doses of 100 to 1000 R on mouse tibia and knee joint. *Br. J. Radiol.* **36**, 505–13. [**85, 87**]

BLAND, M. R., LOUTIT, J. F., SANSOM, J. M., and SMITH, C. (1973). Tumours in bone and bone marrow induced in CBA/H mice by ^{90}Sr and ^{226}Ra, to be published in The Colston Papers No. 24. Bone: Certain Aspects of Neoplasia, ed. C. H. G. Price and F. G. M. Ross. Butterworths, London. [**10**]

BLEANEY, B. (1967). Radiation dose-rates near bone surfaces in rabbits after an injection of plutonium. *Physics Med. Biol.* **12**, 145–60. [**209, 227**]

—— (1969*a*). The radiation dose-rates near bone surfaces in rabbits after intravenous or intramuscular of plutonium 239. *Br. J. Radiol.* **42**, 51–6. [**209, 211, 227**]

—— (1969*b*). Plutonium deposition on bone surfaces and in bone marrow following intravenous and intramuscular injections, pp. 125–35 in *Delayed effects of bone-seeking radionuclides* (ed. Mays, C. W., Jee, W. S., Lloyd, R. D., Stover, B. J., Dougherty, J. H., and Taylor, G. N.). University of Utah Press, Salt Lake City. [**209, 211, 227**]

—— and VAUGHAN, J. (1971). Distribution of ^{239}Pu in the bone marrow and on the endosteal surface of the femur of adult rabbits following injection of ^{239}Pu(NO$_3$)$_4$.

Br. J. Radiol. **44**, 67–73. [**29, 205, 207, 209, 212, 213, 214, 215, 219, 220, 222, 227**]

BLOOM, A. D. and TJIO, J. H. (1964). In vivo effects of diagnostic X-irradiation on human chromosomes. *New Engl. J. Med.* **270**, 1341–4. [**2**]

BOAG, J. W. (1971). Radiation physics in *manual on radiation haematology*, pp. 31–42, Technical Report Series No. 123. International Atomic Energy Agency, Vienna. [**2**]

BODIAN, M. (1963). Aspects of cancer in childhood, pp. 266–93, in *Recent advances in paediatrics*, 3rd ed. (ed. D. Gairdner). J. & A. Churchill Ltd., London. [**48**]

BONFIGLIO, M. (1953). The pathology of fracture of the femoral neck following irradiation. *Am. J. Roentg.* **70**, 449–59. [**88**]

BOOCOCK, G., DANPURE, C. J., POPPLEWELL, D. S., and TAYLOR, D. M. (1970). The subcellular distribution of plutonium in rat liver. *Radiat. Res.* **42**, 381–96. [**221**]

—— and POPPLEWELL, D. S. (1965). Distribution of plutonium in serum proteins following intravenous injection into rats. *Nature, Lond.* **208**, 282–3. [**221**]

BORASKY, R. (1957). Collagen reactivity with plutonium, pp. 36–41, in Biology Research Annual Report 1956. USAEC Report HW 47500, Hanford Atomic Products Operation, General Electric Co. [**30**]

BRILL, A. B., TOMONAGA, M., and HEYSSEL, R. M. (1962). Leukaemia in man following exposure to ionizing radiation. Summary of findings in Hiroshima and Nagasaki and comparison with other human experience. *Ann. intern. Med.* **56**, 590–609. [**66**]

BROOKS, B. and HILLSTROM, H. T. (1933). Effect of roentgen rays on bone growth and bone regeneration. *Am. J. Surg.* **20**, 599–614. [**138, 146**]

BRUES, A. M. (1949) Biological hazards and toxicity of radioactive isotopes. *J. clin. Invest.* **28**, 1286–96. [**8, 222**]

—— (1958). Critique of the linear theory of carcinogenesis. *Science N. Y.* **128**, 693–9. [**51**]

—— (1971). Consideration of carcinogenic mechanisms in the prediction of low-level radiation effects. Conference on the estimation of low-level radiation effects in human populations (ed. G. A. Sacher). ANL 7811, 4–5. Argonne National Laboratory, Argonne. [**1, 5**]

—— AUERBACH, H., de ROCHE, G. M., and GRUBE, D. D. (1969). Mechanisms of carcinogenesis. Argonne National Laboratory Biological and Medical Research Division, Annual Report 1969, ANL 7635. [**5**]

BRYANT, F. J. and LOUTIT, J. F. (1963–64). The entry of strontium 90 into human bone. *Proc. Roy. Soc.* B **159**, 449–65. [**58**]

BUCKTON, K. E., JACOBS, P. A., COURT BROWN, W. M., and DOLL, R. A. (1962). A study of the chromosome damage persisting after X-ray therapy for ankylosing spondylitis. *Lancet* ii, 676–82. [**2**]

BULDAKOV, L. A., LYUBCHANSKII, E. R., MOSKALEV, Y. I., and NIFATOV, A. P. (1970). *Problems of plutonium toxicology.* Atom Publications, Moscow 1969, LF-tr-41, UC-48. [**202, 205**]

BUSTAD, L. K., CLARKE, W. J., GEORGE II, L. A., HORSTMAN, V. G., MCCLELLAN, R. O., PERSING, R. L., SEIGNEUR, L. J., and TERRY, J. L. (1962). Preliminary observations on metabolism and toxicity of plutonium in miniature swine. *Hlth Phys.* **8**, 615–20. [**226**]

—— GOLDMAN, M., ROSENBLATT, L. S., MCKELVIE, D. H., and HERTZENDORF, I. J. (1969). Haematopoietic changes in beagles fed ^{90}Sr, in *Delayed effects of bone-seeking radionuclides* (ed. Mays, C. W., Jee, W. S. S., Lloyd, R. D., Stover, B. J., Dougherty, J. H., and Taylor, G. N.). pp. 279–91. University of Utah Press, Salt Lake City. [**178**]

CAHAN, W. G., WOODARD, H. Q., HIGINBOTHAM, N. L., STEWART, F. W., and COLEY, B. L. (1948). Sarcoma arising in irradiated bone—Report of 11 cases. *Cancer N. Y.* **1**, 3–29. [**79, 83**]

CALABRESI, P. and MEYER, O. O. (1959). Polycythaemia vera. *Ann. int. Med.* **50**, 1182–216. [**240**]

CAMPBELL, A., EMERY, E. W., GODLEE, J. N., and PRANKERD, T. A. J. (1970). Diagnosis and treatment of primary polycythaemia. *Lancet* **1**, 1074–7. [**241, 242, 243**]

CARMON, J. L. (1965). The effects of radiation on growth pattern of mice. *Growth* **29**, 85–95. [**15**]

CASARETT, G. W., TUTTLE, L. W., and BAXTER, R. C. (1962). Pathology of imbibed ^{90}Sr in rats and monkeys, in *Some aspects of internal irradiation* (ed. Dougherty, T. F., Jee, W. S. S., Mays, C. W., and Stover, B. J.), pp. 329–36. Pergamon Press, Oxford. [**178, 183, 184**]

CATER, D. B., BASERGA, R., and LISCO, H. (1959). Studies on the induction of bone and soft tissue tumours in rats by gamma irradiation and the effect of growth hormone and thyroxine. *Br. J. Cancer* **13**, 214–27. [**41, 79, 92**]

—— BASERGA, R., and LISCO, H. (1960). Radiographic appearances of bone lesions in rats exposed to local external irradiation from gamma rays. *Acta radiol.* **54**, 273–88. [**41, 79, 92**]

CHALMERS, J., BARCLAY, A., DAVISON, A. M., MACLEOD, D. A. D., and WILLIAMS, D. A. (1969). Quantitative measurements of osteoid in health and disease. *Clin. Orthop.* **63**, 196–209. [**34**]

CHIEVITZ, E. and THIEDE, T. (1962). Complications and causes of death in polycythaemia vera. *Acta med. scand.* **172**, 513–23. [**240, 241**]

CHIPPERFIELD, A. R. and TAYLOR, D. M. (1968). Binding of plutonium and americium to bone glycoproteins. *Nature, Lond.* **219**, 609–10. [**30**]

—— and TAYLOR, D. M. (1970). Binding of plutonium to glycoproteins *in vitro*. *Radiat. Res.* **43**, 393–402. [**30, 220, 231**]

—— and TAYLOR, D. M. (1971). Personal communication. [**221**]

CLARKE, W. J. (1962). Comparative histopathology of ^{239}Pu, ^{226}Ra, and ^{90}Sr in pig bone. *Hlth Phys.* **8**, 621–7. [**226**]

—— BUSCH, R. H., HACKETT, P. L., HOWARD, E. B., FRAZIER, M. E., MCCLANAHAN, B. J., RAGAN, H. A., and VOGT, G. S. (1972). Strontium-90 effects in swine. A summary up-to-date, in *Biomedical implications of radiostrontium exposure* (ed. Goldman, M. and Bustad. L. K.). Division of Technical Information, U.S. Atomic

Energy Commission, Oak Ridge, Conference CONF. 710201. [6, 46, 180, 183, 197]

—— HOWARD, E. B., and HACKETT, P. L. (1969). ^{90}Sr induced neoplasia in swine, in *Delayed effects of bone-seeking radionuclides* (ed. Mays, C. W., Jee, W. S. S., Lloyd, R. D., Stover, B. J., Dougherty, J. H., and Taylor, G. N.), pp. 263-75. University of Utah Press, Salt Lake City. [180]

CLEVELAND, J. M. (1970). *The chemistry of plutonium*. Science Publishers, Gordon and Breach, New York. [202]

CLUZET. (1910). Action des rayons X sur le développement du cal (Étude macroscopique et radiographique). *Lyon méd.* **114**, 22-4. [138, 146]

COHEN, J. and D'ANGIO, G. J. (1961). Unusual bone tumours after roentgen therapy of children. Two case reports. *Am. J. Roentg.* **86**, 502-12. [79, 81]

COHN, S. H. and GONG, J. K. (1953). Effect of 2000 roentgens local X-irradiation on metabolism and alkaline phosphatase activity of rat bone. *Am. J. Physiol.* **173**, 115-19. [87]

COLE, A. R. C. and DARTE, J. M. M. (1963). Osteochondromata following irradiation in children. *Pediatrics N. Y.* **32**, 285-8. [84]

CONARD, R. A. (1970). Medical survey of the people of Rongelap and Utirik Islands thirteen, fourteen, and fifteen years after exposure to fall-out radiation. (March 1967, March 1968, and March 1969). BNL 50220 (T562). Brookhaven National Laboratory, Upton, New York. [58, 60, 61, 62]

CONTI, E. A., PATTON, G. D., CONTI, J. E., and HEMPELMANN, L. H. (1960). Present health of children given X-ray treatment to the anterior mediastinum in infancy. *Radiology* **74**, 386-91. [74]

COOK, G. M. W. (1958). Chemistry of membranes. *Br. med. Bull.* **24**, 118-23. [221]

COURT BROWN, W. M. and DOLL, R. (1957). Leukaemia and aplastic anaemia in patients irradiated for ankylosing spondylitis. *Spec. Rep. Ser. med. Res. Coun.* (Lond.) 295. [39, 51, 74, 75, 76]

—— and DOLL, R. (1958). Expectation of life and mortality from cancer among British radiologists. *Br. med. J.* **2**, 181-7. [39, 69]

—— and DOLL, R. (1959). Adult leukaemia. Trends in mortality in relation to aetiology. *Br. med. J.* **1**, 1063-9. [48]

—— and DOLL, R. (1965). Mortality from cancer and other causes after radiotherapy for ankylosing spondylitis. *Br. med. J.* **2**, 1327-32. [39, 74, 75]

—— SPIERS, F. W., DOLL, R., DUFFY, B. J., and MCHUGH, M. J. (1960). Geographical variation in leukaemia mortality in relation to background radiation and other factors. *Br. med. J.* **1**, 1753-9. [52, 54]

CRANDALL, J. L. (1968). Survey of applications for Cf 252. CONF. 681032 (U.S.A.E.C. Div. of Tech. Info), 225-56. [246]

CRUZ, M., COLEY, B. L., and STEWART, F. W. (1957). Post-radiation bone sarcoma. Report of 11 cases. *Cancer* **10**, 72-88. [79, 82, 83]

CURTIS, H. J. (1969). Somatic mutations in radiation carcinogenesis, in *Radiation-induced cancer*, pp. 45-55. Atomic Energy Agency, Vienna. [1]

DAHL, B. (1934). Effects des rayons X sur les os longs en développement. Étude radiographique et anatomique. *J. Radiol. Électrol.* **18**, 131–40. [88]

—— (1936). La théorie de l'ostéoclasie et le comportement des ostéoclastes vis à vis du bleu trypan et vis à vis de l'irradiation aux rayons X. *Acta path. microbiol. scand.*, suppl. **26**, 234–9. [88]

DAHLIN, D. C. (1957). *Bone tumours.* C. C. Thomas. Springfield, Illinois. [121]

—— and COVENTRY, M. B. (1967). Osteogenic sarcoma. A study of six hundred cases. *J. Bone Jt Surg.* (Am. ed.) **49**, 101–10. [78, 79, 84]

DALAND, E. M. (1949). Radiation necrosis of the jaw. *Radiology* **52**, 205–15. [89]

DALBY, R. G., JACOX, H. W., and MILLER, N. F. (1936). Fracture of the femoral neck following irradiation. *Am. J. Obstet. Gynec.* **32**, 50–9. [88]

DAMESHEK, W. (1951). Some speculations on the myeloproliferative syndromes. *Blood* **6**, 372–5. [48, 196]

—— (1956). The myeloproiferative disorders. Proc. Third National Cancer Conference. Philadelphia and Montreal, J. B. Lippincott Co. pp. 383–7. [48, 196]

—— (1970). The myeloproliferative disorders, in *Myeloproliferative disorders of animals and man.* U.S. Atomic Energy Commission. [48, 196]

DARLEY, P. J. (1968). Measurements of linear path length distributions in bone and bone marrow using a scanning technique. In *Proceedings of the symposium on microdensitometry*, E.A.E.C. Publication EUR 3747, d-e-f, 509–26. [17]

DAWE, C. J., LAW, L. W., and DUNN, T. B. (1959). Studies of parotid tumour agent in cultures of leukaemic tissues of mice. *J. natn. Cancer Inst.* **23**, 717–97. [7]

DICKSON, R. J. (1969). The late results of radium treatment for benign uterine haemorrhage. *Br. J. Radiol.* **42**, 582–94. [78]

DMOCHOWSKI, L. (1969). Comparison of leukomogenic and sarcomogenic viruses at the ultra-structural level. *Proc. 4th international symposium on comparative leukaemia research.* Cherry Hill, N.J. [8]

DOLL, R. and SMITH, P. G. (1968). The long-term effects of X-irradiation in patients treated for metropathia haemorrhagica. *Br. J. Radiol.* **41**, 362–8. [77]

DOLPHIN, G. W. and EVE, I. S. (1963). The metabolism of strontium in adult humans. *Physics Med. Biol.* **8**, 193–203. [155, 156]

DOUGHERTY, J. H. (1962). Haematological responses to internal irradiation in the beagle in *Some aspects of internal irradiation*, (ed. Dougherty, T. F., Jee, W. S. S., Mays, C. W., and Stover, B. J.), pp. 79–93. Pergamon Press, Oxford. [101, 165, 239]

—— (1971). Early haematologic effects of californium in the beagle. Research in Radiobiology, University of Utah report COO-119-244, pp. 117–25. [246]

—— and ROSENBLATT, L. S. (1969). Leukocyte depression in beagles injected with ^{226}Ra or ^{239}Pu, pp. 457–70, in *Delayed effects of bone-seeking radionuclides*, (ed. Mays, C. W., Jee, W. S. S., Lloyd, R. D., Stover, B. J., Dougherty, J. H., and Taylor, G. N.). University of Utah Press, Salt Lake City. [165, 226]

—— Taylor, G. N., and Mays, C. W. (1972). ^{90}Sr toxicity in adult beagles in *Biomedical implications of radiostrontium exposure*, ed. Goldman, M. and Bustad, L. M. Division of Technical Information, U.S. Atomic Energy Commission, Oak Ridge,

Conference CONF.710201. [165]

DOUGHERTY, T. F. (1962). Incidence of bone cancer in internally irradiated dogs in *Some aspects of internal irradiation*, (ed. Dougherty, T. F., Jee, W. S. S., Mays, C. W., and Stover, B. J.), pp. 47–57. Pergamon Press, Oxford. [138]

—— and MAYS, C. W. (1969). Bone cancer induced by internally-deposited emitters in beagles, pp. 361–7, in *Radiation-induced cancer*. International Atomic Energy Agency, Vienna. [**49, 102, 104, 105, 106, 133, 139, 162, 186, 222, 238, 239**]

DOWNIE, E. D., MACPHERSON, S., RAMSDEN, E. N., SISSONS, H. A., and VAUGHAN, J. (1959). The effect of daily feeding of ^{90}Sr to rabbits. *Br. J. Cancer* **13**, 408–23. [**148, 183**]

DUDLEY, R. A. (1967). A survey of radiation dosimetry in thorium dioxide cases. *Ann. N.Y. Acad. Sci.* **145**, 595–607. [**231, 236, 237**]

DUNGWORTH, D. L., GOLDMAN, M., SWITZER, J. W., and MCKELVIE, D. H. (1969). Development of a myeloproliferative disorder in beagles continuously exposed to ^{90}Sr. *Blood* **34**, 610–32. [**48, 178, 179**]

DUNLAP, C. E., AUB, J. C., EVANS, R. D., and HARRIS, R. S. (1944). Transplantable osteogenic sarcoma induced in rats by feeding radium. *Am. J. Pathol.* **20**, 1–21. [**138, 139, 145, 147, 148**]

DUNNILL, M. S. (1970). Personal communication. [32]

—— ANDERSON, J. A. and WHITEHEAD, R. (1967). Quantitative histological studies on age changes in bone. *J. Path. Bact.* **94**, 275–91. [32]

DURBIN, P. W. (1962). Distribution of the transuranic elements in mammals. *Hlth Phys.* **8**, 665–71. [**154, 231, 247**]

—— (1972). Plutonium in Man: A new look at the old data, pp. 469–530, in *Radiobiology of Plutonium*, ed. B. J. Stover and W. S. S. Jee. J. W. Press, University of Utah. [**203**]

DZIEWIATKOWSKI, D. D. and WOODARD, H. Q. (1959). Effect of irradiation with X-rays on the uptake of ^{35}S sulphate by the epiphyseal cartilage of mice. *Lab. Invest.* **8**, 205–13. [**86**]

EISENBUD, M. (1968). Radionuclides in the environment, in *Diagnosis and treatment of deposited radionuclides* [pp. 3–18.] Proceedings of symposium, Richland 1967 (ed. H. A. Kornberg and W. D. Norwood). Excerpta Medica Foundation, Monographs nuclear medicine No. 2. [**55**]

ELLSASSER, J. C., FARNHAM, J. E., and MARSHALL, J. H. (1969). Comparative kinetics and autoradiography of ^{45}Ca and ^{133}Ba in 10-year old beagle dogs. *J. Bone Jt Surg.* (Am. ed.) **51**, 1397–412. [**97, 98, 111, 115**]

EVANS, R. D. (1966). The effects of skeletally deposited alpha-ray emitters in man (Silvanus Thompson Memorial Lecture). *Br. J. Radiol.* **39**, 881–95. [**13, 39, 114, 115, 119, 123, 124, 126, 130, 131, 132**]

—— HARRIS, R. S., and BUNKER, J. W. M. (1944). Radium metabolism in rats and production of osteogenic sarcoma by experimental radium poisoning. *Am. J. Roentgenol.* **52**, 353–73. [**138**]

—— KEANE, A. T., KOLENKOW, R. J., NEAL, W. R., and SHANAHAN, M. M. (1968). Radiogenic tumours in the radium and mesothorium cases studied at M.I.T. Annual Progress Report on Radium and Mesothorium Poisoning and Dosimetry and Instrumentation Techniques in Applied Radioactivity, MIT-952-5 Part 1, pp. 3–48. Massachusetts Institute of Technology, Massachusetts. [**105**]

—— KEANE, A. T., KOLENKOW, R. J., NEAL, W. R., and SHANAHAN, M. M. (1969). Radiogenic tumours in the radium and mesothorium cases studied at M.I.T., in *Delayed effects of bone-seeking radionuclides*, (ed. Mays, C. W., Jee, W. S. S., Lloyd, R. D., Stover, B. J. Dougherty, J. H., and Taylor, G. N.), pp. 157–94. University of Utah Press, Salt Lake City. [**105, 114, 119, 130, 133, 151, 157**]

—— KEANE, A. T., and SHANAHAN, M. M. (1969). Radium and mesothorium toxicity in human beings. Annual Progress Report on Radium and Mesothorium Poisoning and Dosimetry and Instrumentation Techniques in Applied Radioactivity, MIT-952-6, pp. 1–30. Massachusetts Institute of Technology, Massachusetts. [**105, 114, 119, 121, 123, 126, 130, 132, 134, 193**]

EWING, J. (1926). Radiation osteitis. *Acta radiol.* **6**, 399–412. [**88, 89**]

FABER, M. (1968). Epidemiological experience with thorotrast in Denmark, *in* Dosimetry and Toxicity of Thorotrast, pp. 139–146. International Atomic Energy Agency, Vienna. [**235**]

—— and JOHANSEN, C. (1967). Leukaemia and other haematological disease after Thorotrast. *Ann. N. Y. Acad. Sci.* **145**, 755–8. [**235**]

FABRIKANT, J. I. and SMITH, C. L. D. (1964). Radiographic changes following the administration of bone-seeking radionuclides. *Br. J. Radiol.* **37**, 53–62. [**226**]

FINCH, S. C., HOSHINO, T., TOGA, T., ICHIMARU, M., and INGRAM, R. H., Jr. (1969). Chronic lymphocytic leukaemia in Hiroshima and Nagasaki, Japan, *Blood* **33**, 79–86. [**123**]

FINKEL, A. J., MILLER, C. E., and HASTERLIK, R. J. (1968). Radium-induced malignant tumours in man. Rep. ANL 7461, pp. 5–25. Argonne natn. Lab., Argonne, Ill. [**133**]

—— MILLER, C. E., and HASTERLIK, R. J. (1969a). Radium-induced malignant tumours in man, pp. 195–225, in *Delayed effects of bone-seeking radionuclides*, (ed. Mays, C. W., Jee, W. S. S., Lloyd, R. D., Stover, B. J., Dougherty, J. H., and Taylor, G. N.). University of Utah Press, Salt Lake City. [**105, 119, 121, 123, 130, 133**]

—— MILLER, C. E., and HASTERLIK, R. J. (1969b). Radiobiological parameters in human cancers attributable to long-term radium deposition in *Radiation-induced cancer*. International Atomic Energy Agency, Vienna. [**105, 119, 121, 130, 133**]

FINKEL, M. P. (1953). Relative biological effectiveness of radium and other alpha emitters in CFI female mice. *Proc. Soc. Exper. Biol. Med.* **83**, 494–8. [**138, 222**]

—— (1956). Relative biological effectiveness of internal emitters. *Radiology* **67**, 665–72. [**138, 222**]

—— (1959a). Late effects of internally deposited radioisotopes in laboratory animals. *Radiat. Res. Suppl.* **1**, 265–79. [**138, 222**]

—— (1959b). Induction of tumours with internally administered isotopes, pp. 322–35, in *Radiation biology and cancer*, No. 12. University of Texas Press, Austin. [**138, 172, 173, 222**]

—— and BISKIS, B. O. (1959). The induction of malignant bone tumours in mice by radioisotopes. *Acta. Un. int. Cancr.* **15**, 99–106. [**138**]

—— —— (1960). Radium damage in mouse bones in *The relation of radiation damage to radiation dose*. International Atomic Energy Agency, Vienna. pp. 27–8. [**138**]

—— —— (1962). Toxicity of plutonium in mice. *Hlth Phys.* **8**, 565–79. [**222, 224**]

—— —— (1968). Experimental induction of osteosarcomas. *Prog. exp. Tumour Res.* **10**, 72–111. [**8, 90, 91, 107, 148**]

—— —— (1969). Osteosarcomas induced in mice by FBJ virus and strontium-90, pp. 417–35, in *Delayed effects of bone-seeking radionuclides*, (ed. Mays, C. W., Jee, W. S. S., Lloyd, R. D., Stover, B. J., Dougherty, J. H., and Taylor, G. N.). University of Utah Press, Salt Lake City. [**8, 9, 172**]

—— —— and FARRELL, C. (1967). Pathogenic effects of extracts of human osteosarcomas. *Arch. Path.* **84**, 425–8. [**8**]

—— —— —— (1968). Osteosarcomas appearing in Syrian hamsters after treatment with extracts of human osteosarcomas. *Proc. Nat. Acad. Sci.* **60**, 1223–30. [**8**]

—— —— —— (1969). Non-malignant and malignant changes in hamsters inoculated with extracts of human osteosarcomas. *Radiology* **92**, 1546–52. [**8**]

—— —— GRECO, J. L., and CAMDEN, R. W. (1972). ^{90}Sr toxicity in dogs. Status of Argonne study on influence of age and dose pattern, in *Biomedical implications of radiostrontium exposure* (ed. M. Goldman and L. K. Bustad). Division of Technical Information, U. S. Atomic Energy Commission, Oak Ridge, Conference CONF. 710201. [**48, 49, 165, 171, 186**]

—— —— and JINKINS, P. B. (1966). Virus induction of osteosarcomas in mice. *Science* **151**, 698–701. [**7**]

—— —— —— (1969). Toxicity of radium 226 in mice, pp. 369–391, in *Radiation-induced cancer*. International Atomic Energy Agency, Vienna. [**107, 138, 140, 141, 142, 143**]

—— —— and SCRIBNER, G. M. (1958). The influence of ^{90}Sr upon life span and neoplasms of mice. *2nd Proc. Int. Conf. peaceful uses atom energy.* Geneva **22**, 65–70. [**173**]

—— —— —— (1959). The influence of ^{90}Sr upon life span and neoplasms of mice. *Progress in Nuclear energy*, Series VI. Vol 2. Biological sciences. pp. 199–209. Pergamon Press, Oxford. [**49, 196**]

—— —— TOLLE, J. and BISKIS, B. O. (1966). Serial radiography of virus-induced osteosarcomas in mice. *Radiology* **87**, 333–9. [**7**]

—— LISCO, H., and BRUES, A. M. (1954). *Toxicity of ^{89}Sr in mice. Tumours among the control animals*. Argonne National Laboratory, Quarterly Report on Biological and Medical Research Division, ANL 5288, pp. 23–34. [**8**]

—— REILLY, Jr., C. A., BISKIS, B. O. and CAMDEN, R. W. (1971). Bone tumours caused by oncogenic viruses. *J. Bone & Joint Surg.* **43A**, 806. [**8**]

FOLLEY, J. H., BORGES, W. H., and YAMAWAKI, T. (1952). Incidence of leukaemia in survivors of the atomic bomb in Hiroshima and Nagasaki. *Am. J. Med.* **13**, 311–21. [62]

FOREMAN, J. H., MOSS, W., and LANGHAM, W. (1960). Plutonium accumulation from long-term occupational exposure. *Hlth Phys.* **2**, 326–33. [204, 206]

FORREST, A. W. (1961). Tumours following radiation about the eye. *Trans. Am. Acad. Opthal. Oto-lar.* **65**, 694–717. [80]

FRANCISCO, F. B., PUSITZ, M. E., and GERUNDO, M. (1936). Malignant degeneration of a benign bone cyst. *Arch. Surg.* **32**, 669–78. [83]

FRANCONI, VON G. and ILLIG. R. (1959). Beinverkurzung und Enstehung einer solitären cartilaginären Exostose nach gelenknaher Bestrahlung eines Hämangioms in Sänglingsalter. *Helv. paediat. Acta* **14**, 425–9. [84]

FRANTZ, C. H. (1950). Extreme retardation of epiphyseal growth from roentgen irradiation. A case study. *Radiology* **55**, 720–4. [84]

—— (1968b). Induction of bone tissue by transitional epithelium. *Clin. Orthop.* **59**, 21–37. [28, 29]

FRIEDENSTEIN, A. J. and KURALESOVA, A. I. (1970). Osteogenic precursor cells of bone marrow in radiation chimaeras. To be published. [29]

FRIEDENSTEIN, A. J., PETRAKOVA, K. V., KUROLESOVA, A. J., and FROLOVA, G. P. (1968). Heterotopic transplants of bone marrow. Analysis of precursor cells for osteogenic and haematopoietic tissues. *Transplantation* **6**, 230–47. [28]

FRIEDENSTEIN, A. J., PIATETZKY-SHAPIRO, I. I., and PETRAKOVA, K. V. (1966). Osteogenesis in transplants of bone marrow cells. *J. Embryol. exp. Morph.* **16**, 381–90. [28, 29]

FRITZ, T. E., NORRIS, W. P., REHFELD, C. E., and POOLE, C. M. (1970). Myeloproliferative disease in beagle dogs given protracted whole-body irradiation of single doses of ^{144}Ce, pp. 219–214, in *Myeloproliferative disorders of animals and man* (ed. Clarke, W. J., Howard, E. B., and Hackett, P. L.). U.S. Atomic Energy Commission. [91]

FRITZSCHE, K. and BAHNEMANN, H. (1969). Further studies of avian osteopetrosis virus 2. *Poultry Science* **48**, 2123–9. [7]

FUJINAGA, S., POEL, W. E., and DMOCHOWSKI, L. (1970). Light and electron microscope studies of osteosarcomas induced in rats and hamsters by Harvey and Moloney sarcoma viruses. *Cancer Res.* **30**, 1698–1708. [7]

FURTH, J. and FURTH, O. B. (1936). Neoplastic disease produced in mice by general irradiation with X-rays. *Am. J. Cancer* **28**, 54–65. [51]

GERBER, C., HAMBURGER, R., and HULL, E. W. S. (1966). *Plowshare*, U.S. Atomic Energy Commission, Division of Technical Information. [56]

GLIMCHER, M. J. and KRANE, S. M. (1964). The incorporation of radioactive inorganic orthophosphate as organic phosphate by collagen fibrils *in vitro*. *Biochem.* **3**, 195–202. [29]

GLUCKSMANN, A. (1963). Carcinogenesis, in *Cellular basis and aetiology of late somatic effects of ionizing radiation*, (ed. Harris, R. J. C.). Academic Press, London. pp. 121–33. [11]

GOLDENBERG, R. R., CAMPBELL, C. J., and BONFIGLIO, M. (1970). Giant cell tumour of bone. An analysis of 218 cases. *J. Bone Jt Surg.* (Am ed.) **52**, 619–64. [**82**]

GOLDMAN, M., DELLA ROSA, R. J., and MCKELVIE, D. H. (1969). Metabolic, dosimetric, and pathologic consequences in the skeletons of beagles fed ^{90}Sr, in *Delayed effects of bone-seeking radionuclides* (ed. Mays, C. W., Jee, W. S., Lloyd, R. D., Stover, B. J., Dougherty, J. H. and Taylor, G. N.). University of Utah Press, Salt Lake City. pp. 61–77. [**138, 178**]

—— DUNGWORTH, D. L., BULGIN, M. S., ROSENBLATT, L. S. RICHARDS, W. P. C., and BUSTAD, L. K. (1969). Radiation-induced neoplasma in beagles after administration of ^{90}Sr and ^{226}Ra in *Radiation-induced cancer*. International Atomic Energy Agency, Vienna. pp. 345–60. [**144, 145, 146, 148, 178, 179**]

GOODMAN, A. H. and SHERMAN, M. S. (1963). Postirradiation fractures of the femoral neck. *J. Bone Jt Surg.* (Am ed.), **45**, 723–30. [**88, 89**]

GOWGIEL, J. M. (1965). A sarcoma observed in the irradiated monkey. *Radiat. Res.* **24**, 446–51. [**93**]

GRATZEK, F. R., HOLMSTROM, E. G., and RIGLER, L. G. (1945). Post-irradiation bone changes. *Am. J. Roentg.* **53**, 62–76. [**88**]

GRILLMAIER, R., MUTH, H., and OBERHAUSER, E. (1968). Untersuchungen zum Stoffwechsel des Radium 244 (ThX) und Abschätzung der Strahlendosen bei therapeutischer Anwendung. *Biophysik* **4**, 266–82. [**112**]

GROSS. L. (1951). 'Spontaneous' leukaemia developing in C3H mice following inoculation in infancy with AK-leukaemia extracts or AK-embryos. *Proc. Soc. exp. Biol. Med.* **76**, 27–32. [**5**]

—— (1958). Attempt to recover filterable agent from X-ray-induced leukaemia. *Acta haemat.* **19**, 353–61. [**6**]

GRUCA, A. (1934). Case of sarcoma developing after roentgenotherapy of tuberculosis of bones. *Chirurja narz. ruchu Ortop. pol.* **7**, 187 (*Abstr. Z. Orthop.* (1936) **65**, 89. [**81**]

GRUNEBERG, H., BAINS, G. S., BERRY. R. J., RILES, L., SMITH, C. A. B., and WEISS, R. A. (1966). A search for genetic effects of high natural radioactivity in South India. *Spec. Rep. Ser. med. Res. Coun.* 307. Her Majesty's Stationery Office, London. [**54**]

GUIMARAES, J. P. and LAMERTON, L. F. (1956). Further experimental observations on the late effects of thorotrast administration. *Br. J. Cancer* **10**, 527–32. [**236**]

HADFIELD, E. H. (1970). A study of adenocarcinomas of the paranasal sinuses in woodworkers in the furniture industry. *Ann. R. Coll. Surg.* **46**, 301–19. [**50, 123**]

HALNAN, K. E. and RUSSELL, M. H. (1965). Polycythaemia vera. Comparison of survival and causes of death in patients managed with and without radiotherapy. *Lancet* **2**, 760–3. [**240**]

HAM, A. W. (1969). *Histology*, 6th edn. Pitman Publishing Co. Ltd., London. [**28**]

HAMILTON, J. G. (1947). The metabolism of the fission products and the heaviest elements. *Radiology* **49**, 325–348. [**94, 118**]

HAMMARSTROM, L. and NILSSON, A. (1970a). Radiopathology of Americium 241. (i) Distribution of americium in adult mice. *Acta radiol.* **9**, 433–42. [**231**]

—— —— (1970b). Radiopathology of Americium 241. (ii) Uptake in the developing teeth of rats. *Acta radiol.* **9**, 609–17. [**231**]

HARMAN, J. B. and LEDLIE, E. M. (1967). Survival of polycythaemia vera patients treated with radioactive phosphorus. *Br. med. J.* **ii**, 146–8. [**240**]

HARRISON, G. E., CARR, T. E. F., and SUTTON, A. (1967). Distribution of radioactive calcium, strontium, barium, and radium following intravenous injection into a healthy man. *Int. J. Radiat. Biol.* **13**, 235–47. [**37**]

—— —— —— and RUNDO, J. (1966). Plasma concentration and excretion of calcium-47, strontium-85, barium-133 and radium-223 following successive intravenous doses to a healthy man. *Nature, Lond.* **209**, 526–7. [**35, 36, 37**]

—— HOWELLS, G. R., and POLLARD, J. (1967). Comparative uptake and elution of ^{45}Ca, ^{85}Sr, ^{133}Ba and ^{223}Ra in bone powder. *Calc. Tiss. Res.* **1**, 105–13. [**31, 156**]

HASTERLIK, R. J. (1967). Personal communication. [**122**]

—— (1968). Radium-induced malignant tumours in man. Twelfth Radiobiology Forum. Medical Research Council, London. [**120**]

—— FINKEL, A. J. and MILLER, C. E. (1964). The cancer hazards of industrial and accidental exposure to radioactive isotopes. *Ann. N. Y. Acad. Sci.* **114**, 832–7. [**39, 124, 127**]

—— MILLER, C. E., and FINKEL, A. J. (1969). Radiographic development of skeletal lesions in man many years after acquisition of radium burden. *Radiology* **93**, 599–603. [**122**]

HATCHER, C. H. (1945). The development of sarcoma in bone subjected to roentgen or radium irradiation. *J. Bone Jt Surg.* **27**, 179–95. [**81**]

—— and CAMPBELL, J. C. (1951). Benign chondroblastoma of bone: its histologic variations and a report of late sarcoma in the site of one. *Bull. Hosp. Jt Dis.* **12**, 411–30. [**83**]

HATFIELD, P. M. and SCHULZ, M. D. (1970). Postirradiation sarcoma. Including 5 cases after X-ray therapy of breast carcinoma. *Radiology* **96**, 593–602. [**78, 79, 80, 81, 82**]

HELLER, M. (1948). *Histopathology of irradiation from external and internal sources*, (ed. Bloom, W.), p. 70. McGraw-Hill, New York. [**145**]

HEMPELMANN, L. H., PIFER, J. W., BURKE, G. J., TERRY, R., and AMES, W. R. (1967). Neoplasms in persons treated with X-rays in infancy for thymic enlargement. A report of the third follow-up survey. *J. natn. Cancer Inst.* **38**, 317–41. [**45, 74, 84**]

HERRING, G. M. (1970). A review of recent advances in the chemistry of calcifying cartilage and bone matrix. *Calc. Tiss. Res. Suppl.* **4**, 17–23. [**29**]

—— (1972). The organic matrix of bone in *The biochemistry and physiology of bone*, 2nd ed., (ed. G. H. Bourne). Academic Press, New York. **1**, 128–189. [**29, 30, 31**]

—— ANDREWS, A. T. de B., and CHIPPERFIELD, A. R. (1971). Chemical structure of bone sialoprotein and a preliminary study of its calcium-binding properties, in

BIBLIOGRAPHY

Cellular mechanisms for calcium transfer (ed. G. Nichols, Jr & R. H. Wasserman). Academic Press, New York and London. pp. 64–72. [**30**]

—— VAUGHAN, J., and WILLIAMSON, M. (1962). Preliminary report on the site of localization and possible binding agents for yttrium, americium, and plutonium in cortical bone. *Hlth Phys.* **8**, 717–24. [**30, 210, 211**]

HEWITT, D. (1955). Some features of leukaemia mortality. *Br. J. prev. soc. med.* **9**, 81–8. [**69**]

HIGHT. D. (1941). Spontaneous fracture of the femoral neck following roentgen-ray therapy over the pelvis. *J. Bone Jt Surg.* **23**, 676–81. [**89**]

HINDMARSH, M., OWEN, M., and VAUGHAN, J. (1959). A note on the distribution of radium and a calculation of the radiation dose nonuniformity factor for radium 226 and strontium 90 in the femur of a luminous dial painter. *Br. J. Radiol.* **32**, 183–7. [**128, 129, 135, 136**]

—— —— —— LAMERTON, L. F., and SPIERS, F. W. (1958). The relative hazards of ^{90}Sr and ^{226}Ra. *Br. J. Radiol.* **31**, 518–33. [**135, 136**]

—— and VAUGHAN, J. (1957). The distribution of radium in certain bones from a man exposed to radium for 34 years. *Br. J. Radiol. Suppl.* **7**, 71–80. [**129, 135, 136**]

HOECKER, F. E. and ROOFE, P. G. (1951). Studies of radium in human bone. *Radiology* **56**, 89–98. [**136**]

HOFER, R. and KUHBOCK, J. (1970). Late results of radiophosphorus therapy in polycythemia vera. *Blut* **20**, 237–41. [**241**]

HOLM, N. W. and BERRY, R. J. (eds) Manual on Radiation Dosimetry. Marcel Dekker, Inc., New York 1970. [**vii**]

HOLMES, J. R. (1961*a*). Post-mortem findings in avian osteopetrosis. *J. Comp. Path.* **71**, 20–7. [**7**]

—— (1961*b*). Radiological lesions in avian osteopetrosis. *Br. J. Radiol.* **34**, 368–77. [**7**]

—— (1964). Avian osteopetrosis. *Natn. Cancer Inst. Monogr.* **17**, 63–79. [**7**]

HOWARD, E. B. and CLARKE, W. J. (1970). Induction of haematopoietic neoplasms in miniature swine by chronic feeding of strontium 90. *J. natn. Cancer Inst.* **44**, 21–9. [**48, 180, 181**]

—— —— KARAGIANES, M. T., and PALMER, R. F. (1969). ^{90}Sr-induced bone tumours in miniature swine. *Radiat. Res.* **39**, 594–607. [**41, 46, 165, 186**]

—— FRAZIER, M. E., and JANNKE, C. C. 1970. Viral studies in a miniature swine. BNWL 1306, Part 1, UC-48, pp. 39–50. Battelle North West Laboratory, Richland. Washington. [**6, 182**]

—— JANNKE, C. C., FRAZIER, M. E. and ADEE, R. R. (1970). Strontium-90 swine tissue culture, studies. BNWL 1050, Part 1, UC-48. Battelle North West Laboratory, Richland, Washington. [**182**]

—— USHIJIMA, R. N., HACKETT, P. L. and FRAZIER, M. E. (1970). Corollary studies in ^{90}Sr induced leukaemogenesis in swine in *Myeloproliferative disorders of animals and man*. U.S. Atomic Energy Commission Division of Technical Information. [**6**]

HUEBNER, R. J. (1970*a*). Viruses and cancer. *Proc. natn. Cancer Conf.* **6**, 263–4. [**10**]

—— (1970*b*). Identification of leukaemogenic viruses: specifications for vertically transmitted, mostly 'switched off' RNA tumour viruses as determinants of the generality of cancer, in *Comparative leukaemia research* (1969). S. Karger Basel. pp. 22–44. [10]

—— HARTLEY, J. W., LANE, W. T., TURNER, H. C., and KELLOFF, G. (1969). Unpublished data quoted Rhim, J. S. *et al.* 1969. [8, 140]

—— KELLOFF, G. J., SARMA, P. S., LANE, W. T., and TURNER, H. C. (1970). Group-specific antigen expression during embryogenesis of the genome of the C-type RNA tumour virus: implications for ontogenesis and oncogenesis. *Proc. natn. Acad. Sci. U.S.A.* **67**, 366–76. [6]

HUG, O., GOSSNER, W., MULLER, W., LUZ, A., and HINDRINGER, B. (1969). Production of osteosarcomas in mice and rats by incorporation of radium 224, in *Radiation-induced cancer* (ed. A. Ericson). International Atomic Energy Agency, Vienna. [14, 111, 112, 151, 152]

—— KELLERER, A., and ZUPPINGER, A. (1966). 'Der Zeitfaktor', *Handbach der Medizinischen Radiologie*, Vol. 2. 1. Strahlenbiologie (ed. Diethelm, L. *et al.*). Springer Verlag. pp. 271–354. [152]

HULSE, E. V. (1969). Osteosarcomas, fibrosarcomas, and basal-cell carcinomas in rabbits after irradiation with gamma-rays or fission neutrons: an interim report on incidence, site of tumours and R.B.E. *Int. J. Radiat. Biol.* **16**, 27–30. [90, 91]

HUTCHISON, G. B. (1968). Leukaemia in patients with cancer of the cervix uteri treated with radiation. A report covering the first five years of an international study. *J. natn. Cancer Inst.* **40**, 951–82. [77]

HUXLEY, J. (1958). *Biological aspects of cancer*. Allen and Unwin, London. [1]

INTERNATIONAL COMMISSION ON RADIATION UNITS AND MEASUREMENTS. (1970). ICRU Report 16, Linear Energy Transfer. [3, 12]

INTERNATIONAL COMMISSION ON RADIOLOGICAL PROTECTION (1960). I.C.R.P. Publication No. 2. Pergamon Press, Oxford. [106]

—— (1966). *The evaluation of risks from radiation*. I.C.R.P. Publication No. 8. Pergamon Press, Oxford. [13, 68, 106]

—— (1968). *A review of the radiosensitivity of the tissues in bone*. I.C.R.P. Publication No. II. Pergamon Press, Oxford. [38, 121]

—— (1969). *Radiosensitivity and spatial distribution of dose*. I.C.R.P. Publication No. 14, Pergamon Press, Oxford. [184]

—— (1972). *Metabolism of plutonium and related elements and their compounds*. Report of a Task Group of I.C.R.P. in preparation. Pergamon Press, Oxford. [226]

IRVING, J. T. and WUTHIER, R. E. (1968). Histochemistry and biochemistry of calcification with special reference to the role of lipids. *Clin. Orthop.* **56**, 237–60. [31]

ISHIHARA, T. and KUMATORI, T. (1965). Chromosome aberrations in human leukocytes irradiated *in vivo* and *in vitro*. *Acta haemat. jap.* **28**, 291–307. [61]

ISHIMARU, T., HOSHINO, T., ICHIMARU, M., OKADA, H., TOMIYASU, T., TSUCHIMOTO, T., and YAMAMOTO, T. (1971). Leukaemia in atomic bomb survivors, Hiroshima and Nagasaki, 1 October 1950–30 September 1966. *Radiat. Res.* **45**, 216–33. [**62, 64, 66, 67**]

JABLON, S., ISHIDA, M., and YAMASAKI, M. (1965). Studies of the mortality of atom bomb survivors. 3. Description of the sample and mortality 1950–1960. *Radiat. Res.* **25**, 25–52. [**62**]

—— TACHIKAWA, K., BELSKY, J. L., and STEER, A. (1971). Cancer in Japanese children exposed to atomic bombs. *Lancet* **i**, 927–31. [**40, 62, 63**]

JACOB, F. and MONOD, J. (1961). Genetic regulatory mechanisms in the synthesis of proteins. *J. molec. Biol.* **3**, 318–56. [**6**]

JAFFE, H. L. (1958). *Tumours and tumorous conditions of bone and joints.* Kimpton, London. [**44, 82, 83, 88, 89**]

JEE, W. S. S. and ARNOLD, J. S. (1960). Effect of internally deposited radioisotopes upon blood vessels of cortical bones. *Proc. Soc. exp. Biol. Med.* **105**, 351–6. [**129, 138, 147**]

—— BARTLEY, M. H., DOCKUM, N. L., YEE, J., and KENNER, G. H. (1969). Vascular changes in bones following bone-seeking radionuclides, pp. 437–55, in *Delayed effects of bone-seeking radionuclides* (ed. Mays, C. W., Jee, W. S., Lloyd, R. D., Stover, B. J., Dougherty, J. H., and Taylor, G. N.). University of Utah Press, Salt Lake City. [**200**]

—— DOCKUM, N. L., MICAL, R. S., ARNOLD, J. S., and LOONEY, W. B. (1967). Distribution of thorium daughters in bone. *Ann. N. Y. Acad. Sci.* **145**, 660–73. [**231, 234, 235**]

—— PARK, H. Z., and BURGGRAAF, R. (1969). Estimates of residence time of ^{239}Pu in trabecular bones of beagles. *Research in Radiobiology*, COO-119, 240, pp. 188–98. [**212**]

—— STOVER, B. J., TAYLOR, G. N., and CHRISTENSEN, W. R. (1962). The skeletal toxicity of ^{239}Pu in adult beagles. *Hlth Phys.* **8**, 599–607. [**222, 223**]

JONES, A. (1953). Irradiation sarcoma. *Br. J. Radiol.* **26**, 273–84. [**79, 81**]

JOWSEY, J. (1963). Microradiography of bone resorption, pp. 447–469, in Mechanisms of Hard Tissue Destruction, ed. Sognnaes, R. F., published No. 75. American Association for the Advancement of Science, Washington, D.C. [**22**]

—— KELLY, P. J., RIGGS, B. L., BIANCO, A. J. Jr., SCHOLZ, D. A., and GERSHON-COHEN, J. (1965). Quantitative microradiographic studies of normal and osteoporotic bone. *J. Bone Jt Surg.* (Am. ed.) **47**, 785–806. [**21, 33, 216**]

—— OWEN, M., TUTT, M., and VAUGHAN, J. (1955). Retention and excretion of 90-Sr by adult rabbits. *Br. J. exp. Path.* **36**, 22. [**158**]

—— RAYNER, B., TUTT, M., and VAUGHAN, J. (1953). The deposition of ^{90}Sr in rabbit bones following intravenous injection. *Br. J. exp. Path.* **34**, 384–91. [**157**]

—— and ROWLAND, R. E. (1960). Point-source beta irradiation of bone. Argonne National Laboratory, Radiological Physics Division, Semiannual report, January through June 1960, ANL 6199, pp. 21–35. [**15**]

—— SISSONS, H. A., and VAUGHAN, J. (1956). The site of deposition of Y^{91} in the bones of rabbits and dogs. *J. nucl. Energy* **2**, 168–76. [**154**]

JUNGHERR, E. and LANDAUER, W. (1938). Studies on fowl paralysis. A condition resembling osteopetrosis (marble bone) in the common fowl. *Agr. exp. State Bull.* No. 222, pp. 5–33. [**7**]

KAPLAN, H. S. (1966). Interaction of occult leukaemogenic viruses with ionizing radiation and other external leukaemogenic agents in the induction of thymic lymphosarcoma in the mouse in *Ciba Foundation symposium on the thymus*, pp. 310–24, (ed. Wolstenholme, G. E. W. and Porter, R.). J. & A. Churchill, London. [**6, 139**]

—— (1967). On the natural history of the murine leukaemias. *Cancer Res.* **27**, 1325–40. [**6**]

—— and BROWN, M. P. (1954). Development of lymphoid tumours in non-irradiated thymic grafts in thymectomised irradiated mice. *Science* **119**, 439–46. [**5**]

—— BROWN, M. P., HIRSCH, B. B., and CARNES, W. H. (1956). Indirect induction of lymphomas in irradiated mice. (ii) Factor of irradiation of the host. *Cancer Res.* **16**, 426–8. [**5**]

KATZMAN, A., WAUGH, T., and BERDON, W. (1969). Skeletal changes following irradiation of childhood tumours. *J. Bone Jt Surg.* (Am. ed.) **51**, 825–42. [**80, 81, 83, 84, 85**]

KELLOFF, G. J., LANE, W. T., TURNER, H. C., and HUEBNER, R. J. (1969). *In vivo* studies of the FBJ murine osteosarcoma virus. *Nature, Lond.* **223**, 1379–80. [**8, 172**]

KEMBER, N. F. (1960). *Quantitative studies on radiation injury to bone in experimental animals*. Thesis, London University. [**85**]

—— (1965). An *in vivo* cell survival system based on the recovery of rat growth cartilage from radiation injury. *Nature, Lond.* **207**, 501–3. [**85, 88**]

—— (1967). Cell survival and radiation damage in growth cartilage. *Br. J. Radiol.* **40**, 496–505. [**85, 86, 87, 88**]

—— (1969). Clones in bones. *J. Bone Jt Surg.* (Br. ed.) **51**, 385. [**85**]

—— and COGGINS, J. (1967). Changes in the vascular supply to rat growth cartilage during radiation injury and repair. *Int. J. Radiat. Biol.* **12**, 143–51. [**88**]

—— SADEK, M. (1970). Mitotic suppression in gut and growth cartilage by X-irradiation *in vivo*. *Int. J. Radiat. Biol.* **17**, 19–23. [**88**]

KHODYREVA, M. A. (1965). Penetration of ^{239}Pu through the skin. *Med. Rad.* **10**, Oct., 42–6 (from National Lending Library, Science and Technology RTS 5223). [**206**]

KIDMAN, B., RAYNER, B., TUTT, M. L., and VAUGHAN, J. (1952). Autoradiographic studies of the deposition of Sr^{89} in rabbit bones. *J. Path. Bact.* **64**, 453–9. [**99**]

—— TUTT, M. L., and VAUGHAN, J. M. (1951*a*). The retention of radioactive strontium and yttrium (^{89}Sr, ^{90}Sr, and ^{90}Y) in pregnant and lactating rabbits and their offspring. *J. Path. Bact.* **63**, 253–68. [**97, 99**]

—— —— —— (1951*b*). Excretion of ^{91}yttrium in rabbits. *Nature, Lond.* **167**, 858. [**97**]

KIRSTEN, W. H., ANDERSON, D. G., PLATZ, C. E., and CROWELL, E. B., Jr. (1962). Observations on the morphology and frequency of polyoma tumours in rats. *Cancer Res.* **22**, 484–91. [7]

KLIGERMAN, M. M. (1950). The effect of radioactive phosphorus on the growth of albino rats. *Am. J. Roentg.* **63**, 380–95. [243, 244]

KOLAR, J. (1960). Osteochondrome and Exostosen bei bestrahlten Kindern. *Arch. orthop. Unfallchir.* **51**, 631–4. [84]

KOLETSKY, S. and GUSTAFSON, G. E. (1955). Whole-body radiation as a carcinogenic agent. *Cancer Res.* **15**, 100–4. [90]

KOK, G. (1953). Spontaneous fractures of the femoral neck after the intensive irradiation of carcinoma of the uterus. *Acta radiol.* **40**, 511–27. [88]

KSHIRSAGAR, S. G., LLOYD, E., and VAUGHAN, J. (1966). Discrimination between strontium and calcium in bone and transfer from blood to bone in the rabbit. *Br. J. Radiol.* **39**, 131–40. [42, 156]

—— VAUGHAN, J., and WILLIAMSON, M. (1965). The occurrence of squamous carcinoma and osteosarcoma in young rabbits injected with ^{90}Sr (50–100 µc/kg). *Br. J. Cancer* **19**, 777–86. [40]

KULP, J. L. and SCHULERT, A. R. (1962). Strontium-90 in man. V. *Science N.Y.* **136**, 619–32. [58]

—— —— and HODGES, E. J. (1960). ^{90}Sr in man. Iv. *Science, N.Y.* **132**, 448–54. [58]

KUZMA, J. F. and ZANDER, G. (1957a). Cancerogenic effects of Ca^{45} and Sr^{89} in Sprague-Dawley rats. *A.M.A. Archs Path.* **63**, 198–206. [49, 154, 171, 195]

—— —— (1957b). The histogenesis of osteogenic sarcoma as induced by radioactive calcium and strontium. *Am. J. Path.* **33**, 607–8. [49, 154, 171]

LACASSAGNE, A. and VINZENT, R. (1929). Sarcomes provoqués chez les lapins par l'irradiation d'abcès a *streptobacillus caviae*. *C. r. Sianc. Soc. Biol.* **100**, 249–51. [79, 92]

LAGERQUIST, C. R., BOKOWSKI, D. L., HAMMOND, S. E., and HYLTON, D. B. (1968). Plutonium content of several internal organs following occupational exposure, in *Proceedings of the 13th annual bio-assay and analytical chemistry meeting* held at Berkeley, California, on October 12–13, 1967. CONF.671048, p. 103 (quoted Mays *et al.* 1970). [204]

—— —— —— —— (1969). Plutonium content of several internal organs following occupational exposure. *Am. Industr. Hyg. Ass. J.* **30**, 417–21. Cit. No. 4244435. [204, 205]

LAMERTON, L. F. (1958). An examination of the clinical and experimental data relating to the possible hazard to the individual of small doses of radiation. *Br. J. Radiol.* **31**, 229–39. [13, 51]

—— (1961). Somatic effects of radiation at low levels of dosage, with particular reference to protraction of radiation exposure. *Trans. IXth Int. Congr. Radiol.* **2**, 1203–11. [90]

—— (1966). The response of mammalian cell populations to continuous irradiation

Radiation Research, North Holland Publishing Company, Amsterdam (1967), pp. 643–58. [90]

—— (1966). Personal communication. [90]

—— (1968). Radiation biology and cell population kinetics. *Physics Med. Biol.* **13**, 1–14. [90]

LANGHAM, W. H. (1959). Physiology and toxicology of plutonium-239 and its industrial medical control. *Hlth Phys.* **2**, 172–85. [203, 204]

—— (1970). Biological considerations of non-nuclear incidents involving nuclear warheads. UCRL 50639. University of California Radiation Laboratory. Berkeley. [221]

—— BASSETT, S. H., HARRIS, P. S., and CARTER, R. E. (1950). Distribution and excretion of plutonium administered intravenously to man. Los Alamos Scientific Laboratory Report LA 1151. [203]

—— and CARTER, R. E. (1951). Los Alamos Scientific Laboratory Report LA 1309. [225, 226, 231]

—— LAWRENCE, J. N. P., McCLELLAND, J., and HEMPELMANN, L. H. (1962). The Los Alamos Scientific Laboratory experience with plutonium in man. *Hlth Phys.* **8**, 753–60. [203, 221]

LATOURETTE, H. B. and HODGES, F. J. (1959). Incidence of neoplasia after irradiation of thymic region. *Am. J. Roentg.* **82**, 667–77. [74]

LAWRENCE, J. H. (1955). Polycythemia: physiology, diagnosis, and treatment based on 303 cases. *Modern medical monograph* No. 13, New York. [240]

—— WINCHELL, H. S., and DONALD, W. G. (1969). Leukaemia in polycythaemia vera. Relationship to splenic myeloid metaplasia and therapeutic radiation dose. *Am. J. Int. Med.* **70**, 763–71. [241]

LEAVER, A. G. and SHUTTLEWORTH, C. A. (1968). Studies on the peptides, free amino acids and certain related compounds isolated from ox bone. *Archs oral Biol.* **13**, 509–25. [29]

LEDERER, C. M., HOLLANDER, J. M., and PERLMAN, I. (1967). *Table of isotopes.* 6th ed. J. Wiley & Sons, New York. [201]

LEDLIE, E. M. (1960). The incidence of leukaemia in patients with polycythaemia vera treated by ^{32}P. *Clin. Radiol.* **11**. 130–33. [240]

—— MYNORS, L. S., DRAPER, G. J., and GORBACH, P. D. (1970). Natural history and treatment of Wilm's tumour: an analysis of 335 cases occurring in England and Wales 1962–6. *Brit. med. J.* **4**, 195–200. [85]

LEE, W. R., MARSHALL, J. H., and SISSONS, H. A. (1965). Calcium accretion and bone formation in dogs. *J. Bone Jt Surg.* (Brit. ed.) **47**, 157–80. [21, 216]

LEVDIK, T. I., LEMBERG, V. K., BULDAKOV, L. A., LYUBCHANSKII, E. R., and PESTERNIKOV, V. M. (1972). Biological effectiveness of ^{237}Np. *Hlth. Phys.* **22**, 643–645. [247]

LEWIS, E. B. (1957). Leukaemia and ionising radiation. *Science, N.Y.* **125**, 965–72. [51, 68]

LICHTENSTEIN, L. (1951). Giant-cell tumor of bone; current status of problem in diagnosis and treatment. *J. Bone Jt Surg.* (Am. ed.) **33**, 143–50. [82]

—— (1953). Aneurysmal bone cyst—further observations. *Cancer, N.Y.* **6**, 1228–37. **[82]**

LIEBERMAN, M. and KAPLAN, H. S. (1959). Leukaemogenic activity of filtrates from radiation-induced lymphoid tumours of mice. *Science, N.Y.* **130**, 387–8. **[6]**

LINDENBAUM, A., LUND, C., SMOLER, M., and ROSENTHAL, M. W. (1968). Preparation, characterization, and distribution in mouse tissues of graded polymeric and monomeric plutonium. Radiochemical and autoradiographic studied in *Diagnosis and treatment of deposited radionuclides*. Proceedings of Symposium. Richland 1967 (ed. H. A. Kornberg and W. D. Norwood). Excerpta Medica Foundation, Monographs Nuclear Medicine No. 2. pp. 56–64. **[202, 219]**

—— ROSENTHAL, M. W., RUSSELL, J. J., MORETTI, E. S., and SMYTH, M. A. (1969). Metabolic and therapeutic studies of plutonium V. ANL 7635, pp. 186–90. **[202]**

LISCO, H. (1956). Bone as a critical organ for the deposition of radioactive materials. *In Ciba foundation symposium on bone structure and metabolism* (ed. G. E. W. Wolstenholme and C. M. O'Connor), p. 272–82. Churchill, London. **[11]**

—— and CONARD, R. A. (1967). Chromosome studies on Marshall Islanders exposed to fall-out radiation. *Science. N.Y.* **157**, 445–7. **[61]**

—— ROSENTHAL, M. W., and VAUGHAN, J. (1971). Observations on skeletal pathology in female CF1/ANL mice possibly related to virus infection. Personal communication. To be published. **[8, 107, 139, 152, 172, 224]**

LISTER, B. A. J., MORGAN, A., and SHERWOOD, R. J. (1963). Excretion of plutonium following accidental skin contamination. *Hlth. Phys.* **9**, 803–15. **[202]**

LITVINOV, N. N. (1957). Morphological changes of bone tissue in rats in chronic intoxication by radioactive strontium. *Arkh. Patol.* **19**, Part 1. 26–31. **[40]**

LLOYD, E. (1961*a*). The distribution of radium in human bone. *Br. J. Radiol.* **34**, 521–8. **[109, 114, 129, 137]**

—— (1961*b*). The relative distributions of radioactive yttrium and strontium and the secondary deposition of ^{90}Y built up from ^{90}Sr. *Int. J. Radiat. Biol.* **3**, 475–92. **[154]**

—— (1968). Relative binding of strontium and calcium in protein and non-protein fractions of serum in the rabbit. *Nature, Lond.* **217**, 355–6. **[156]**

—— and HODGES, D. (1971). Quantitative characterization of bone: a computer analysis of microradiographs. *Clin. Orthop.* **78**, 230–50. **[20, 32, 152]**

—— and MARSHALL, J. H. (1972). Toxicity of ^{239}Pu relative to ^{224}Ra in man and dog, pp. 377–383, *in* Radiobiology of Plutonium (ed. Stover, B. J. and Jee, W. S. S.). The J. W. Press, Salt Lake City. **[21]**

LLOYD, R. D., MAYS, C. W., TAYLOR, G. N., and ATHERTON, D. R. (1970). Americium-241 studies in beagles. *Hlth Phys.* **18**, 149–56. **[230]**

—— —— —— and WILLIAMS, J. (1972). Californium excretion and retention by beagles injected with ^{269}Cf or ^{252}Cf. Health Physics **22**, 667–673. **[246]**

LOGUE, G. L., GUTTERMAN, J. U., McGINN, T. G., LAZLO, J., and RUNDLES, R. O. (1970). Mephalan therapy of polycythemia vera. *Blood* **36**, 70–86. **[241]**

LOONEY, W. B. (1955). Late effects (twenty-five to forty years) of the early medical and industrial uses of radioactive materials. Their relation to the more accurate establishment of maximum permissible amounts of radioactive elements in the body. Part I. *J. Bone Jt Surg.* (Am. ed.) **37**, 1169–87. [**123, 124**]

—— (1956a). Late effects (25–40 years) of the early medical and industrial use of radioactive materials. Their relation to the more accurate establishment of maximum permissible amounts of radioactive elements in the body. Part II. *J. Bone Jt Surg.* (Am. ed.) **38**, 175–218. [**123, 124**]

—— (1956b). Late skeletal roentgenographic, histopathological, autoradiographic and radio-chemical findings following radium deposition. *Am. J. Roentg.* **75**, 559–72. [**123, 124**]

—— (1960). An investigation of the late clinical findings following thorotrast (Thorium dioxide) administration. *Am. J. Roentg.* **83**, 163–85. [**235**]

—— HASTERLIK, R. J., BRUES, A. M., and SKIRMONT, E. (1955). A clinical investigation of the chronic effects of radium salts administered therapeutically (1915–1931). *Am. J. Roentg.* **73**, 1006–37. [**123, 124, 127, 129**]

LOUTIT, J. F. (1967). Grafts of haemopoietic tissue: the nature of haemopoietic stem cells. *J. clin. Path.* **20**, suppl. 2, 535–9. [**46, 182, 197**]

—— (1970). Malignancy from radium. *Br. J. Cancer* **24**, 195–207. [**120**]

—— and VAUGHAN, J. M. (1971). The radiosensitive tissues in bone. *Br. J. Radiol.* **44**, 815. [**41**]

LUCAS, H. F. (1961). Correlation of the natural radioactivity of the human body to that of its environment: uptake and retention of Ra^{226} from food and water. Argonne National Laboratory, Radiological Physics Division, Semiannual Report, July through December 1960, ANL-6297, p. 55–66. [**55**]

LUCAS, H. F., Jr., ROWLAND, R. E., MILLER, C. E., HOLTZMAN, R. B., HASTERLIK, R. J., and FINKEL, A. J. (1963). An unusual case of radium toxicity. *Am. J. Roentg.* **90**, 1042–51. [**114**]

MACMAHON, B. and HUTCHINSON, G. B. (1964). Prenatal X-ray and childhood cancer. A review. *Acta Un. int. Cancr.* **20**, 1172–4. [**69**]

MACPHERSON, S. (1961). The stunting of growth in young rabbits injected with ^{90}Sr. *Int. J. Radiat. Biol.* **3**, 515–23. [**14, 85**]

—— OWEN, M., and VAUGHAN, J. (1962). The relation of radiation dose to radiation damage in the tibia of weanling rabbits injected with ^{90}Sr. *Br. J. Radiol.* **35**, 221–34. [**85, 167, 198, 199, 200**]

MAGNO, P. J., KAUFFMAN, P. E., GROULY, P. R. (1969). Plutonium 239 in human tissues and bone. *Radiol. Hlth Data*, Rep. **10**, 47–50 (quoted Mays *et al.* 1970). [**204**]

—— —— and SHLEIEN, B. (1967). Plutonium in environmental and biological media. *Hlth Phys.* **13**, 1325–30. [**204**]

MALETSKOS, C. J., KEANE, A. T., TELLES, N. C., and EVANS, R. D. (1969). Retention and absorption of ^{224}Ra and ^{234}Th and some dosimetric consequences of ^{224}Ra in human beings, in *Delayed effects of bone-seeking radionuclides* (ed. Mays, C. W., Jee, W. S., Lloyd, R. D., Stover, B. J., Dougherty, J. H., and Taylor, G. N.) pp. 29–48. University of Utah Press, Salt Lake City. [**111, 113, 117, 118, 238**]

BIBLIOGRAPHY 271

MARCH, H. C. (1944). Leukaemia in radiologists. *Radiology* **43**, 275–8. [**68**]

—— (1950). Leukaemia in radiologists. *Am. J. med. Sci.* **220**, 282–6. [**68**]

MARCIAL-ROJAS, R. A. (1960). Primary haemangiopericytoma of bone. Review of the literature and report of the first case with metastases. *Cancer N.Y.* **13**, 309–11. [**28**]

MARIE, P., CLUNET, J., and RAULOT-LAPOINTE, G. (1910). Contribution à l'étude du développement des tumeurs malignes sur les ulcères de roentgen. *Bull. Ass. fr. Étude Cancer.* **3**, 404. [**92**]

MARINELLI, L. D. (1958). Radioactivity and the human skeleton. *Am. J. Roentg.* **80**, 729–39. [**55**]

—— (1964). The doses from thorotrast and migrated descendents: status, prospects and implications. Argonne National Laboratory, Radiological Physics Division, Annual Report (July 1964–June 1965). ANL-7060, 37–45. [**233**]

—— and LUCAS, H. F. (1962). Translocation of thorium daughters to bone. In *Some aspects of internal irradiation,* (ed. T. F. Dougherty), Symposium, Heber, Utah 1961. Pergamon Press, Oxford. 499–516. [**233**]

MARSCH, E. (1922). Tuberkulose und Sarkom (Rontgensarkom). *Zentbl. Chir.* **49**, 1057–60. [**81**]

MARSHAK, A. (1949). Radioactive phosphorus in studies of the metabolism of normal and neoplastic tissues. *J. clin. Invest.* **28**, 1324–9. [**108**]

MARSHALL, J. H. (1964). Theory of alkaline metabolism, *J. theor. Biol.* **6**, 386–412. [**32, 34**]

—— (1969*a*). The retention of radionuclides in bone, pp. 7–23, in Delayed effects of bone-seeking radionuclides (ed. Mays, C. W., Jee, W. S., Lloyd, R. D., Stover, B. J., Dougherty, J. H. and Taylor, G. N.). University of Utah Press, Salt Lake City. [**94**]

—— (1969*b*). Measurements and models of skeletal metabolism in *Mineral metabolism III*, pp. 2–122. [**32, 34, 156**]

—— and FINKEL, M. P. (1960). Comparison of microdosimetry and tumour production ^{45}Ca, ^{90}Sr and ^{226}Ra in mice, pp. 40–4, in *The relation of radiation damage to radiation dose in bone.* International Atomic Energy Agency, Venna [**138**]

—— ROWLAND, R. E., and JOWSEY, J. (1959). Microscopic metabolism of calcium in bone. (v) The paradox of diffuse activity and long-term exchange. *Radiat. Res.* **10**, 258–70. [**35**]

—— RUNDO, J., and HARRISON, G. E. (1969). Retention of radium in man. *Radiat. Res.* 445–51. [**113**]

MARTELL, E. A. (1969). Plowing a nuclear furrow. *Environment* **11**, 2–15. [**56**]

MARTLAND, H. S. (1926). Microscopic changes of certain anemias due to radioactivity. *Archs Path.* **2**, 465–72. [**118**]

—— (1931). The occurrence of malignancy in radioactive persons. *Am. J. Cancer* **15**, 2435–516. [**119, 120**]

—— CONLON, P., and KNEF, J. P. (1925). Some unrecognized dangers in the use and handling of radioactive substances: with special reference to the storage of insoluble products of radium and mesothorium in the reticulo-endothelial system.

J. Am. med. Ass. **85**, 1769–76. [**118, 119**]

—— and HUMPHRIES, R. E. (1929). Osteogenic sarcoma in dial painters using luminous paint. Archs Path. **7**, 406–17. [**118**]

MATHEWS, J. L. (1971). Personal communication. [**27**]

MAYS, C. W. (1969). In Radiation-induced cancer. Proceedings of a symposium, Athens, International Atomic Energy Agency, Vienna. p. 360 (in discussion). [**109**]

—— DOUGHERTY, T. F., TAYLOR, G. N., LLOYD, R. D., STOVER, B. J., JEE, W. S. S., CHRISTENSEN, W. R., DOUGHERTY, J. H., and ATHERTON, D. R. (1969). Radiation-induced bone cancers in beagles, pp. 387–408, in Delayed effects of bone-seeking radionuclides (ed. Mays, C. W., Jee, W. S., Lloyd, R. D., Stover, B. J., Dougherty, J. H., and Taylor, G. N.). University of Utah Press, Salt Lake City. [**103, 104, 105, 106, 138, 157, 162, 186**]

—— —— —— STOVER, B. J., JEE, W. S. S., CHRISTENSEN, W. R., DOUGHERTY, J. H., STEVENS, W., Jr., and NABORS, C. J., Jr. (1970). Bone cancer induction by radionuclides: incidence vs dose. COO–119, 242. pp. 385–401. [**162, 165**]

—— HALDIN, R., and VAN DILLA, M. A. (1958). Thoron exhalation in radiothorium-burdened beagles. Radiat. Res. **9**, 438–44. [**112**]

—— and LLOYD, R. D. (1972). Bone sarcoma risk from ^{90}Sr in Biomedical implications of radiostrontium exposure, (ed. Goldman, M. and Bustad, L. K.). Division of Technical Information, U.S. Atomic Energy Commission, Oak Ridge, Conference CONF.710201. [**162, 189, 190, 191, 192, 193, 194**]

—— and TAYLOR, G. N. (1964). Low natural incidence of osteosarcomas in beagles. Research in radiobiology, University of Utah report, COO–119–231, pp. 70–75. [**139**]

—— —— JEE, W. S. S., and DOUGHERTY, T. F. (1970). Speculated risk to bone and liver from ^{239}Pu. Hlth Phys. **19**, 601–10. [**203, 204, 205, 206, 222, 226**]

—— and TUELLER, A. B. (1964). Determination of localized alpha-dose in soft tissue near radioactive bone. In Research in radiobiology, University of Utah, College of Medicine, Department of Anatomy, Radiobiology Laboratory, Annual report of work in progress on the chronic toxicity program COO–119–229, 199–205. [**239**]

—— VAN DILLA, M. A., FLOYD, R. L., and ARNOLD, J. S. (1958). Radon retention in radium-injected beagles. Radiat. Res. **8**, 480–9. [**109**]

MCCLANAHAN, B. J., HACKETT, P. L. and BEAMER, J. L. (1970). Chromosomal effects of ^{90}Sr in miniature swine. Battelle North-West Laboratory Annual Report 1050, Part 1, UC-48, pp. 217–218. [**182**]

MCCLELLAN, R. O., BOECKER, B. B., JONES, R. K., BARNES, J. E., CHIFFELLE, T. L., HOBBS, C. H. and REDMAN, H. C. (1972). Toxicity of inhaled radiostrontium in experimental animals, in Biomedical Implications of Radiostrontium Exposure, (eds M. Goldman and L. Bustad, Division of Technical Information, U.S. Atomic Energy Commission, Oak Ridge, Conference CONF.710201. [**49, 165, 166, 195**]

MCKENNA, R. J., SCHWINN, C. P., SOONG, K. Y., and HIGINBOTHAM, N. L. (1966). Sarcomata of the osteogenic series (osteosarcoma, fibrosarcoma, chondrosarcoma, parosteal osteogenic sarcoma and sarcomata arising in abnormal bone. J. Bone Jt Surg. (Am. Ed.) **48A**, 1–26. [**79, 82**]

MCKENZIE, A., COURT BROWN, W. M., DOLL, R., and SISSONS, H. A. (1961). Mortality from primary tumours of bone in England and Wales. *Br. med. J.* 1, 1782–90. [**39, 75**]

MCLEAN, F. C. and ROWLAND, R. E. (1963). Internal remodelling of compact bone, pp. 371–85, *in* Mechanisms of Hard Tissue Destruction, ed. R. F. Sognnaes. American Association for the Advancement of Science, Washington, D.C. [**99**]

Medical Research Council Monitoring Report No.17. (1963). London, Her Majesty's Stationery Office. [**59, 60, 61**]

MENOT, J. C., MASSE, R., MORIN, M. and LAFUMA, J. (1972). An experimental comparative study of the behaviour of ^{237}Np, ^{238}Pu, ^{239}Pu, ^{241}Am, and ^{242}Cm in the bone. *Hlth Phys.*, 22, 657–65. [**247**]

MERWIN, R. M. and REDMON, L. W. (1969). Skeletal and reticular tissue disorders produced in mice by agent (S) from sarcoma 37. *J. natn. Cancer Inst.* 43, 365–76. [**7, 9**]

MILLARD, R. E. (1965). Abnormalities of human chromosomes following therapeutic irradiation. *Cytogenetics* 4, 277–94. [**2, 243, 245**]

—— KAY, H. E. M., and LAWLER, S. D. (1969). Chromosome studies in patients with polycythaemia vera after treatment with ^{32}P. pp. 289–303, in *Radiation-induced cancer.* International Atomic Energy Agency, Vienna. [**241, 242, 243**]

MILLER, C. E., HASTERLIK, R. J., and FINKEL, A. J. (1969). *The Argonne radium studies: summary of fundamental data.* ANL.7531 and ACRH 106. Argonne National Laboratory, Argonne and Argonne Cancer Research Hospital, Chicago. [**119**]

MILLER, R. W. (1964). Radiation, chromosomes, and viruses in etiology of leukaemia, evidence from epidemiological research. *New Engl. J. Med.* 271, 30–6. [**2, 62**]

—— (1969). Delayed radiation effects in atomic-bomb survivors. *Science, N.Y.* 166, 569–74.

MILTON, R. C. and SHOKOJI, T. (1968). Tentative dose estimations for A-bomb survivors, Hiroshima and Nagasaki. ABCC Technical Report 1-68. [**66**]

MODAN, B. and LILIENFELD, A. M. (1965). Polycythaemia vera and leukaemia—The role of radiation treatment. *Medicine, Baltimore* 44, 305–44. [**241**]

—— and MODAN, M. (1968). Benign erythrocytosis. *Br. J. Haemat.* 14, 375–81. [**242**]

—— —— (1970). Treatment of primary polycythaemia. *Lancet* 2, 525.[**242**]

MOERTEL, C. G., DOCKERTY, M. B., and BAGGENSTOSS, A. H. (1961). Multiple primary malignant neoplasms. (i) Introduction and presentation of data. *Cancer N.Y.* 14, 221–30. [**80**]

MOLE, R. H. (1958). The dose-response relationship in radiation carcinogenesis. *Br. med. Bull.* 14, 184–9. [**13, 51**]

—— (1963). Cellular basis and aetiology of late somatic effects of ionizing radiation (ed. Harris, R. J. C.), pp. 3–16, in *Leukaemogenesis, quantitative aspects and cofactors.* Academic Press, London. [**47**]

—— (1966). Bone tumour and leukaemia induction in man by Radium-224 (Thorium X). ICRP/66/C1-5/3. Personal communication. [**150**]

MOORE, M. A. S. and METCALF, D. (1970). Ontogeny of the haemopoietic system.

Yolk sac origin of *in vivo* and *in vitro* colony forming cells in the developing mouse embryo, *Br. J. Haemat.* **18**, 279–96. [28]

MORROW, P. E., GIBB, F. R., DAVIES, H., MITOLA, J., WOOD, D., WRAIGHT, N., and CAMPBELL, H. S. (1967). The retention and fate of inhaled plutonium dioxide in dogs. *Hlth Phys.* **13**, 113–33. [207]

MORTON, D. L., EILBER, F. R., MALMGREN, R. A., and COOKE, K. O. (1970). Evidence for a virus in human osteosarcoma, in *Comparative leukaemia research* (ed. Dutcher, R. M.), pp. 754–60. S. Karger Basle. [10]

MOSKALEV, Y. I., STRELTSOVA, V. N., and BULDAKOV, L. A. (1969). Late effects of radionuclide damage, in *Delayed effects of bone-seeking radionuclides* (ed. Mays, C. W., Jee, W. S., Lloyd, R. D., Stover, B. J., Dougherty, J. H., and Taylor, G. N.), pp. 489–509. University of Utah Press, Salt Lake City. [170, 189, 197, 206, 222, 224, 226]

MOSS, M. L. (1966). In *Histology*, 2nd edn, (ed. R. O. Greep). Blakiston Division, pp. 155–73. McGraw Hill, New York. [17]

MULLER, J., DAVID, A., REJSKOVA, M., and BREZIKOVA, D. (1961). Chronic occupational exposure to ^{90}Sr and ^{226}Ra. *Lancet* **2**, 129–31. [178]

—— KLENER, V., TUSCANY, R., THOMAS, J., BREZIKOVA, D., and HOUSKOVA, M. (1966). Study of internal contamination with strontium-90 and radium-226 in man in relation to clinical findings. *Hlth Phys.* **12**, 993–1006. [178]

—— and THOMAS, J. (1969). Strontium retention in man, in *Delayed effects of bone-seeking radionuclides* (ed. Mays, C. W., Jee, W. S., Lloyd, R. D., Stover, B. J., Dougherty, J. H., and Taylor, G. N.), pp. 51–9. University of Utah Press, Salt Lake City. [178]

MURPHY, F. D. Jr. and BLOUNT, W. P. (1962). Cartilaginous exostoses following irradiation. *J. Bone Jt Surg.* (Am. ed.) **44**, 662–8. [84]

MURPHY, W. R. and ACKERMAN, L. V. (1956). Benign and malignant giant cell tumours of bone. *Cancer N.Y.* **9**, 317–39. [82]

NEARY, G. J. (1970). Track structure in relation to radiobiology, in *Charged particle tracks in solids and liquids.* Proc. Second L. H. Gray Conference, Cambridge 1969 (ed. G. E. Adams, D. K. Bewley, and J. W. Boag). London Institute of Physics and the Physical Society. [2]

NEUHAUSER, E. B. D., WITTENBORG, M. H., BERMAN, C. Z., and COHEN, J. (1952). Irradiation effects of roentgen therapy on the growing spine. *Radiology* **59**, 637–50. [83, 84, 85]

NEUMAN, W. F., HURSH, J. B., BOYD, J., and HODGE, H. C. (1955). On the mechanism of skeletal fixation of radium. *Ann. N.Y. Acad. Sci.* **62**, 123–36. [156]

—— and NEUMAN, M. W. (1958). *The chemical dynamics of bone mineral.* University of Chicago Press, Chicago. [31]

NEWTON, C. E., HEID, K. K., LARSON, H. V., and NELSON, J. C. (1967). Tissue sampling for plutonium through an autopsy program, in *Proceedings of the* 12*th annual bio-assay and analytical chemistry meeting*, at Gatlinburg, Tennessee, Oct. 13–14,

1966. CONF.661018, p. 220. quoted Mays *et al.* 1970. [**221**]

NILSSON, A. (1962). Histogenesis of ^{90}Sr induced osteosarcomas of mice *Acta vet scand.* **3**, 1–16. [**196**]

—— (1968). Pathologic effects of different doses of ^{90}Sr in mice. Development of carcinomas in the mucous membranes of the head. *Acta radiol.* **7**, 27–41. [**175**]

—— (1969). Dose-dependent carcinogenic effect of radiostrontium, in *Radiation-induced cancer*. International Atomic Energy Agency, Vienna. pp. 173–82. [**174, 175, 197**]

—— (1970*a*). Pathologic effects of different doses of radiostrontium in mice. Dose effect relationship in ^{90}Sr induced bone tumours. *Acta radiol.* **9**, 155–76. [**45, 49, 175, 196**]

—— (1970*b*). Pathologic effects of different doses of radiostrontium in mice. Changes in the haematopoietic system. *Acta radiol.* **9**, 528–44. [**49, 175, 176, 177**]

NORMAN, A., SASAKI, M., OTTOMAN, R. E., and VEOMETT, R. C. (1964). Chromosome aberrations in radiation workers. *Radiat Res.* **23**, 282–9. [**2**]

NORRIS, W. P., FRITZ, T. E., REHFELD, C. E., and POOLE, C. M. (1968). The response of the beagle dog to cobalt-60 gamma radiation: determination of the LD$_{50}$(30) and description of associated changes. *Radiat. Res.* **35**, 681–708. [**91**]

NOWELL, P. C. (1965). Unstable chromosome changes in tuberculin-stimulated leucocyte cultures from irradiated patients. Evidence for immunologically committed long lived lymphocytes in human blood. *Blood* **26**, 798–804. [**2**]

OWEN, M. (1962). ^{90}Sr dosimetry in rabbits, in *Some aspects of internal irradiation* (ed. Dougherty, T. F., Jee, W. S. S., Mays, C. W., and Stover, B. J.), pp. 409–20. Pergamon Press, Oxford. [**169, 170, 183**]

—— (1970). The origin of bone cells. *Int. Rev. Cytol.* **28**, 213–38. [**23, 28**]

—— SISSONS, H. A., and VAUGHAN, J. (1957). The effect of a single injection of a high dose of ^{90}Sr (500–1000 µCi) in rabbits. *Br. J. Cancer* **11**, 229–48. [**167, 170**]

—— and TRIFFITT, J. T. (1972). *Plasma glycoproteins and bone*. Proceedings IV Parathyroid Conference, Chapel Hill. Excerpta Medica, Amsterdam. pp. 316–26. [**247**]

—— and VAUGHAN, J. (1959*a*). Dose-rate measurements in the rabbit tibia following uptake of ^{90}Sr. *Br. J. Radiol.* **32**, 714–24. [**160, 167, 184, 185, 186**]

—— —— (1959*b*). Radiation dose and its relation to damage in the rabbit tibia following a single injection and daily feeding of ^{90}Sr. *Br. J. Cancer* **13**, 424–38. [**167, 184, 186**]

PALMER, R. F., THOMAS, J. M., and WATSON, C. R. (1970). Dosimetry aspects of chronically fed ^{90}Sr swine. BNWL-1050, Part 1, UC-48, pp. 2.3–2.7. Battelle North West Laboratory, Richland, Washington. [**183**]

—— —— —— and BEAMER, J. L. (1970). Dosimetry studies. BNWL 1306, Part 1, UC-48, pp. 37–8. Battelle North West Laboratory, Richland, Washington. [**183**]

PARK, J. F., BAIR, W. J., and BUSCH, R. H. (1972). Progress in beagle dog studies with transuranium elements at Battelle North West. *Hlth Phys.*, **22**, 803–810. [**207**]

PARR, R. M. (1967). Information on thorotrast dosimetry supplied by the radiochemical analysis of tissue specimens. *Ann. N.Y. Acad. Sci.* **145**, 644–53. [**237**]

PATERSON, C. R., WOODS, C. G., and MORGAN, D. B. (1968). Osteoid in metabolic bone disease. *J. Path. Bact.* **95**, 449–56. [**34**]

PEACOCKE, A. R. and WILLIAMS, P. A. (1966). Binding of calcium, yttrium and thorium to a glycoprotein from bovine cortical bone. *Nature, Lond.* **211**, 1140–1. [**30, 233**]

PECK, W. S. (1939). Fractures of the femoral neck following pelvic irradiation. *Univ. Mich. med. Bull.* **5**, 33–4. [**88**]

PENDERGRASS, E. P. (1968). Adenocarcinoma of the right breast and osteogenic sarcoma of the right third rib in a patient who did not receive post-operative radiation. *Cancer N.Y.* **22**, 644–9. [**79, 80**]

PENNA FRANCA, E., ALMEIDA, J. C., BECKER, J., EMMERICH, M., ROSER, F. X., KEGEL, G., HAINSBERGER, L. CULLEN, T. L., PETROW, H., DREW, R. T., and EISENBUD, M. (1965). Status of investigations in the Brazilian areas of high natural radioactivity. *Hlth Phys.* **11**, 699–712. [**54**]

PERKINS, J., ISRAELS, M. C. G., and WILKINSON, J. F. (1964). Polycythaemia vera: clinical studies on a series of 127 patients managed without radiation therapy. *O. Jl Med.* **33**, 499–518. [**240, 241**]

PERTHES, G. (1903). Ueber den Einfluss der Rontgenstrahlen auf epitheliale Gewebe, insbesondere auf das Carcinom, *Arch. klin. Chir.* **71**, 955–1000. [**83**]

PHEMISTER, D. B. (1926). Radium necrosis of bone. *Am. J. Roentg.* **16**, 340–8. [**138, 146**]

PHILLIPS, T. L. and SHELINE, G. E. (1963). Bone sarcomas following radiation therapy. *Radiology* **81**, 992–6. [**80**]

PICKREN, J. W. (1963). Cancer often strikes twice. *N. Y. St. J. Med.* **63**, 95–9. [**80**]

PIZZARELLO, D. J. and WITCOFSKI, R. L. (1967). *Basic radiation biology.* Lea and Febiger. [**7**]

PLATZMAN, R. L. (1967). Energy spectrum of primary activations in the action of ionizing radiation, in *Radiation research* (ed. G. Silini), pp. 20–42. Amsterdam-North Holland Publishing Company. [**2**]

POCHIN, E. E. (1969). Carcinogenic effects of radiation in man, in *Radiation induced cancer*, p. 6. International Atomic Energy Agency, Vienna. [**13**]

POOL, R. R., WILLIAMS, R. J. R., and GOLDMAN, M. (1972). ^{90}Sr toxicity in adult beagles, in *Biomedical implications of radiostrontium exposure* (ed. Goldman, M. and Bustad, L. K.). Division of Technical Information, U.S. Atomic Energy Commission, Oak Ridge, Conference CONF.710201. [**178**]

PRITCHARD, J. J. (1956). The osteoblast, in *The biochemistry and physiology of bone* (ed. G. H. Bourne), pp. 179–212. Academic Press Inc., New York. [**25**]

PRITCHARD, D. J., REILLY, C. A. Jr., and FINKEL, M. P. (1971). Evidence for a human osteosarcoma virus. *Nature, Lond.* (New Biology) **234**, 126–7. [**8**]

PUGH, L. P. (1927). Sporadic diffuse osteoperiostitis of fowls. *Vet. Rec.* **7**, 189–90. [**7**]

REESE, A. B., HYMAN, G. A., MERRIAM, G. R., Jr., and FORREST, A. W. (1957). The treatment of retinoblastoma by radiation and triethylene melamine. *Am. J. Opthal.* **43**, 865–72. [**80**]

REGAUD, C. (1922). Sur la sensibilité du tissu osseux normal vis-à-vis des radiations X et γ sur le mécanisme de l'ostéoradio-nécrose. *C. r. Séane Soc. Biol.* **87**, 629–32. [**89**]

REGEN, E. M. and WILKINS, W. E. (1936). The effect of large doses of X-rays on the growth of young bone. *J. Bone Jt Surg.* **18**, 61–8. [**138, 146**]

REISKIN, A. B. (1971) Cytogenetics, in *Conference on the estimation of low-level radiation effects in human populations* (ed. G. A. Sacher), pp. 15–16. ANL 7811. Argonne National Laboratory, Argonne. [**2, 10**]

REPORT OF THE RBE COMMITTEE TO THE INTERNATIONAL COMMISSIONS ON RADIOBIOLOGICAL PROTECTION AND ON RADIOLOGICAL UNITS AND MEASUREMENTS. (1963). *Hlth Phys.* **9**, 357–86. [**11, 107**]

RHIM, J. S., HUEBNER, R. J., LANE, W. T., TURNER, H. C., and RABSTEIN, L. (1969). Neoplastic transformation and derivation of a focus forming sarcoma virus of rat embryo cells infected with a murine osteosarcoma (FBJ) virus. *Proc. Soc. exp. Biol. Med.* **132**, 1091–8. [**8**]

ROSEN, J. C., COHEN, N., and WRENN, M. E. (1971). Short-term metabolism of ^{241}Am in the adult baboon. *Hlth Phys.* **22**, 621–26. [**231**]

ROSENBLATT, L. S. and GOLDMAN, M. (1967). The use of probit analysis to estimate dose effects on post irradiation leukocyte depressions. *Hlth Phys.* **13**, 795–8. [**144**]

ROSENTHAL, M. and GRACE, E. J. (1936). Experimental radium poisoning. I. Bone marrow and lymph-node changes in rabbits, produced by oral administration of radium sulphate. *Am. J. med. Sci.* **191**, 607–18. [**138, 146**]

ROSENTHAL, M. W. and LINDENBAUM, A. (1969). Osteosarcomas as related to tissue distribution of monomeric and polymeric plutonium in mice, in *Delayed effects of bone-seeking radionuclides* (ed. Mays, C. W., Jee, W. S., Lloyd, R. D., Stover, B. J., Dougherty, J. H., and Taylor, G. N.), pp. 371–84. University of Utah Press, Salt Lake City. [**219**]

—— MARSHALL, J. H., and LINDENBAUM, A. (1968). Autoradiographic and radiochemical studies of the effects of colloidal state of intravenously injected plutonium on its distribution in bone and marrow, in *Diagnosis and treatment of deposited radionuclides*, pp. 73–80. Proceedings of Symposium, Richland 1967 (ed. H. A. Kornberg and W. D. Norwood). Excerpta Medica Foundation, Monographs nuclear medicine No. 2. [**218**]

Ross, J. M. (1936). The carcinogenic action of radium in the rabbit. The effect of prolonged irradiation with screened radium. *J. Path. Bact.* **43**, 267–76. [**138**]

ROUGET, C. (1873). Mémoire sur le développement de la tunique contractile des vaisseaux. *C. r. hebd. Séanc. Acad. Sci.*, Paris **79**, 559. [**28**]

ROWLAND, R. E. (1959a). Late observations of the distribution of radium in the human skeleton. Argonne National Laboratory, Radiological Physics Division, Semiannual Report, (July-December 1959), ANL-6104, 16–29. [**136**]

—— (1959b). The radium distribution in the tooth of a dial painter. Argonne National Laboratory, Radiological Physics Division, Semiannual Report, (July-December (1959), ANL-6104, 30–3. [**128**]

—— (1960a). Microscopic metabolism of Ra^{226} in canine bone. Part II. Argonne National Laboratory, Radiological Physics Division, Semiannual Report, (January-June 1960), ANL-6199, 16–20. [**35, 129**]

—— (1960b). Plugged haversian canals in a radium case. Argonne National Laboratory, Radiological Physics Division, Semiannual Report, (January-June 1960), ANL-6199, 36–43. [**129, 200**]

—— (1960c). Bone studies on an exhumed radium patient. ANL-6297, 8. [**129**]

—— (1960d). Radioisotopes in the skeleton: late observations of the distribution of radium in the human skeleton, in *Radioisotopes in the biosphere* (ed. R. S. Caldecott and L. A. Snyder), pp. 339–53. Symposium, University of Minnesota 1959. [**129, 136, 137, 138, 139**]

—— (1962). Skeletal retention of the alkaline earth radioisotopes and bone dosimetry, in *Some aspects of internal irradiation, Heber, Utah*, pp. 455–69. Pergamon Press, Oxford. [**129**]

—— (1966). Exhangeable bone calcium. *Clin. Orthop.* **49**, 233–48. [**34, 35, 111, 156**]

—— FAILLA, P. M., KEANE, A. T., and SLEHNEY, A. F. (1969–70). Some dose-response relationships for tumour incidence in radium patients. Argonne National Laboratory, Radiological Physics Division, Annual Report. ANL-7760, Part II, pp. 1–17. [**119, 120, 121, 123, 130, 133, 151**]

—— —— —— —— (1970–71). Tumour incidence for the radium patients. Argonne National Laboratory, Radiological Physics Division, Annual Report. ANL-7860, Part. II. [**119, 130, 133**]

—— and MARSHALL, J. H. (1959). Radium in human bone: The dose in microscopic volumes of bone. *Radiat. Res.* **11**, 299–313. [**109, 129, 137**]

—— —— and JOWSEY, J. (1959). Radium in human bone. The microradiographic appearance. *Radiat. Res.* **10**, 323–34. [**129**]

RUBIN, P., DUTHIE, R. B., and YOUNG, L. W. (1962). The significance of scoliosis in post irradiated Wilm's tumour and neuroblastoma. *Radiology* **79**, 539–59. [**83**]

RUSHTON, M. A., OWEN, M., HOLGATE, W., and VAUGHAN, J. (1961). The relation of radiation dose to radiation damage in the mandible of weanling rabbits. *Archs oral Biol.* **3**, 235–46. [**90, 169, 186**]

SABANAS, A. O. DAHLIN, D. C., CHILDS, D. S., Jr., and IVINS, J. C. (1956). Post irradiation sarcoma of bone. Cancer *N. Y.* **9**, 528–42. [**79, 82, 83**]

SABIN, F. R., DOAN, C. A., and FORKNER, C. E. (1932). The production of osteogenic sarcomata and the effects on lymph nodes and bone marrow of intravenous injections of radium chloride and mesothorium in rabbits. *J. exp. Med.* **56**, 267–89. [**138**]

SAENGER, E. L., SILVERMAN, F. N., STERLING, T. D., and TURNER, M. E. (1960). Neoplasia following therapeutic irradiation for benign conditions in childhood. *Radiology* **74**, 889–904. [**74**]

SCHAJOWICZ, F. and GALLARDO, H. (1970). Epiphyseal chondroblastoma of bone: a clinico-pathological study of 69 cases. *J. Bone Jt Surg.* (Brit. ed.) **52***b*, 205–26. [**45**]

SCHMIER, H., SEELENTAG, W., and WALDESKOG, B. (1971). The radiation environment of human beings, in *Manual on radiation haematology*, pp. 7–29. International Atomic Energy Agency. [**54**]

SCOTT, K. G. and LAWRENCE, J. H. (1941). Effect of radio phosphorus on blood of monkeys. *Proc. Soc. exp. Biol. Med.* **98**, 155–8. [**243**]

SEABORG, G. T. (1968). Californium 252: radioisotopes with a future; in CONF.681032 (U.S.A.E.C. Div. of Tech. Info.), 1–9. [**246**]

SEELENTAG, W. (1968). On the importance of the radiation burden on a population with special reference to the genetically significant dose from application of radiation in medicine. *Prog. nucl. Energy. Series XII. Hlth Phys.* **2**, 125–55. Pergamon Press, Oxford. [**54**]

SEGI, M., KURIHARA, M., and MATSUYAMA, T. (1965). Cancer mortality in Japan (1899–1962), p. 125, Sendai. Tohoku University School of Medicine, Department of Public Health. pp. 14, 23, and 45. [**63**]

SELTSER, R. and SARTWELL, P. E. (1965). The influence of occupational exposure to radiation on the mortality of American radiologists and other medical specialists. *Am. J. Epidemiology* **81**, 2–22. [**39, 40, 68**]

SHARPE, W. D. (1971). Radium osteitis with osteogenic sarcoma. The chronology and natural history of a fatal case. *Bull. N.Y. Acad. Med.* **47**, 1059–82. [**118**]

SILVA HORTA, J. (1967). Late effects of Thorotrast on the liver and spleen and their efferent lymph nodes. *Ann N Y Acad. Sci.* **145**, 676–99. [**235**]

—— ABBATT, J. D., MOTTA, L. C., and RORIZ, M. L. (1965). Malignancy and other late effects following administration of thorotrast. *Lancet* **2**, 201–5. [**235**]

SIMON, N., BRUCER, M. and HAYES, R. (1960). Radiation and leukaemia in carcinoma of the cervix. *Radiology* **74**, 905–11. [**77**]

SIMPSON, C. F. and SANGER, V. L. (1968). A review of avian osteopetrosis. Comparisons with other bone diseases. *Clin. Orthop.* **58**, 271–81. [**7**]

SINCLAIR, W. K., ROWLAND, R. E., and SACHER, G. A. (1971). Overview and critique of the conference, pp. 41–42, *Conference on the estimation of low-level radiation effects in human populations*. ANL-7811. [**12, 13**]

SISSONS, H. A. (1966). Tumours of bone, pp. 1396–428, in *Systematic pathology* (eds G. Payling Wright and W. St. Clair Symmers). Longmans, London. [**44, 46, 121**]

—— (1970). Dimensions of cells covering bone surfaces. Medical Research Council (London) Subcommittee on Permissible Levels. PIRC/PL/70/4. [**27, 40**]

—— HOLLEY, K. J., and HEIGHWAY, J. (1967). Normal bone structure in relation to osteomalacia, pp. 19–37, in *l'ostéomalacie* (ed. D. J. Hioco), Symposium organisé par le Centre du Métabolisme Phospho-Calcique, Tours, 1965. [**21, 32, 33, 216**]

SJOGREN, H. O. and RINGERTZ, N. (1962). Histopathology and transplantability of polyoma-induced tumours in strain A/Sn and three coisogenic resistant (IR) substrains. *J. natn. Cancer Inst.* **28**, 859–95. [**7**]

SKOLNIK, E. M., FORNATTO, E. J., and HEYDEMANN, J. (1965). Osteogenic sarcoma of the skull following irradiation. *Ann. Otol. Rhinol. Lar.* **65**, 915–36. **[81]**

SKORYNA, S. C. and KAHN, D. S. (1959). The late effects of radioactive strontium on bone. *Cancer N.Y.* **12**, 306–22. **[49, 183, 184]**

—— —— and WEBSTER, D. R. (1958). Histogenesis of bone tumours produced by radioactive strontium in rats. *Proc. Am. Ass. Cancer Res.* **2**, 347. **[183, 184]**

SLAUGHTER, D. P. (1942). Radiation osteitis and fractures following irradiation. *Am. J. Roentg.* **48**, 201–12. **[88]**

SMITHERS, D. W. and RHYS-LEWIS, R. D. S. (1945). Bone destruction in cases of carcinoma of the uterus. *Br. J. Radiol.* **18**, 359–62. **[88]**

SOEHNER, R. L. and DMOCHOWSKI, L. (1969). Induction of bone tumours in rats and hamsters with murine sarcoma virus and their cell-free transmission. *Nature, Lond.* **224**, 191–2. **[7]**

SOLOWAY, H. B. (1966). Radiation-induced neoplasms following curative therapy for retinoblastoma. *Cancer N.Y.* **19**, 1984–8. **[81]**

SPIERS, F. W. (1960). The dose to tissues of the body from natural background radiation. *Trans. IXth Int. Congr. Radiol.* **2**, 1133–40. Georg Thieme Verlag, Stuttgart. **[52]**

—— (1966). Dose to bone from strontium 90: implications for the setting of the maximum permissible body burden. *Radiat. Res.* **28**, 624–2. **[43, 187, 188]**

—— (1968a). *Radioisotopes in the human body*. Academic Press, New York and London. **[vii, 3, 108, 156, 245]**

—— (1968b). Dose to trabecular bone from internal beta emitters. *Proc. First. Int. Congress of Radiation Protection* (ed. Snyder W.S.). Pergamon Press, Oxford. pp. 165–72. **[19, 108, 245]**

—— (1969). Beta particle dosimetry in trabecular bone, pp. 95–107, in *Delayed effects of bone-seeking radionuclides* (ed. Mays, C. W., Jee, W. S., Lloyd, R. D., Stover, B. J., Dougherty, J. H., and Taylor, G. N.). University of Utah Press, Salt Lake City. **[19, 108]**

—— (1971). Biophysical basis for radiation haematology. in *Manual on radiation haematology*, pp. 45–69. Technical Report Series No. 123. International Atomic Energy Agency, Vienna. **[3]**

—— WHITWELL, J. R. and DARLEY, P. J. (1971). Dose in bone marrow cavities from radium-226. To be published, personal communication. **[108, 109, 110, 135, 152]**

—— ZANELLI, G. D., DARLEY, P. J., WHITWELL, J. R., and GOLDMAN, L. M. (1972). Beta particle dose rates in human and animal bone. in *Biomedical implications of radiostrontium exposure* (ed. Goldman, M. and Bustad, L. K.). Division of Technical Information, U.S. Atomic Energy Commission, Oak Ridge, Conference CONF. 710201. **[17, 19, 108, 187, 188, 189]**

SPIESS, H. (1950). Discussion on 'Die Wirkung spezifischer Medikamente bei der Knochen-und Gelenktuberkulose im Tierversuch.' *Verh. dt. orthop. Ges.* **38**, 204–6. **[149]**

—— (1952). Über Anwendung und Wirkung des Peteosthor bei pulmonaler und extrapulmonaler Tuberkulose im Kindesalter. Zugleich eine allgemeine Stellungnahme zur Thorium X- und Peteosthor- Therapie. *Z. Kinderheilk.* **70**, 213–52. [**149**]

—— (1956). Schwere Strahlenschäden nach der Peteosthorbehandlung von Kindern. *Dt. med. Wschr.* **81**, 1053–4. [**149**]

—— (1957). Exostotische Dysplasia durch Strahlenwirkung. *Dt. med. Wschr.* **82**, 1483–4. [**149**]

—— (1969). ^{224}Ra induced tumours in children and adults, pp. 227–46, in *Delayed effects of bone-seeking radionuclides* (ed. Mays, C. W., Jee, W. S., Lloyd, R. D., Stover, B. J., Dougherty, J. H., and Taylor, G. N.). University of Utah Press, Salt Lake City. [**45, 84, 149, 150, 151**]

—— (1970). Late effects of ^{224}Ra injections in man. *Hlth Phys.* **19**, 98. [**149**]

—— and MAYS, C. W. (1970). Some cancers induced by ^{224}Ra (ThX) in children and adults. *Hlth Phys.* **19**, 713–29. [**112, 130, 149, 151**]

STAMPFLI, W. P. and KERR, H. D. (1947). Fractures of the femoral neck following pelvic irradiation. *Am. J. Roentg.* **57**, 71–83. [**88**]

STARA, J. F., NELSON, N. S., DELLA ROSA, R. J., and BUSTAD, L. K. (1971). Comparative metabolism of radionuclides in mammals—a review. *Hlth Phys.* **20**, 113–37. [**113, 114, 138, 155, 203**]

STEPHENSON, W. H. and COHEN, B. (1956). Post-irradiation fractures of the neck of the femur. *J. Bone Jt Surg.* (Brit. ed.). **38**, 830–45. [**88**]

STEVENS, W. and BRUENGER, F. W. (1972). Interaction of ^{249}Cf and ^{252}Cf with constituents of dog and human blood. *Hlth Phys.* **22**, 679–683. [**246**]

—— —— and STOVER, B. J. (1968). *In vivo* studies for the interactions of PuIV with blood constituents. *Radiat. Res.* **33**, 490–500. [**221**]

STEWART, A. M. (1970). Personal communication. [**70, 71**]

STEWART, A. M. (1972*a*). Tissue ageing as a factor in juvenile cancers. *Proc. R. Soc. Med.* **65**, 245–246. [**51**]

STEWART, A. M. (1972*b*). Epidemiology of Leukaemia, pp. 3–22, *in* Clinics in Haematology, ed. Stuart Roath. Saunders Co. Ltd., London, Philadelphia, Toronto. [**51**]

—— and BARBER, R. (1971). Epidemiological importance of childhood cancers. *Br. med. Bull.* **27**, 64–70. [**40, 62, 69, 71, 72**]

—— and KNEALE, G. W. (1968). Changes in the cancer risk associated with obstetric radiography. *Lancet* **i**, 104–7. [**69, 70, 71**]

—— —— (1969). Role of local infections in the recognition of haemopoietic neoplasms. *Nature, Lond.* **223**, 741–2. [**69, 71, 72**]

—— —— (1970). Age-distribution of cancers caused by obstetric X-rays and their relevance to cancer latent periods. *Lancet* **ii**, 4–8. [**69, 71**]

STEWART, A., PENNYBACKER, W., and BARBER, R. (1962). Adult leukaemias and diagnostic X-rays. *Br. med. J.* **2**, 882–90. [**73**]

STOUT, A. P. (1949). Hemangiopericytoma: study of 25 new cases. *Cancer N.Y.* **2**, 1027–54. [**28**]

STOVER, B. J., ATHERTON, D. R., BUSTER, D. S., and BRUENGER, F. W. (1965). The Th^{228} decay series in adult beagles: Ra^{224}, Pb^{212}, and Bi^{212} in selected bones and soft tissues. *Radiat. Res.* **26**, 132–45. [**112, 235**]

—— —— —— (1971). Protracted hepatic, splenic and renal retention of ^{239}Pu in the beagle. *Hlth Phys.* **20**, 369–74. [**201, 221**]

—— —— —— and KELLER, N. (1965). The Th^{228} decay series in adult beagles: Ra^{224}, Pb^{212}, and Bi^{212} in blood and excreta. *Radiat. Res.* **26**, 226–43. [**235**]

—— —— KELLER, N., and BUSTER, D. S. (1960). Metabolism of Th^{228} decay series in adult beagle dogs. I. Th^{228} (RdTh). *Radiat. Res.* **12**, 657–71. [**96, 202**]

—— —— and MAYS, C. W. (1962). Studies on the retention and distribution of ^{226}Ra, ^{239}Pu, ^{228}Ra (MsTh$_1$), ^{228}Th (RdTh) and ^{90}Sr in adult beagles, pp. 7–25, in *Some aspects of internal irradiation* (ed. Dougherty, T. F., Jee, W. S. S., Mays, C. W. and Stover, B. J.). Pergamon Press, Oxford. [**101**]

—— BRUENGNER, F. W., and STEVENS, W. (1970). Association of americium with ferretin in canine liver. *Radiat. Res.* **43**, 173–86. [**221**]

STROEBEL, C. F. (1954). Current status of radiophosphorus therapy. *Proc. Mayo Clinic* **29**, 1–4. [**240**]

STUART, B. O. (1970). Comparative distribution of ^{238}Pu and ^{239}Pu in rats following inhalation of the oxide. BNWL 1050, Part 1, UC 48, pp. 3.19–3.20. Battelle North West Laboratory, Richland, Washington. [**201**]

SUNDELIN, P. and NILSSON, A. (1968). Cytoplasmic ultraviolet extinction of strontium-90 induced fibroblastic osteosarcomas correlated to histologic appearance and ultrastructure. *Acta Radiol.* **7**, 161–70. [**196**]

SZUR, L. and LEWIS, S. M. (1966). The haematological complications of polycythaemia vera and treatment with radioactive phosphorus. *Br. J. Radiol.* **39**, 122–30. [**240, 243**]

—— and SMITH, M. D. (1961). Red cell production and destruction in myelosclerosis. *Br. J. Haemat.* **7**, 147–68. [**243**]

TAKAHASHI, S., KITABATAKE, T., YAMAGATA, S., MIYAKAWA, T., MASUYAMA, M., MORI, T., TANAKA, T., HIBINO, S., MIYAKAWA, M., KANEDA, H., OKAJIMA, S., KOMIYAMA, K., KOGA, Y., ADACHI, T., HASHIZUME, T., and HASHIMOTO, Y. (1965). Statistical study on thorotrast-induced cancer of the liver. *Tohoku J. exp. Med.* **87**, 144–54. [**235**]

TALIAFERRO, W. H., TALIAFERRO, L. G. and JAROSLOW, B. N. (1964). *Radiation and immune mechanisms*. Academic Press Inc., New York. [**61**]

TASK GROUP REPORT. (1972). *Report of task group on metabolism of plutonium and related elements and their compounds*. 1st draft to ICRP Committee 2, March 1971. [**226**]

TAYLOR, D. M. (1962). Some aspects of the comparative metabolism of plutonium and americium in rats. *Hlth Phys.* **8**, 673–677. [**228, 229**]

—— The metabolism of plutonium in adult rabbits. *Br. J. Radiol.* **42**, 44–50. [**205, 207, 211, 220, 221**]

—— (1970). Personal communication. [**30**]

—— (1973). Plutonium chemical and physical properties. To be published in *Handbook of experimental pharmacology Series. Uranium, plutonium and transplutonic elements*, (ed. H. C. Hodge and J. N. Stannard). Springer Verlag, Berlin. [**30, 202**]

—— and BENSTED, J. P. M. (1969). Long-term biological damage from plutonium 239 and americium 241 in rats, pp. 357–70, in *Delayed effects of Bone-seeking radionuclides* (ed. Mays, C. W., Jee, W. S., Lloyd, R. D., Stover, B. J., Dougherty, J. H., and Taylor, G. N.). University of Utah Press, Salt Lake City. [231]

—— and CHIPPERFIELD, A. R. (1970). The mode of fixation of plutonium 239 and americium 241 in bone: a possible explanation of their different carcinogenicity, *Symposium Ossium* 1968. E. & S. Livingstone. London. [30]

—— and —— (1971). The binding of transplutonium elements to proteins of bone. *Proceedings of Euratom/ENEA seminar on radiation protection problems relating to transuranic elements*. EUR 4612 d-f-e pp. 187–204, Luxembourg. [30]

—— SOWBY, F. D., and KEMBER, N. F. (1961). The metabolism of americium and plutonium in the rat. *Physics Med. Biol.* **6**, 73–86. [230]

TAYLOR, G. N., CHRISTENSEN, W. R., JEE, W. S. S., REHFELD, C. E., and FISHER, W. (1962). Anatomical distribution of fractures in beagles injected with ^{239}Pu. *Hlth Phys.* **8**, 609–13. [225]

—— DOUGHERTY, T. F., SHABESTARI, L., and DOUGHERTY, J. H. (1969). Soft tissue tumours in internally irradiated beagles, pp. 323–36, in *Delayed effects of bone-seeking radionuclides* (ed. Mays, C. W., Jee, W. S., Lloyd, R. D., Stover, B. J., Dougherty, J. H., and Taylor, G. N.). University of Utah Press, Salt Lake City. [224]

—— JEE, W. S. S., CHRISTENSEN, W. R., REHFELD, C. E., and NEBEKER, N. (1965). Thorium-228 induced fractures in beagles. University of Utah, College of Medicine, Department of Anatomy, Radiobiology Division, *Annual report of progress in the internal irradiation program*, COO-119-232, pp. 74–87. [239]

TOUGH, I. M., BUCKTON, K. E., BAIKIE, A. G., and COURT BROWN, W. M. (1960). X-ray induced chromosome damage in man. *Lancet* **ii**, 849–51. [2]

TURNER, G. A. and TAYLOR, D. M. (1968). The transport of plutonium americium and curium in the blood of rats. *Physics Med. Biol.* **13**, 535–46. [221, 229]

TURNER, R. C., RADLEY, J. M. and MAYNEORD, W. V. (1958). Alpha-ray activities of humans and their environment. *Nature, Lond.* **181**, 518–21. [55]

—— —— —— (1961). Naturally occurring alpha-activity of drinking waters. *Nature, Lond.* **189**, 348. [55]

TUTT, M., KIDMAN, B., RAYNER, B. and VAUGHAN, J. (1952). The deposition of ^{89}Sr in rabbit bones following intravenous injection. *Br. J. exp. Path.* **33**, 207–15. [97]

TWENTYMAN, P. R. and BLACKETT, N. M. (1970). Red cell production in the continuously irradiated mouse. *Br. J. Radiol.* **43**, 898–902. [90]

UNITED NATIONS. (1962). *UNSCEAR Report*. Report of the United Nations Scientific Committee on the Effects of Atomic Radiation. General Assembly: Official Records, Seventeenth Session, Supplement No. 16 (A/5216). United Nations, New York. [54, 55]

—— (1964). *Radiation carcinogenesis in man*. Report of the United Nations Scientific Committee on the Effects of Atomic Radiation, General Assembly, Official Records,

Nineteenth session, Supplement No. 14 (A/15814). United Nations, New York. [54]

VAETH, J. M., LEVITT, S. H., JONES, M. D., and HOLTFRETER, C. (1962). Effects of radiation therapy in survivors of Wilm's tumour. *Radiology* 79, 560–68. [83, 85]

VAN PUTTEN, L. M. and DE VRIES, M. J. (1962). ^{90}Sr toxicity in mice. *J. natn. Cancer Inst.* 28, 587–603. [178]

VAUGHAN, J. (1961). The relation of radiation dose to skeletal damage from bone-seeking isotopes. *Lectures on Scientific Basis of Medicine*, pp. 47–62. Athlone Press, London. [129]

—— (1962a). Bone disease induced by radiation. *Int. Rev. exp. Path.* 1, 244–369. [88, 100, 139]

—— (1962b). The effect of internal irradiation from ^{90}Sr on the bone marrow and peripheral blood picture of young rabbits. *Some aspects of internal irradiation*, (ed. T. F. Dougherty). Symposium, Heber, Utah 1961. pp. 361–77. Pergamon Press, New York. [171, 172]

—— (1970a). *The physiology of bone*, pp. 51–60. Clarendon Press, Oxford. [21, 23, 29, 34, 35, 97, 155, 156, 158, 216, 217]

—— (1970b). Radiation and myeloproliferative disorders in man. *Symposium on myeloproliferative disorders of animals and man.* pp. 489–500. International Atomic Energy Agency. [48, 186, 196, 235, 240]

—— (1970c). Note on character of cells on trabecular bone surfaces in adult human vertebrae. Medical Research Council (London) Subcommittee on Protection against Ionizing Radiation, PIRC/PL/70/1. [40]

—— Distribution, excretion and effects of plutonium as a bone seeker. *Handbook of experimental pharmacology series, volume on uranium, plutonium and transplutonic elements* (ed. H. C. Hodge and J. N. Stannard). Springer Verlag, Berlin. To be published. [202, 205, 206, 227]

—— BLEANEY, B. and WILLIAMSON, M. (1967). The uptake of plutonium in bone marrow—a possible leukaemic risk. *Br. J. Haemat.* 13, 492–402. [209, 219, 221, 238]

—— LAMERTON, L. F. and LISCO, H. (1960). *The relation of radiation damage to radiation dose in bone*. pp. 7–20. International Atomic Energy Agency, Vienna. [11, 19, 122]

—— and WILLIAMSON, M. (1967). Variation in 'turnover rates' in different parts of the skeleton in relation to tumour incidence due to ^{90}Sr deposition. *Proceedings international symposium on some aspects of strontium metabolism*. Chapelcross, Scotland. 5–6 May 1966. pp. 195–206. Academic Press, New York and London. [35, 157]

—— —— (1969). ^{90}Sr in the rabbit: the relative risks of osteosarcoma and squamous cell carcinoma, *Delayed effects of bone-seeking radionuclides* (ed. Mays, C. W., Jee, W. S., Lloyd, R. D., Stover, B. J., Dougherty, J. H. and Taylor, G. N.). pp. 337–55. University of Utah Press, Salt Lake City. [135, 157, 159, 160, 165, 166, 168, 186]

VIDEBAEK, A. (1950). Polycythaemia vera. *Acta med. Scand.* 138, 179–87. [241]

VOLF, V. (1972). Strontium 90 effects in man, *Biomedical implications of radiostrontium exposure* (ed. Goldman, M. and Bustad, L. K.). Division of Technical Information, U.S. Atomic Energy Commission, Oak Ridge, Conference CONF.710201. [178]

WALSER, M. (1969). Renal excretion of alkaline earths, *Mineral metabolism III*, (ed. C. L. Comar and F. Bronner). pp. 235–320. Academic Press, New York. [156]

WANEBO, C. K., JOHNSON, K. G., SATO, T., and THORSLUND, T. W. (1968*a*). Breast cancer after exposure to the atomic bombings of Hiroshima and Nagasaki (ABCC). *N. Engl. J. Med.* **279**, 667–71. [63]

—— —— SATO, K., and THORSLUND, T. W. (1968*b*). Lung cancer following atomic radiation. *Am. Rev. resp. Dis.* **98**, 778–87. [63]

WARREN, S. (1956). Longevity and causes of death from irradiation in physicians. *J. Am. med. Ass.* **162**, 464–68. [68]

—— and CHUTE, R. N. (1963). Radiation-induced osteogenic sarcoma in parabiont rats *Lab. Invest.* **12**, 1041–5 [90]

—— and LOMBARD, O. M. (1966). New data on the effects of ionizing radiation on radiologists. *Archs envir. Hlth* **13**, 415–21. [68, 69]

—— and MEISNER, L. (1965). Chromosomal changes in leucocytes of patients receiving irradiation therapy. *J. Am. med. Ass.* **193**, 351–8. [2]

WASSERMAN, L. R. (1954). Polycythemia vera—its course and treatment. Relation to myeloid metaplasia and leukaemia. *Bull. N.Y. Acad. Med.* **30**, 343–75. [240]

WASSERMAN, R. H. and COMAR, C. L. (1961). The parathyroids and the intestinal absorption of calcium, strontium, and phosphate ions in the rat. *Endocrinology* **69**, 1074–9. [156]

—— and TAYLOR, A. N. (1969). Some aspects of the intestinal absorption of calcium with special reference to Vitamin D, in *Mineral metabolism III* (ed. C. L. Comar and F. Bronner), pp. 221–402. [156]

WATKINS, P. J., HAMILTON FAIRLEY, G., and BODLEY SCOTT, R. (1967). Treatment of polycythaemia vera. *Br. med. J.* **2**, 664–6. [240]

WENGER, P. and CASSIMATIS, D. (1962). 92. Recherches sur l'accumulation et la toxicité du radium et du radiostrontium dans le corps humain. I. Determination du strontium 90 dans quelques cas de personnes contaminées. *Helv. chim. Acta* **45**, 783–9. [178]

—— and MILLER, C. E. (1962). Recherches sur l'accumulation et la toxicité du radium et du radiostrontium dans le corps humain. II. L'anthropogammamètre 'Whole-body counter) de Genève. *Helve. chim. Acta* **46**, 467–79. [178]

—— and SOUCAS, K. (1963). 50. Recherches sur l'accumulation et la toxicité du radium et du radiostrontium dans le corps humain. III. Courbes d'élimination du strontium 90. *Helve. chim. Acta* **46**, 479–82. [178]

—— —— (1965). La contamination et l'accumulation du radium et du radiostrontium chez les horologers suisses. *Radiologia clin.* **34**, 67–71. [178]

WHITEHOUSE, W. M. and LAMPE, I. (1953). Osseous damage in irradiation of renal tumours in infancy and childhood. *Am. J. Roentg.* **70**, 721–9. [83, 84, 85]

WHITMORE, G. F. (1971). *In vitro* studies of genetic and carcinogenic effects of radiation, in, *Conference on the estimation of low-level radiation effects in human populations*, p. 19. ANL-7811. [10]

WHITWELL, J. R. and SPIERS, F. W. (1971). The dosimetry of β emitters in bone. 'Theoretical methods'. In *Proceedings of the Fifth Congress of the French Society for Radioprotection*, Grenoble, France, 1971. To be published. [135, 188]

WICK, O. J. (1967). ed. *Plutonium handbook. A guide to the Technology.* Vols. I and II. New York, Gordon and Breach Science Publishers. [202]

WILLIAMS, P. A. and PEACOCKE, A. R. (1967). The binding of calcium and yttrium ions to a glycoprotein from bovine cortical bone. *Biochem. J.* **105**, 1177–85. [30]

WILLIAMS, R. R., DAHLIN, D. C., and GHORMLEY, R. K. (1954). Giant-cell tumour of bone. *Cancer N.Y.* **7**, 764–73. [82]

WILLIAMSON, M. (1963). The sites of deposition of certain radioactive isotopes which concentrate in bone. B.Sc., Thesis, Oxford. [207, 208]

—— and VAUGHAN, J. (1967). Histochemistry of the mucosaccharides in the epiphyseal plate of young rabbits. *Nature, Lond.* **215**, 711–4. [31, 221]

WILLIS, R. A. (1967). *Pathology of tumours*, 4th edn, pp. 658–9, 682–5. Butterworths, London. [23]

WILSON, C. W. (1956a), The uptake of phosphorus-32 by the knee joint and tibia of six week old mice and the effect of X-rays upon it. Variation of uptake with time after a dose of 2000 rad of 200 kV X-rays. *Br. J. Radiol.* **29**, 86–91. [85, 87]

—— (1956b). The effect of X-rays on the uptake of phosphorus-32 by the knee joint and tibia of six-week old mice: relation of depression of uptake to X-ray dose. *Br. J. Radiol.* **29**, 571–3. [85, 87]

—— (1958). The effect of X-rays upon the uptake of ^{32}P by the knee joint of the mouse. Relation between the depression of ^{32}P uptake and the age of the animal. *Br. J. Radiol.* **31**, 384–6. [87]

—— (1959). Effect of X-rays on uptake of ^{32}P by the mouse knee joint when the X-ray dose is given in two carefully spaced fractions. *Br. J. Radiol.* **32**, 547–51. [87]

WISEMAN, B. K., ROHN, R. J., BOURONCLE, B. A., and MYERS, W. G. (1951). The treatment of polycythaemia with radioactive phosphorus. *Ann. intern Med.* **34**, 311–30. [240]

WOLFE, J. J. and PLATT, W. R. (1949). Post-irradiation osteogenic sarcoma of the nasal bone. *Cancer N.Y.* **2**, 438–46. [81]

WOODARD, H. Q. and LAUGHLIN, J. S. (1957). The effect of X-rays of different qualities on the alkaline phosphatase of living mouse bone. *Radiat. Res.* **7**, 236–52. [85]

—— and SPIERS, F. W. (1953). The effect of X-rays of different qualities on the alkaline phosphatase of living mouse bone. *Br. J. Radiol.* **26**, 38–46. [85]

WOOD, J. W., TAMAGAKI, H., NERIISHI, J., SATO, T., SHELDON, W. F., ARCHER, P. G., HAMILTON, H. B., and JOHNSON, K. G. (1969). Thyroid carcinoma in atomic bomb survivors Hiroshima and Nagasaki. *Am. J. Epidemiology* **89**, 4–14. [63]

WUTHIER, R. E. (1968). Lupids of mineralizing epiphyseal tissues in the bovine foetus. *J. Lipid Res.* **9**, 68–78. [29]

YUMOTO, T., POEL, W. E., KODAMA, T., and DMOCHOWSKI, L. (1970). Studies on FBJ virus induced bone tumours in mice. *Tex. Rep. Biol. Med.* **28**, 145–65. [8]

YOUNG, R. W. (1962). Cell proliferation and specialization during endochondral osteogenesis in young rats. *J. cell Biol.* **14**, 357–70. [**40**]

ZANELLI, G. D. (1968). Thermoluminescence methods applied to dosimetry in trabecular bone, in *Proceedings of the symposium on microdosimetry.* E.A.E.C. Publication EUR-3747, d-e-f, 527–49. [**188**]

—— DARLEY, P. J., and GOLDMAN, M. (1971). Marrow absorbed dose rates in bones of beagle dogs raised on diets containing ^{90}Sr, in *Proceedings of conference of the fifth congress of the French Society for Radioprotection*, Grenoble, France (1971), to be published. [**188**]

ZIMMERMAN, K. W. (1923). Der feinere Bau der Blut-capillaren. *Z. Anat. EntwGesch.* **68**, 29–109. [**28**]

Subject Index

Aberdeen, 53, 54
Accretion, 21, 34, 58, 156, 198
Accumulated dose, 160, 167, 169, 184
Acoustic neuroma, 120
Adenocarcinoma, 50, 123
Adeno-virus, 7, 182
Adrian Committee, 68
Age, 14
Alkaline earths, 21
   discrimination between, 35
   distribution in bone, 20, 96, 97, 99, 100
   excretion, 35, 36, 37
   plasma levels, 37
   retention, 37
Alkaline phosphatase, 85, 86, 139
Alpha particles, 2, 4, 40, 42, 110, 153
^{241}Americium, 95, 100
   binding, 231
   chemistry, 228, 229
   distribution,
      bone, 230
      dental pulp, 231
      liver, 230
      marrow, 231
      thyroid, 230
   metabolism, 228, 229, 230
   toxic effects
      leukaemia, 231
      osteosarcoma, 231
Anaemia
   aplastic, 39, 48, 74, 75, 123, 180
   associated with
      ^{242}Cf, 246
      ^{252}Cf, 246
      ^{60}Co, 91
      ^{239}Pu, 226
      ^{224}Ra, 151
      ^{226}Ra, 119, 144, 145
      ^{90}Sr, 165, 170
      ^{228}Th, 238, 239
      ^{232}Th, 236

Anaemia—*continued*
   Leucoerythroblastic, 48
Anaplastic unclassified carcinomas, 50
Angiosarcoma, 44, 49, 165, 183, 196
Ankylosing spondylitis, 38, 39, 50, 74, 149, 150, 151
Antibiotics, 69
Antibodies, 10
Aplasia, 170, 180, 183, 239
Apposition, 20, 21
Astrocytoma, 120
Atomic bomb
   casualty Commission, 62, 64
   chromosome effects, 2
   dose commitments, 55
   Hiroshima, 62, 63, 64, 65, 67, 68
   Japanese fisherman, 61
   Nagasaki, 62, 63, 64, 65, 67, 68
   Rongelap Islanders, 48, 61, 62
   survivors, 38, 39
Autoradiographic techniques, 136, 137, 186
Average skeletal dose, 103, 107, 186, 190, 249
Avian petrosis, 7

Baboon, 230
Background radiation, 50
Barium, 96
   bone distribution, 96, 97, 98, 99
   metabolism, 35, 36, 37
Beagle dogs, 101, 139, 162, *see also* dogs
Benign erythrocytosis, 242
Beta particles, 2, 3, 5, 17, 40, 42, 110, 135, 188
Bile duct tumours, 201, 226
Bone
   cavity, *see* marrow cavity
   cyst, 83
   matrix, 29
   marrow, 23, 27, 28, 41, 46, 50, 60, 63, 75, 77, 123, 129, 169, 170, 175,

SUBJECT INDEX 289

Bone—*continued*
  176, 177, 187, 195, 196, 197, 216, 217, 220, 224, 226, 231, 235, 236, 237, 238, 239
  mineral, 31
  physiology, 23
  powder, 31, 156
  remodelling, 20, 21, 216
  resorption, 20, 21
  structure, 16
  surfaces, 21, 23, 29, 211
    area, 21, 32, 33
    character, 23, 24, 29, 33, 211
    endosteal, 23, 40, 41, 208, 231, 232
      area, 32
      growing, 21
      resorbing, 21, 208, 231
      resting, 21
    periosteal, 21, 23, 32, 40, 41, 208, 231, 232
      area, 32
    uptake
      ^{241}Am, 207, 208
      ^{133}Ba, 97, 98, 115
      ^{45}Ca, 97, 207, 208
      ^{237}Np, 247
      ^{239}Pu, 207, 208, 211
      ^{244}Ra, 111, 116, 151
      ^{228}Th, 232
      ^{232}Th, 232
  trabecular, 16, 21
  trabeculation, 10, 16
    dog, 16, 19
    man, 16, 19
    mouse, 10, 20
    pig, 16, 19
  'turnover', 21, 135, 227
  turnover rate, 35, 157, 160, 167
Brazil, 54

Caesium, 56, 61
Caisson disease, 124
Calcium, 95
  accretion, 97
  appatite crystals, 97
  binding, 29, 30
  exchange, 34
  kinetics, 34
  metabolism, 35, 36, 37, 156
Calcium apatite, 31, 221
^{45}Calcium, 42, 87, 154, 171, 172

Californium, 100, 246
^{249}Californium, 100, 246
^{252}Californium, 100, 246
Cancers other than skeletal, 38, 63, 68, 74, 75, 78, 149
^{14}C, 52, 56
Carbohydrate protein complexes, 29, 30
Carboxyl groups, 233
Carcinogenesis, 1
Carcinogens, chemical, 5, 10
Carcinoma
  breast, 79
  cervix, 77–9
  intraoral, 50
  penis, 88
  sinuses of skull, 38, 39, 42, 50, 121, 134, 157, 165, 166, 167, 173, 244
  ^{239}Pu, 224
  ^{226}Ra, 42, 121, 122, 123
  ^{228}Ra, 116, 134
  ^{90}Sr, 162, 165, 166, 173
  thyroid, 74
  uterus, 78
Cardiac pacemakers, 56, 202
Cartilage cells hypertrophic, 31
Cartilagenous exostosis *see* osteochroma benign
Central nervous system tumours, 120
Cerium, 100, 246
Chelating agents, 227
Chimaeras, 23, 29, 44
Chondroblastoma, 83
Chondroitin sulphate protein complex, 221
Chondrosarcomas, 44, 45, 79, 80, 81, 83, 122
Chromosomes, 2, 10, 13, 60, 61, 172, 243
  Japanese fisherman, 61
  ^{32}P, 243
  Philadelphia, 172
  Rongelap islanders, 60, 61
Clavicle, 81, 82
Coconut crab, 62
Cofactors, 1, 5, 10
^{60}Co source, 91
Collagen, 29, 221
Cosmic rays, 2, 52
Cretinism, 83
Cumulative rad CR, 130
Cumulative rad years CRY, 130
Curium, 30, 100, 246, 247

De Guglielmo's disease, 48, 235
DDTA, 227
Delta rays, 2
Dental pulp, 231
Dentine, 129
Dial painters, 39, 117, 178
  ^{226}Ra, 39, 117, 118, 119
  ^{90}Sr, 178
Diet, 14
DNA, 108
Dogs
  anaemia, 165
  angiosarcoma, 165, 178, 179
  bone trabeculation, 194
  carcinoma, 106, 162, 178, 179
  chondrosarcoma, 165, 178, 179
  dosimetry in, 19, 101, 103, 104, 105, 106, 144, 179, 188, 189, 191
  fibrosarcoma, 165, 178, 179
  haemangiosarcoma, 162
  leukaemia, 165, 179
  myeloproliferative dyscrasia, 178
  osteosarcoma, 102, 104, 105, 162, 165, 178
DOPC determined osteoprogenitor cell, 28, 29
Dose equivalent, 12
Dose rate, 134, 160, 167, 170, 184
Dose response curve, 12, 13, 14, 70, 71, 75, 76, 133, 134, 190, 191, 192, 193, 194
Dosimetry techniques, 129, 130, 133
  average skeletal dose, 190, 191, 192, 193, 194
  autoradiographic, 136, 137, 186
  bone mensuration, 135, 136, 187, 188
  thermoluminescence, 188, 189, 190
Double primary, 79
Dundee, 53, 54
Dysplasia, 1, 77, 83
  Radiation, external
    exostosis, 44, 45, 74, 81, 84
    fracture, 88, 225, 238
    growth retardation, 83, 226
    necrosis, 83, 88, 225, 226
    scoliosis, 83, 84
    osteomyelitis, 89
    pathology, 85, 86, 87, 88, 89
  Radiation, internal
    ^{239}Pu, 225
    ^{224}Ra, 150

Dysplasia—*continued*
  ^{226}Ra, 123, 124, 138, 146
  ^{99}Sr, 197
  ^{228}Th, 238, 239

Ear bone, 159, 160
Edinburgh, 53, 54
Einsteineum, 246
Embryomas, 51, 70
Endothelial cells, 28
Epidemiological studies, 38
Epithelium applied to bone, 38, 41, 42, 50, 224
Epiphyseal plate, 156, 160, 198, 208, 221, 231
  radiation injury, 85, 198
Epiphysis, 160
Ewing's tumour, 44, 50, 79
Exchange
  long term, 32, 35, 129
  short term, 35
Excitation, 2

Fall out, 56, 58, 61, 154, 162, 202
  ^{137}Cs, 56, 61, 62
  ^{131}I, 56, 58, 62
  ^{238}Pu, 202
  ^{239}Pu, 204
  ^{89}Sr, 56
  ^{90}Sr, 56, 60
Fat cells, 28, 43, 46
Foetus, 48, 71
Fibroblasts, 23, 24, 45, 46
  endosteal, 23
  periosteal, 23
  stem cell, 23, 46
Fibrosarcoma
  definition, 45, 46
  origin, 9, 44
  radiation, external
    man, 80, 81
    monkey, 93
    rabbit, 90
    rat, 92
    ^{239}Pu, 222
    ^{226}Ra, 121, 122, 149
    ^{90}Sr, 163, 164, 165
Fracture
  radiation, external, 88
  radiation, internal, 123, 146, 225, 239
Food chains, 56, 154

SUBJECT INDEX

Gamma rays, 2, 3, 40, 52, 53, 67, 90
  definition, 2
  effects, 60, 66, 90
Giant cells, 44, 46
Giant cell tumour, 44, 46, 82, 142, 165
  treatment, 46, 82
Glasgow, 54
Glioblastoma, 120
Glioma, 120
Glycoproteins, 30, 31, 220, 221, 231, 233
  Metal binding, 30, 31, 220, 221, 231, 233, 247
Granulocytes, 41
Growth retardation
  radiation external, 83
  radiation internal, 150, 198
Growth hormone, 92

Haemangioendothelioma, 49, 94, 142, 173, 196
Haemangioma, 84
Haemangiosarcoma, 49, 162
Haemocytoblastosis, 171
Haemoglobin, 151, 170, 172
Haemopoiesis extramedullary, 144
Haemopoietic death, 165
Haemosiderin, 221
Hamsters, 7, 8
Hanford plutonium registry, 227
Haversian canals, 99, 146, 208, 231
  plugged, 129, 200
  surfaces, 23
Hiroshima, 62, 64, 65, 66, 67
Histiocytes, 28
Hot line, 239
Hot spot, 21, 99
Hurlers disease, 83
Hydroxyapatite, 97, 156

Immunofluorescence, 10
Immunological deficiency, 6
Inflammation, 78, 92
Inhalation, 61, 117, 165, 202, 206, 219
^{131}Iodine, 56, 58
Ionization, 2, 3
IOPC, inducable osteoprogenitor cell, 28, 29
Iridium source, 92
Iron, 221

Japanese fisherman, 61

Jaw, 90, 119, 167, 169, 231

Kerala, 54

Lacunae
  plugged, 129, 200
Latent period
  carcinoma sinuses, 38, 123
  leukaemia, 38, 51, 71, 182, 184, 186, 236
  osteosarcoma, 38, 51, 70, 81, 120, 140, 175
LET, 3, 4, 11, 12, 193
Leucopaenia, 91, 144, 165, 170, 183
Leukaemia
  definition, 47, 48
  latent period, 38, 51, 71, 182, 184, 186, 236
  aleukaemic, 48, 235
  blast cell, 48, 71
  lymphatic
    animals, 48, 180
    man, 48, 67, 71
  myeloid, *see also* myeloproliferative dyscrasia, 41, 48
  animals
    ^{60}Co, 91
    ^{239}Pu, 223, 224
    ^{90}Sr, 165, 175, 180, 182, 197
    ^{228}Th, 238
    ^{232}Th, 235
  man
    ankylosing spondylitis, 38, 39, 74, 75
    atom bomb casualties, 63, 64, 65, 67, 68
    diagnostic radiology, 69, 70, 71, 72, 73, 74
    foetal irradiation, 69, 71
    pelvic irradiation, 77, 78
    radiologists, 68, 69
    therapeutic radiology, 73, 74
    ^{226}Ra, 123
    ^{224}Ra, 150
    ^{232}Th, 235
Leukaemogenic virus, 6
Life shortening, 12
Linear energy transfer *see* LET
Lipids, 29, 31
Liposarcoma, 49
Liver, 201, 202, 203, 204, 205, 226, 230, 235, 236, 246
  tumours, 200, 231, 235

Lymphocytes, 41
Lymphomas, 5, 9, 98, 182, 224
Lymphosarcoma, 48, 70, 73
Lysosomes, 29

Macrophage, 28, 206, 219, 221, 233, 235
Malignancy, 1
Malignant transformation, 2, 5, 13, 28, 38, 82, 120, 227
Mandible, 142, 165
Manhattan project, 221
Marrow *see* bone marrow
Marrow cavity, 4, 16, 135, 188
Mastectomy, 51, 79
Mean absorbed dose, 187, 188
Mean endosteal dose, 188
Mean marrow dose, 43, 136, 188
Mean path length, 17, 19, 135
  bone trabeculae, 17, 19, 135
  marrow cavities, 17, 19, 135
Megakaryocyte, 179
Megakaryocytic leukaemia, 48
Menorrhagia, 78
Mesenchyme, 7, 41, 46, 195
Mesenchyme cell, 25, 44, 49, 221
Mesenchymal marrow elements, 41
Mesons, 2
Mesothorium, *see* ^{228}Ra
Metabolism
  americium, 228, 229, 230
  barium, 35, 36, 37
  calcium, 35, 36, 37, 156
  plutonium, 202, 203
  radiation effects on, 15
  radium, 35, 36, 37, 111
  strontium, 35, 36, 37, 154, 155, 156, 178
  thorium, 232, 238
Metal binding, 29, 30
  glycoproteins, 30, 31, 220, 221, 231, 233, 247
  osteogenic tissue, 220, 221, 233
Metastases, 145, 148, 149
Methylcholanthrene, 6, 10
Metropathia haemorrhagica, 77
Mice, 6
  bone dyscrasia, 8, 9
  bone trabeculation, 152
  dose response curve, 142, 143, 144, 190
  external radiation, 90
  radiation effects
    anaemia, 177

Mice—*continued*
  carcinoma, 173, 176
  chondrosarcoma, 173
  fibrosarcoma, 176
  haemangio endothelioma, 173
  leukaemia, 6, 175, 176, 197
  myeloproliferative dyscrasia, 177–8
  osteosarcoma, 90, 91, 172, 173, 175, 176, 178
  strains
    AKR, 5
    BALB/C, 8, 9
    C 57 BL, 90
    CBA, 10, 172, 175, 196
    CF1, 7, 8
    CF1/ANL, 7, 8, 90, 172
    $C_3H/131$, 5
    ICR, 8
  trabeculation, 20, 152
Milk, 56
Microcateract, 13
Miniature swine, *see* pig
Mitochondria, 29
Monkey, 93, 178
Monosite sands, 54
Mutation, 1
Myelofibrosis, 8, 165, 178, 242
Myeloid metaplasia, 178, 197, 242, 243
Myeloproliferative disorders, 41, 47, 179, 242, *see also* leukaemia
Myeloproliferative dyscrasia, 6
  definition, 47
  dogs, 91, 178, 179
  pigs, 180, 181, 182
Myelosclerosis, 10, 48, 120, 240, 243

^{24}Na, 56
Nagasaki, *see* Hiroshima
^{237}Neptunium, 246, 247
Neuroblastoma, 70, 84, 85
Neutrons, 2, 40, 66, 67, 90
Non teratomatous tumour, 71
Non uniformity, 128, 135, 136

Osteoblast, 7, 23, 24, 40, 89, 139
Osteochondroma benign, 44, 45, 74, 81, 84, 150, 151, 172
Osteochondrosarcoma, 7, 165
Osteoclast, 23, 24, 40, 89, 216
Osteoclastoma, *see* giant cell tumour
Osteocyte, 1, 23, 40, 54, 89, 200

SUBJECT INDEX 293

Osteogenic tissue, 23, 24, 25, 29, 40, 43, 211, 216
  metal binding, 29, 211, 216, 221
  old bone, 24, 25, 26, 27
  young bone, 24
Osteoid, 33, 34
Osteoma, 8, 142
Osteomyelitis, 89
Osteoporosis, 88, 225
Osteoprogenitor cells, 25
Osteosarcoma, 44
  embryonic, 70
  latent period, 38, 70
  natural incidence, 80, 139, 172, 175
  radiation, external
    dog, 91
    man, 78, 79, 80, 81
    mouse, 90, 91
    rabbit, 90
  radiation, internal
    ^{241}Am, 231
    ^{237}Np, 247
    ^{32}P, 244
    ^{239}Pu, 216, 222
    ^{224}Ra, 149, 151, 152
    ^{226}Ra, 102, 105, 121, 122, 139, 140, 141, 142, 143, 144, 145
    ^{228}Ra, 102, 105
    ^{90}Sr, 41, 162, 165, 166, 167, 171, 172, 173, 175, 178, 179, 182, 183
    ^{228}Th, 102, 105, 238
    ^{232}Th, 236
  virus in, 8, 10
Ovarian radiation, 78

Pancytopaenia, 238
Parietal bone, 135
Path length, 16, 20, 135
Pelvic irradiation, 77, 78
Penis, 88
Peptides, 29, 31
Pericytes, 28, 46
Periodic acid Schiff stain, 231
Periosteum, 7
Periosteal surface, 21, 23, 32, 40, 41, 208, 231, 232
Periostitis, 89
Peru, 54
Peteosthor, 149
^{32}Phosphorus, 48, 108, 240
  Chromosome effects, 243

^{32}Phosphorus—*continued*
  Distribution, 108
  dosimetry
    marrow, 244–5
  physical characteristics, 108
  therapeutic use
    polycythaemia vera, 240, 241, 242, 243
  toxic effects
    anaemia, 243, 244
    leukaemia, 241, 242, 243
    acute myeloid, 241
    myeloproliferative malignancy, 48
    myeloid metaplasia, 242
    myelosclerosis, 241, 242, 243
    osteosarcoma, 244
Pig, 6, 180, 182
  dosimetry, 182, 183, 192, 193
  myeloproliferative disorder, 6, 180, 182, 187
  osteosarcoma, 182
  ^{90}Sr, 6, 165, 180, 181, 182
  trabeculation, 19
  viruses in, 6, 7, 9, 182
Pink teeth, 127
Plasma proteins, 204, 247
Ploughshare, 56
Plugged canals, 129, 200
Plugged lacunae, 129, 200
Plutonium isotopes, 95, 96, 100, 201
  ^{237}plutonium, 201
  ^{238}plutonium, 201
  binding, 30, 204, 220
  chemistry, 202
  distribution, 101
    aggregate, 216, 217, 218, 221
    diffuse, 216, 217, 218, 221
    liver, 201, 203, 204, 206, 221
    marrow, 216, 219, 220, 221, 222, 224, 226
    osteogenic cells, 221, 216
    skeleton, 203, 204, 206, 207
  dosimetry, 226, 227
  entry route
    lungs, 202, 206, 219
    gut, 202, 206
    skin, 206
    wound, 202, 207, 212, 220
  maximum permissible body burden, 202, 226
  metabolism, 202, 203, 204

Plutonium isotopes—*continued*
  monomeric, 202, 205, 219, 220
  physical characteristics, 201
  polymeric, 202, 206, 219, 220
  toxic effects, 222
    anaemia, 226
    bile duct tumours, 201, 226
    carcinoma, 224
    chondrosarcoma, 224
    dysplasia, 225
    fibrosarcoma, 222
    leukaemia, 223
    myeloproliferative dyscrasia, 223, 224
    osteosarcoma, 216, 222
  translocation, 202, 205, 206, 207, 212
Polycythaemia vera, 48, 240, 241, 242, 243
Preosteoblast, 2, 24, 27, 40
Proliferating cells, 38, 39, 40, 41, 42
Proliferative potential, 38
Protons, 2
Pure radium equivalent, PRE, 118

Quality factor, 12

Rabbits, 156, 157, 158, 159, 160, 161, 162, 165, 167, 168, 169, 170, 171
  aplasia, 182
  carcinoma, 90, 157, 165
  external radiation, 90, 92
  fibrosarcoma, 90
  osteosarcoma, 166, 167
Radiation
  chimaera, 23, 29, 44
  dose, 61, 75, 77, 80, 135, 136, 137
  low level, 13, 71
  necrosis, 83
  osteitis, 89
Radiography obstetric, 70
Radiologists, 39, 68
Radiothor, 118
Radium isotopes, 95, 96, 109
  normal bone content, 55
  water content, 55
^{224}Radium, 94, 233
  decay chain, 111
  distribution skeletal, 111, 112, 116
  dosimetry, 151
  metabolism, 35, 36, 37, 111
  physical characteristics, 111

^{224}Radium—*continued*
  toxic effects
    anaemia, 151
    cartilaginous exostosis, 45, 150, 151
    chondrosarcoma, 149
    dysplasia, 84, 85
    fibrosarcoma, 11
    leukaemia, 150
    osteosarcoma, 149, 151
  therapeutic use
    ankylosing spondylitis, 149
    tuberculosis, 45, 149
^{226}Radium, 94, 109
  dogs
    single injection, 102
    multiple injections, 144
    toxic effects
      anaemia, 144
      bone dysplasia, 145, 146
      leucopaenia, 144
      osteosarcoma, 102, 145, 146
  man
    chemists, 39, 117
    dial painters, 117, 118, 119
    distribution
      diffuse, 136
      hot spot, 21, 136, 137
      teeth, 128
    dosimetry, 130, 131, 133, 134, 135
    dose response curve, 132, 133
    metabolism, 35, 36, 37, 113
      retention, bone, 113
      soft tissues, 113
    physical characteristics, 109
    therapeutic use, 39, 117, 118
    toxic effects
      acute, 119, 120
      anaemia, 119, 120, 123
      dysplasia, 123, 124, 125, 128, 129
      carcinoma sinuses, 120, 157
      central nervous system tumours, 120
      chondrosarcoma, 122
      fibrosarcoma, 121, 122
      fracture, 123
      leukaemia, 123
      osteosarcoma, 120
  mice, 139
    dose response curve, 143, 144
    osteosarcoma, 140, 141, 142
^{228}Radium, 94, 110, 233

SUBJECT INDEX 295

^{228}Radium—*continued*
  decay chain, 110
  distribution, skeletal, 116
  dial paint, 117
Radon, 42, 94, 109, 134
Rats
  angiosarcoma, 183
  chondrosarcoma, 183
  dose response curve ^{90}Sr, 191
  external radiation, 90
  fibrosarcoma, 92, 183
  leukaemia, 171, 197
  osteosarcoma, 92, 171, 183
  reticulum cell tumour, 183
  reticulosis, 171
Reactors, 56, 202
Relative biological efficiency, RBE, 11, 105, 107, 108, 133
  gamma rays, 67
  fast neutrons, 67
  ^{226}Ra, 105, 133
  ^{90}Sr, 105
  ^{239}Pu, 105
  ^{228}Th, 105, 239
  Utah experiment, 101, 102, 103, 104, 105, 106, 133
Rem, 12
Remodelling, 20, 21, 160
Resorption, 20
  cavities, 129, 146, 231
Reticuloendothelial cells, 28
Reticuloendothelioma, 50
Reticulosis, 171
Reticulum cell sarcoma, 44, 49, 50, 183, 196
Retinoblastoma, 80, 81
Ribs, 225, 239
Rongelap islanders, 48, 58, 61, 62

Sarcoma, 7
Satellite, 56, 202
Scapula, 82
Scoliosis, 83, 84
Secondary deposits, *see* metastases
Sex differences, 14, 152
Sialomucin, 31
Sialoprotein (bone), 31, 221, 233
Skull, 21
  chondrosarcoma, 81
  dysplasia, 124, 135
  fibrosarcoma, 81

Skull—*continued*
  osteosarcoma, 81, 82
Somatic mutation, 1
Species difference, 14, 17, 90, 113, 155, 202
Spleen, 177, 179, 237, 242
Squamous carcinoma, 50, 123
Stem cell, 27, 182, 197
  fibroblast, 23, 46
  haemopoietic, 28, 41, 46, 170
  leukaemia, 48
  osteogenic, 28, 29
Sternum, 81
Strontium isotopes, 95, 96, 154
^{85}Strontium, 154, 195, 196
^{87}Strontium, 154
^{89}Strontium, 62, 154, 171, 172
^{90}Strontium, 43, 94, 154
  dial painters, 161, 178
  discrimination against, 156
  distribution skeletal, 156
    single administration, 156, 158, 159, 160, 162
    continuous intake, 156, 158
  Fall out, 56, 154, 162
    bone, 57, 58, 60
    food, 56, 57
    milk, 56, 57
    Rongelap islanders, 58, 60, 61, 62
  dose response curves, 190, 191, 192
  dosimetry, 184, 185, 186
  metabolism, 35, 36, 37, 154, 155, 156, 178
  physical characteristics, 154
  RBE, 105
  toxic effects
    anaemia, 165, 170, 183
    angiosarcoma, 49, 165, 195, 196
    carcinoma, 157, 162, 165, 166, 173, 176, 178
    chondrosarcoma, 165, 173, 178, 194
    dysplasia, 198, 199, 200
    fibrosarcoma, 165, 176, 178, 195, 196
    giant cell tumour, 165
    haemangiosarcoma, 49, 162, 173, 178, 195, 196
    leukaemia, 48, 178, 184, 197
    myeloproliferative dyscrasia, 48, 165, 177, 178, 197
    osteosarcoma, 162, 165, 166, 167, 172, 172, 173, 175, 176, 178, 182,

^{90}Strontium—*continued*
 183, 194, 195, 196
  reticulum cell sarcoma, 50, 183
  reticuloendothelioma, 50
  reticulosarcoma, 50
^{32}Sulphur, 108
^{35}Sulphur, 86
Surface seekers, 94, 100, 101, 231
Surfaces, *see* bone
Swine, *see* pigs

Teeth, 89, 127, 128, 129, 145
Thermoluminescence dosimetry, 188
Thermoluminescent dosimeters, 182, 188
Thorium isotopes, 52, 100
Thorium X, *see* ^{224}Ra
^{228}Thorium, 233
 absorption from gut, 117, 238
  dial paint, 117
 distribution
  bone, 238
  marrow, 238
  metabolism, 238
 dosimetry, 239
 RBE, 105, 239
 toxic effects,
  anaemia, 238, 239
  haemangiosarcoma, 49, 238
  osteosarcoma, 238
  pancytopaenia, 238
  skeletal dysplasia, 238, 239
^{232}Thorium, 233, 235
Thoron, 116, 134
Thorotrast, 233, 235, 236, 237
 distribution
  bone, 233, 234, 235
  liver, 235, 236
  lung, 236
  marrow, 235, 236
  spleen, 235, 236
 dosimetry, 236, 237
 ionium enriched thorotrast, 236
 toxic effects
  anaemia aplastic, 235
  De Guglielmo, 235
  leukaemia acute, 235
   aleukaemic, 235
   myeloid, 235, 236
  liver neoplasms, 235
  myelosclerosis, 235

Thorotrast—*continued*
  osteosarcoma, 235
Threshold, 13, 75, 133, 190, 193
Thrombocytopaenia, 165, 183
Thymic irradiation, 51, 74
Thymic lymphomas, 5
Thymidine, 42
Thymus, 74, 176, 177
Thyroid, 74, 230
 cancer, 74
 deficiency, 58
Thyroxine, 92
Trabeculae, 16, 19
 dog, 19
 man, 19, 135, 152
 mouse, 20, 152
 pig, 17
Transferrin, 30, 31
Transitional epithelium, 28
Triethylene-tetramine-hexaacetic Acid, TTPA, 227
Tuberculosis, 78, 149
Tumour expectancy, 142, 143
Turnover rate, *see* bone

Uranium, 52
^{238}Uranium, 200
Urethane, 10
Utah dogs, 49, 101, 102, 103, 104, 105, 106, 162
Utirik islands, 46

Vascular damage, 85, 88, 198, 200
Vertebrae, 18, 20, 140, 222
Virus
 avian osteopetrosis, 7
 bone dyscrasias, 7
 co-factors, 5
 leukaemia, 6
 osteosarcoma, 7
 types
  adenovirus, 7, 182
  C type RNA, 5, 6
  FBJ, 7, 8, 172
  Harvey sarcoma, 7
  Molony, 7
  parotid tumour agent, 7
  polyoma, 7
  Raucher, 7
  sarcoma, 37, 9
Volume seekers, 20, 95, 96, 99

SUBJECT INDEX

Weather station, 56
Weight loss, 14
Wilm's tumour, 70, 81, 84, 85
Woodworkers, 50, 123
Wound, 202, 207, 212

X-rays, 2, 3, 5, 40, *see* also external irradiation
  diagnostic
    adult, 73, 74
    foetal, 69, 70, 71, 72

X-rays—*continued*
  occupational, 68, 69
  therapeutic
    adults, 73, 74, 75, 76, 77, 78, 79, 80, 82, 241
    children, 74, 80

Yolk sac, 28
Yttrium, 100
^{90}Y, 43, 94, 154
^{91}Y, 154

RA
1231
.R2
V36